Lecture Notes in Co

The series Lecture Notes in Computer Science (LNCS), including its subseries Lecture Notes in Artificial Intelligence (LNAI) and Lecture Notes in Bioinformatics (LNBI), has established itself as a medium for the publication of new developments in computer science and information technology research, teaching, and education.

LNCS enjoys close cooperation with the computer science R & D community, the series counts many renowned academics among its volume editors and paper authors, and collaborates with prestigious societies. Its mission is to serve this international community by providing an invaluable service, mainly focused on the publication of conference and workshop proceedings and postproceedings. LNCS commenced publication in 1973.

Robert Wrembel · Silvia Chiusano ·
Gabriele Kotsis · A Min Tjoa · Ismail Khalil
Editors

Big Data Analytics and Knowledge Discovery

26th International Conference, DaWaK 2024
Naples, Italy, August 26–28, 2024
Proceedings

Springer

Editors
Robert Wrembel
Poznan University of Technology
Poznan, Poland

Silvia Chiusano
Polytechnic University of Turin
Turin, Italy

Gabriele Kotsis
Johannes Kepler University Linz
Linz, Austria

A Min Tjoa
Vienna University of Technology
Vienna, Austria

Ismail Khalil
Johannes Kepler University Linz
Linz, Austria

ISSN 0302-9743 ISSN 1611-3349 (electronic)
Lecture Notes in Computer Science
ISBN 978-3-031-68322-0 ISBN 978-3-031-68323-7 (eBook)
https://doi.org/10.1007/978-3-031-68323-7

This Springer imprint is published by the registered company Springer Nature Switzerland AG
The registered company address is: Gewerbestrasse 11, 6330 Cham, Switzerland

Preface

DAWAK was established in 1999 as an International Conference on Data Warehousing and Knowledge Discovery. It was run continuously under this name until its 16th edition in 2014 (Munich, Germany). In 2015 (Valencia, Spain) it was renamed to International Conference on Big Data Analytics and Knowledge Discovery, but retained its DAWAK acronym. This name change was to reflect new research directions in the broad and dynamically developing areas of big data analytics and building knowledge from massive data.

Since the very beginning, the DAWAK conference has served as a high-quality forum for researchers, practitioners, and developers in the fields of: data integration, data warehousing, standard and big data analytics, and recently in artificial intelligence. The main objectives of the conference are to explore, disseminate, and exchange knowledge in these fields through scientific and industry talks. Big data analytics and artificial intelligence are hot research areas for both academia and the IT industry. They are continuously evolving, fueled by advances in new algorithms, data processing architectures, and by specialized and high-performance hardware.

Important research topics associated with these major areas include: data lakes (schema-free repositories of heterogeneous data), data lakehouses, conceptual/logical/physical database design, theoretical foundations for data engineering, data integration (especially the fusion of structured and semi-structured data sources), big data management (mixing relational tables, text, and any types of files), query languages (beyond SQL), scalable analytical algorithms, parallel storage and computing systems (cloud, parallel database systems, Spark, MapReduce, HDFS), graph processing, stream and time series processing, IoT architectures, artificial intelligence/machine learning algorithms, specialized hardware for data processing and machine learning (GPU, FPGA, TPU, NPU), and applications of these solutions in industry.

The proceedings include papers from the 26th DAWAK edition, which took place on August 26–28, 2024 in Naples (Italy).

DAWAK 2024 received 83 paper submissions. Out of these, the Program Committee selected 16 as regular papers, resulting in an acceptance rate of 19%. Additionally, 20 papers were accepted as short papers, aimed at showcasing pioneering research and innovative projects across various disciplines. These short papers highlight early-stage research, novel ideas, and preliminary findings, fostering meaningful discussions and potential collaborations.

Accepted papers cover a variety of research topics on both theoretical and practical aspects. The program includes among others the following topics: (1) machine learning, (2) data storage and architectures, (3) standard and big data analytics, (4) data matching, (5) time series and stream processing, (6) natural language processing and text analytics, (7) similarity computation, (8) graph processing.

Thanks to the reputation of DAWAK, selected best papers of DAWAK 2024 will be invited for a special issue of the Data & Knowledge Engineering (DKE, Elsevier) journal. Therefore, the PC-chairs would like to thank the DKE Editor-in-Chief, Carson Woo, for his approval of the special issue.

We would like to express our sincere gratitude to all the Program Committee members and the external reviewers who reviewed the papers profoundly and in a timely manner. Finally, we are thankful for the continuous support and guidance from the DEXA conference organizers, especially, Ismail Khalil for his invaluable help in all tasks leading to having these proceedings published.

August 2024

Robert Wrembel
Silvia Chiusano
Gabriele Kotsis
A Min Tjoa
Ismail Khalil

Organization

Program Committee Chairs

Robert Wrembel	Poznań University of Technology; Interdisciplinary Centre for Artificial Intelligence and Cybersecurity, Poland
Silvia Chiusano	Polytechnic University of Turin, Italy

Steering Committee

Gabriele Kotsis	Johannes Kepler University Linz, Austria
A Min Tjoa	Vienna University of Technology, Austria
Lukas Fischer	Software Competence Center Hagenberg, Austria
Bernhard Moser	Software Competence Center Hagenberg, Austria
Ismail Khalil	Johannes Kepler University Linz, Austria
Nicola Mazzocca	University of Naples Federico II, Italy
Elio Masciari	University of Naples Federico II, Italy

Program Committee Members

Abhishek Santra	University of Texas at Arlington, USA
Alberto Abello	Universitat Politècnica de Catalunya, Spain
Alejandro Maté	University of Alicante, Spain
Andrea Kő	Corvinus University of Budapest, Hungary
Anton Dignös	Free University of Bozen-Bolzano, Italy
Benkrid Soumia	École Nationale Supérieure d'Informatique, Algeria
Besim Bilalli	Universitat Politècnica de Catalunya, Spain
Boris Novikov	National Research University, Finland
Carlos Ordonez	University of Houston, USA
Carson Leung	University of Manitoba, Canada
Christos Doulkeridis	University of Piraeus, Greece
Claudia Diamantini	Università Politecnica delle Marche, Italy
Cristina Aguiar	Universidade de São Paulo, Brazil
Darja Solodovnikova	University of Latvia, Latvia
Dimitris Sacharidis	Université libre de Bruxelles, Belgium

Dimitri Theodoratos	New Jersey Institute of Technology, USA
Ela Pustulka	University of Applied Sciences and Arts Northwestern Switzerland, Switzerland
Elisa Quintarelli	Università di Verona, Italy
Emanuele Storti	Università Politecnica delle Marche, Italy
Enrico Gallinucci	University of Bologna, Italy
Fadila Bentayeb	Université Lyon 2, France
Franck Ravat	IRIT, Université de Toulouse, France
Frank Höppner	Ostfalia University of Applied Sciences, Germany
Genoveva Vargas-Solar	CNRS, LIRIS, France
Günther Specht	University of Innsbruck, Austria
Isabelle Comyn-Wattiau	ESSEC Business School, France
Jaroslav Pokorný	Charles University in Prague, Czech Republic
Jens Lechtenbörger	University of Münster, Germany
Jérôme Darmont	Université Lyon 2, France
Jorge Bernardino	Polytechnic Institute of Coimbra, Portugal
Jun Miyazaki	Tokyo Institute of Technology, Japan
Kamel Boukhalfa	USTHB, Algeria
Karen Davis	Miami University, USA
Kazuo Goda	University of Tokyo, Japan
Ladjel Bellatreche	LIAS/ENSMA, France
Lars Ailo Bongo	University of Tromsø, Norway
Laurent D'Orazio	University Rennes, CNRS, IRISA, France
Lena Wiese	Fraunhofer Institute for Toxicology and Experimental, Medicine, Germany
Leonidas Fegaras	University of Texas at Arlington, USA
Luca Gagliardelli	University of Modena and Reggio Emilia, Italy
Maik Thiele	HTW Dresden, Germany
Marcin Gorawski	Silesian University of Technology, Poland
Marco Angelini	Sapienza University of Rome, Italy
Markus Endres	University of Passau, Germany
Matteo Francia	University of Bologna, Italy
Matteo Golfarelli	University of Bologna, Italy
Mirjana Ivanovic	University of Novi Sad, Serbia
Olivier Teste	IRIT, France
Omar Boussaid	Université Lyon 2, France
Oscar Romero	Universitat Politècnica de Catalunya, Spain
Patrick Marcel	University of Orléans, France
Pawel Boinski	Poznań University of Technology, Poland
Pedro Furtado	University of Coimbra, Portugal
Sana Sellami	Aix-Marseille University, France
Sandro Bimonte	INRAE, France

Selma Khouri	École Nationale Supérieure d'Informatique, Algeria
Sofian Maabout	University of Bordeaux, France
Soror Sahri	Université Paris Cité, France
Soumyava Das	Teradata Lab, USA
Stephane Bressan	National University of Singapore, Singapore
Stephane Jean	University of Poitiers, ISAE-ENSMA, LIAS, France
Sven Groppe	University of Lübeck, Germany
Syed Muhammad Fawad Ali	Accenture, Germany
Sylvio Barbon Junior	University of Trieste, Italy
Szymon Wilk	Poznań University of Technology, Poland
Witold Andrzejewski	Poznań University of Technology, Poland
Yinuo Zhang	Teradata, USA
Yong Hwan-Seung	Ewha Womans University, South Korea

External Reviewers

Eros Fabrici	Universitat Politècnica de Catalunya, Spain
Ingi Lee	Ewha Womans University, South Korea
Moncef Garouani	IRIT-Université Toulouse Capitole, France

Organizers

Web Applications Society
@WAS
www.iiwas.org

DIETI.
UNIVERSITA' DEGLI STUDI DI NAPOLI FEDERICO II
DIPARTIMENTO DI INGEGNERIA ELETTRICA E DELLE TECNOLOGIE DELL'INFORMAZIONE

Abstracts of Keynote Talks

Multimodal Deep Learning in Medical Imaging

Carlo Sansone

Department of Electrical Engineering and Information Technology,
University of Naples Federico II, Italy

Abstract. In this talk, we will consider how Deep Learning (DL) approaches can profitably exploit the presence of multiple data sources in the medical domain.

First, the need to be able to use information from multimodal data sources is addressed. Starting from an analysis of different multimodal data fusion techniques, an innovative approach will be proposed that allows the different modalities to influence each other.

However, in medical applications it is often very difficult to obtain high quality and balanced labelled datasets due to privacy and sharing policy issues. Therefore, several applications have leveraged DL approaches in data augmentation techniques, proposing models that can create new realistic and synthetic samples. Consequently, a new data source can be identified, namely a synthetic data source. In this context, a data augmentation method based on deep learning, specifically designed for the medical domain, will be presented. It exploits the biological characteristics of images by implementing a physiologically aware synthetic image generation process.

Digital Humanism as an Enabler for a Holistic Socio-Technical Approach to the Latest Developments in Computer Science and Artificial Intelligence

A Min Tjoa

TU Wien (Vienna University of Technology), Austria

Abstract. The rapid development of computer science and artificial intelligence (AI) has brought about transformative changes, but not without significant ethical, social, and technical challenges. As early as 2017, Tim Berners-Lee, the inventor of the World Wide Web, warned of the "nasty storm" threatening the future of the web, including the proliferation of fake news, propaganda, and increasing polarization. These issues highlight the urgent need for a paradigm that ensures technology serves the best interests of humanity.

This keynote will explore the foundational principles of Digital Humanism and its role in guiding the development of computer science and AI to align with human values and societal well-being.

In December 2023, the United Nations Advisory Panel on AI released its interim report, "Governing AI for Humanity," which highlights the need for AI governance to address challenges and harness AI's potential in an inclusive way, ensuring that no one is left behind. A key measure of AI's success will be its contribution to achieving the SDGs.

The keynote will illustrate how Digital Humanism can be operationalized to create technologies that enhance human capabilities and societal well-being. It will highlight the need for interdisciplinary research and development to harness the potential of computer science and AI for the benefit of humanity.

Digital Humanism offers a vital pathway for navigating the complexities of modern technological advancements. By taking a holistic socio-technical approach, it can be ensured that developments in computer science and AI are aligned with our core human values, thereby fostering a more just, ethical, and sustainable digital future.

Deep Entity Processing in the Era of Large Language Models: Challenges and Opportunities

Toshiyuki Amagasa

University of Tsukuba, Japan

Abstract. Handling entities has long been a critical task in data analytics and integration, with over 80% of time and effort often devoted to data preprocessing. Improving the performance of this task has been a persistent challenge. Recently, transformer-based pre-trained language models and large language models (LLMs) have emerged as key tools for entity processing tasks such as named entity recognition (NER) and entity matching. However, these models introduce new challenges, including significant demands for computational resources and high-quality training data. In this talk, we will review recent advances in deep entity processing and explore the associated challenges and research opportunities.

Contents

Modeling and Design

LiteSelect: A Lightweight Adaptive Learning Algorithm for Online Index Selection

Xiaoying Wu[1], Senyang Wang[1], Xin Liu[1], Dimitri Theodoratos[2](✉), and Md Rakibul Hasan[2]

[1] School of Computer Science, Wuhan University, Wuhan, China
{xiaoying.wu,whuwsy,xliu1997}@whu.edu.cn
[2] NJ Institute of Technology, Newark, NJ, USA
{dth,mh629}@njit.edu

Abstract. Using appropriately selected indexes can dramatically improve the performance of query workloads in database systems. Typically, the access patterns of the workloads in real-world applications change frequently. This poses the challenge of automatically adapting the indexes to the changing workload. An effective approach to solve this problem is an online index selection process, which does not assume prior knowledge of the workload pattern but adapts the index configuration based on the history of the workload.

In this paper, we address the Online Index Selection problem. Our study on recent learning-based solutions shows that their methods incur significant tuning overhead, making them unsuitable for online-tuning in real-world systems. To address this limitation, we model online index selection as a problem of sequential decision making under uncertainty, and we design a lightweight adaptive learning algorithm called *LiteSelect*. At the core of *LiteSelect* is an exponential smoothing method which takes a sequence of observations to estimate index benefits for future queries with unknown distribution. *LiteSelect* enjoys a fast convergence rate and has low memory cost. We further design optimizations for *LiteSelect* to control the online tuning overhead and to enhance the solution quality. Our extensive experiments demonstrate that *LiteSelect* effectively performs online index tuning on different kinds of workloads under widely used benchmarks and greatly outperforms index tuning algorithms using sophisticated learning methods.

1 Introduction

Indexes are important physical structures for the efficient processing of database workloads. Index selection is the problem of finding an optimal set of indexes for a

X. Wu was supported by CCF-Huawei Database System Innovation Research Plan No. CCF-HuaweiDBIR003A and the National Natural Science Foundation of China under Grant No. 61872276.

R. Wrembel et al. (Eds.): DaWaK 2024, LNCS 14912, pp. 3–18, 2024.
https://doi.org/10.1007/978-3-031-68323-7_1

given workload and dataset while considering constraints, e.g., a storage budget. Workloads from real-world applications typically change frequently. For example, monitoring the performance of HTAP applications often involves looking for anomalies and trends in data. The set of tables and predicates of interest for this kind of ad-hoc, exploratory analysis often shifts over time. An offline approach, i.e., optimizing the design for a representative workload, assuming the workload does not change much over time and an index is globally useful over the entire workload, does not work in this setting. Instead, an online approach which does not assume a prior knowledge of the workload but automatically adapts the design based on the history of the workload is needed.

The Framework. In this paper we address the Online Index Selection (OIS) problem. This problem assumes a sequence of queries (and/or updates) and can be modeled by a system which evolves over a finite number of stages. Each stage corresponds to a query to be evaluated and is associated to a *configuration*, which is a set of indexes for queries that have not yet arrived. The configuration transition cost is the cost of creating and dropping indexes between two stages. The goal of the problem is to find an optimal sequence of configurations that minimizes the total cost of query execution, configuration selection and configuration transition.

Online-tuning solutions can be broadly classified as either non-learning- or learning-based. A solution is learning-based if it takes information that is known to predict or estimate something that is unknown [8]. Earlier proposals all belong to the non-learning category [2,6,9]. More recent works are learning-based. They model the online tuning as a Markov decision process and study the problem under the setting of reinforcement learning (RL) [1] or multi-armed bandit (MAB) learning [7]. We are interested in designing a learning-based solution.

Challenges. An online algorithm must address the following challenges that are unique to continuous tuning [2,10]: (1) it must have very low overhead and not interfere with the normal functioning of the DBMS; and (2) the tuning process must react quickly to *real* changes in the workload but not overreact to temporary variations of the query distribution to avoid unwanted oscillations. Appropriately handling these challenges are key to the incorporation of online-tuning in real-world systems.

Two sub-problems need to be addressed to solve the online-tuning problem. First, we need to quantify the benefit of an index to the present query, i.e., the cost-saving when the index is utilized for executing the query. Most index selection algorithms exploit a what-if optimizer which returns estimated costs without physically creating indexes (called hypothetical indexes). Optimizer-based cost estimations can result in significant estimation errors due to cardinality misestimations or inaccurate cost models [5]. To improve the accuracy of estimation, a couple of recent works propose a learning-based method that models the index benefit estimation as a regression task [11,12]. These solutions need to gather numerous historical index statistics in order to train a regression model. They are unable to do the "on-the-go" learning and estimation, thus not applicable to the online index tuning task.

Second, we need to estimate online the utility of a configuration for queries in the near future. This is the key component that affects critically the online-tuning performance. For this, online-tuning solutions build and/or compute a model for candidate index configurations that summarizes historical observations and/or index configuration features, and update the model overtime when new observations are available. For instance, the model in the learning-based solutions was represented as value functions of a Bellman Equation [1], or a regression model of index scores [7]. As we discuss in Sect. 5 and experimentally verify in Sect. 4, the model building and maintenance in existing solutions incur significant computational overhead.

Contribution. In order to address the aforementioned challenges of online tuning and the limitations of existing solutions, we model online index selection as a sequential decision making problem under uncertainty. In this context, we make the following additional contributions to provide an efficient solution to this problem:

- We use the exponential smoothing method to iteratively build and update a model for estimating index benefit for future queries with unknown distribution. The model summarizes all the history observations of the tuning system. It is lightweight yet effective.
- We design *LiteSelect*, a lightweight learning-based algorithm for the OIS problem. A salient feature of *LiteSelect* is that its runtime behavior can be easily and flexibly tuned through a couple of parameters.
- To control the tuning overhead of *LiteSelect*, we design a mechanism to budget what-if calls during the tuning process of *LiteSelect*.
- To reduce the processing cost of expensive queries, we design a "restart" mechanism which increases the chances of finding promising configurations when *LiteSelect* explores the search space.
- We run extensive experiments to evaluate the performance of *LiteSelect* on different types of evolving workloads on TPC-H and TPC-C benchmarks. We also compare *LiteSelect* with recent learning-based solutions which use sophisticated learning methods. Our results show that *LiteSelect* can effectively perform online index tuning, and it greatly outperforms the state-of-the-art.

2 The Online Index Selection Problem

We formally define now the problem we address. Let Q be a workload query trace, defined as a sequence of N queries. Query q_k represents the kth query in Q executed at time k. We model q_k as a random variable, whose generating distribution may not be known a priori. This means that q_k is only observable at time k. Let I denote the set of secondary indices of the DBMS. A configuration $s \subseteq I$ is the set of indexes available at some point in time. We introduce a configuration constraint by assuming that $s \in S$, where S is a set of admissible configurations. Popular configuration constraints are the storage space available for storing indexes or an upper bound on the number of indexes.

We model the problem as a sequential decision problem under uncertainty, where the system evolves over a finite number N of time steps (also called stages) determining a configuration at each stage for queries that have not yet arrived.

States and decisions of the system are index configurations and configuration changes respectively. The state at time k of the system is denoted by s_k. Configuration s_0 is some initial index configuration which is in place before the workload Q is executed. A *policy* is a selection strategy for the next configuration given the present configuration and query. It is formally defined as a function $\pi : S \times Q \rightarrow S$. Given a configuration s_{k-1} and a query q_k, a policy π determines the next configuration $s_k := \pi(s_{k-1}, q_k)$.

The following events happen in order at any time k $(0 < k \leq N)$ during the sequential decision making:

1. Query q_k arrives and is executed by the DBMS under configuration s_{k-1}.
2. A configuration $s_k \in C$ is determined by policy π based on the workload $(q_1, q_2, \dots, q_k)$ and s_{k-1}.
3. The configuration changes from s_{k-1} to s_k by materializing the indexes in s_k which do not exist yet. If no configuration change occurs at time k, $s_k = s_{k-1}$.
4. The system state is updated (non-randomly) from time k to time $k+1$.

The cost per stage is denoted by $g_k(s_{k-1}, s_k, q_k)$, and consists of three parts: (1) the cost of executing query q_k when optimized under configuration s_{k-1}, (2) the cost of index creation and dropping in transitioning from configuration s_{k-1} to s_k, and (3) the recommendation cost for s_k. Note that the system is deterministic as each state s_k is determined solely by its preceding state and the decision made at time $k-1$. On the other hand, costs are stochastic, as they depend on the query which is a random variable. Given an initial state x_0 and a policy π, the expected cost of π starting at s_0 is

$$J_\pi(s_0) = \sum_{k=1}^{N} \mathbb{E}(g_k(s_{k-1}, s_k, q_k)) \tag{1}$$

where $s_k = \pi(s_{k-1}, q_k)$.

An optimal policy π^* is one that minimizes this cost; i.e.,

$$J_{\pi^*}(s_0) = \min_{\pi \in \Pi} J_\pi(s_0) \tag{2}$$

where Π is the set of all policies.

The goal of online index selection is to find an optimal policy π^* that generates an optimal configuration sequence $(s_0, s_1, \dots, s_N)$, thereby obtaining the optimal value $J^*(s_0) = J_{\pi^*}(s_0)$.

Main Challenges. This problem is challenging for several reasons. First, the probability distribution of queries is not known a priori, making it impossible to compute the expectation of query execution cost (and hence index benefit). Even with a complete knowledge of the workload trace, it is practically impossible to do the exact computation for Eq. (1) because of an exponential explosion of the number of states (i.e., configurations) as time progresses. We therefore have to resort to approximation techniques to find suboptimal solutions for the online tuning. Second, the tuning process runs concurrently with normal query processing, where an online tuner has to continuously monitor changes in workloads and adapt the index configuration during different workload phases, we need to ensure that it has very low overhead. Finally, to be effective, an online tuner has to balance the cost of transitioning between configurations and the potential benefits of such design changes for future queries.

3 LiteSelect: An Lightweight Online Index Tuner

To address the aforementioned computational challenges for online index selection, we design a learning-based algorithm. The algorithm is called *LiteSelect* and is presented in Algorithm 1. *LiteSelect* is adaptive to time-evolving workloads with controllable overheads while resilient to noise (one-off queries) in workloads. One salient feature of *LiteSelect* is that its runtime behavior can be easily and flexibly tuned through a couple of knobs it exposes. It can also be integrated to any DBMS to perform the continuous index tuning.

3.1 Algorithm *LiteSelect*

Algorithm Overview. Recall that the goal of online index selection is to find a sequence of configurations that minimizes the end-to-end workload cost, or equivalently, maximizes the end-to-end workload cost reduction. This amounts to find an optimal policy which determines, at each decision stage, a potentially beneficial configuration for future queries. Here, the *benefit* of n index configuration for a query is defined as the cost reduction for the query using the configuration. The core issue is how to adjust the configuration for each incoming query, without knowing the remaining sequence of queries in the workload.

To tackle that issue, *LiteSelect* adopts an online learning approach. It makes a decision based on the history of observations on index benefits and updates its estimates of configuration utility overtime as new queries arrive. Its decision making policy considers balancing the following two trade-offs. One is between the query execution cost and the configuration transition cost. Note that it is not enough for the online approach to select the best configuration for each query at present, since the configuration transition cost also contributes to the overall online tuning cost. The other trade-off is between the exploration of new configurations and the exploitation of promising configurations already known.

For the policy implementation, *LiteSelect* makes two important approximations. First, it approximates the benefit of a configuration for a query by aggregating the benefits of the individual indexes in the configuration for that query. Second, it approximates the mean benefit of an index for the future queries with the exponentially weighted moving average of benefit on the query history.

Query Cost and Index Benefit. To get the execution cost $g_{exe}(q, s)$ for query q under configuration s, we invoke the query optimizer to produce a *what-if* plan

[4], which is the query execution plan under an index configuration without actually building the indexes.

The benefit $b(q, x, s)$ of index x for query q under index configuration s is then measured as the difference in the query cost between the configuration that does not contain x and the one that contains it. More formally, $b(q, x, s) = g_{exe}(q, s - \{x\}) - g_{exe}(q, s \cup x)$. The benefit of x for a workload w under configuration s is defined as the sum of the benefits of x for the queries in w under s.

Adaptive Learning of Expectation of Index Benefit. Let $b_k(x)$ denote the benefit of index x for query q_k at stage k, defined as $b_k(x) = b(q_k, x, s_{k-1})$. Let also $B_k(x)$ denote the estimate of the mean benefit of x after k queries are seen. After k steps, the sequence $(s_0, q_1, b_1, s_1, q_2, b_2, \ldots, s_{k-1}, q_k, b_k)$ of states and benefits is obtained. We use this data to compute $B_k(x)$ for each index x. This estimate is used to make a decision for s_k. The new query q_{k+1} is evaluated using s_k at which point we observe the new benefit $b_{k+1}(x)$ of each index x for q_{k+1}. We need to use the prior estimate $B_k(x)$ along with the new information $b_{k+1}(x)$ to produce the new estimate $B_{k+1}(x)$. We apply *exponential smoothing* [8], a popular method for adaptive learning, to estimate B_{k+1} as follows:

$$B_{k+1}(x) = \lambda_k \times B_k(x) + (1 - \lambda_k) \times b_{k+1}(x) \tag{3}$$

where λ_k $(0 \leq \lambda_k \leq 1)$ is the smoothing factor that determines how much the current observation affects the current estimation. Factor λ_k controls the weight of benefit that the index can get from past queries in the cumulative benefit statistics. Intuitively, a less recently made estimation should have less impact on measuring the index benefit for future use. Following a widely used strategy, we set λ_k to a constant chosen in advance. Designing an adaptive smoothing factor which depends on the observation b_k is left as future work. The cumulative benefit of configuration s_k is defined as $B_k(s_k) = \sum_{x \in s_k} B_k(x)$.

Putting Everything Together. *LiteSelect* consists of a cycle of three phases: (a) profiling, (b) forecasting, and (b) reaction. In the phase of profiling, the algorithm gathers performance statistics of candidate indexes for the past queries and computes a hypothetical benefit for index candidates on the current query. These statistics are maintained and adjusted dynamically. Then, in the forecasting phase, the algorithm predicts the benefit of candidate indexes for the future queries based on the gathered performance statistics. In the reaction phase, the algorithm uses a greedy strategy to recommend the best candidate configuration for materialization. This iterative process which alternates among profiling, forecasting and reaction enables the algorithm to continuously track the dynamic workload trace and adapt the index configuration accordingly.

As shown in Algorithm 1, *LiteSelect* performs the index profiling in lines 2–5. In order to maintain statistical information about indexes, the set of all materialized and hypothetical indexes considered so far is collected in the index catalog I. At stage k, *LiteSelect* computes, for each index $x \in I$ relevant to query q_k, its benefit b under configuration s_{k-1} (line 3). An index is *relevant* to a query if it is derived from the WHERE, GROUP BY, ORDER BY, or SELECT clause

Algorithm 1 *LiteSelect*

Input: $Q = (w_1, w_2, ..., w_N)$: query workload, I: index catalog, λ: smoothing factor, β: the transition threshold, K: maximum number of indexes in a configuration, s_0: an initial index configuration;

Output: A configuration sequence $(s_0, s_1, \ldots, s_N)$;

1: **for** time $k = 1, 2, \ldots, N$ **do**
2:    **for** each index $x \in I$ **do**
3:       $b \leftarrow \sum_{q \in w_k} b(q, x, s_{k-1})$;
4:       $B_k(x) \leftarrow \lambda \times B_{k-1}(x) + (1 - \lambda) \times b$;
5:    **end for**
6:    $s_k \leftarrow$ the top-K most beneficial indexes ranked by their current B_k values;
7:    $B \leftarrow \sum_{x \in s_k}(B_k(x)) - \sum_{x \in s_k - s_{k-1}}(g_{mat}(x))$;
8:    $B' \leftarrow \sum_{x \in s_{k-1}}(B_k(x))$;
9:    **if** $B/B' < \beta$ **then**
10:       $s_k \leftarrow s_{k-1}$;
11:    **end if**
12:    Append s_k to the sequence;
13: **end for**

of the query. Benefit b is set to 0 if x is irrelevant to q_k. Subsequently, at the forecasting phase, *LiteSelect* updates $B_k(x)$ using Eq. 3 (line 4).

Following the forecasting phase is the phase of reaction (lines 6–12), which decides on a new configuration. In this phase, a greedy strategy is applied to select the top-K indexes based on B_k in order to form the new configuration s_k (line 6). The materialization of a new index configuration implies creating new indexes and dropping those that were found less beneficial. The transition cost is then measured as the cost $g_{mat}(x)$ of materializing the new indexes and is subtracted from the cumulative benefit of configuration (line 7).

To avoid frequently repeated alternations of creation and drop of the same indexes, we replace the current configuration s_{k-1}, only if the ratio between the benefit of the newly recommended configuration s_k (after discounting the transition cost) and the benefit of the current configuration s_{k-1} is above a threshold β (lines 9–11). Parameter β can be tuned to control the balance between the workload execution cost and the configuration transition cost.

Complexity. At each stage, for each index x, the computation of b involves at most two calls to the optimiser, one for the case where x is materialized and one where it is not (line 3). Updating $B_k(x)$ takes $O(1)$ (line 4), and getting the top-K elements from B_k takes $O(|I| \times log|I|)$ (line 6). Since K is a constant, the cost of determining a configuration at each stage is bounded by $O(m \times |I| \times log|I|)$, where m denotes the cost for the physical plan analysis by the optimizer. The space complexity is $O(|I|)$.

3.2 Fine Tuning *LiteSelect*

We now discuss a couple of refinements and optimizations for *LiteSelect*.

Budget What-If Calls. The exact computation of the index benefit would require a prohibitive cost in terms of what-if calls to the query optimizer to produce query plans and estimate the query cost. To obtain reasonable estimates of index benefit at moderate cost, we constrain the number of what-if calls per query that is allowed to be made during the profiling phase. Specifically, when a new query q_k arrives, we first use the recorded benefit B_k value to select the most promising candidate indexes relevant to q_k, the number of which is determined by the parameter α (a constant factor between 0 and 1) times the total number of indexes in the catalog I relevant to q_k. Only for these indexes, *LiteSelect* uses what-if optimization calls to compute their benefit for q_k under s_{k-1}. For the rest of the relevant indexes in I, *LiteSelect* approximates their benefits for q_k using directly the recorded B_k values. The parameter α can be adjusted to optimize the trade-off between exploration and exploitation.

Profiling Restart. Finding a promising index configuration to optimize expensive queries is critical in online index tuning. *LiteSelect* can benefit from a "restart" mechanism for this issue: each time the dbms executes a new query, the query execution time is recorded by *LiteSelect*. If that time exceeds a threshold, *LiteSelect* will restart the profiling process for gathering performance statistics. The "restart" process puts *LiteSelect* to its original state by resetting the recorded cumulative index benefits. This "restart" mechanism clears the caches of performance statistics so as to increase the chances of the algorithm to find more promising configurations while exploring the search space.

4 Experimental Evaluation

We conduct extensive performance studies to evaluate the effectiveness and efficiency of our proposed online index selection approach.

4.1 Experimental Setup

Algorithms in Comparison. We evaluate the performance of our approach, abbreviated as *Lite*, with the following three index selection algorithms. The first is a state-of-the-art offline tuning algorithm called *Anytime* [3]. We used the *Anytime* source code from the offline index selection algorithm evaluation platform[1]. The other two are recently developed learning based online index selection algorithms: *rCOREIL* [1] and *MAB* [7]. We used their respective source code from GitHub: *rCOREIL*[2] and *MAB*[3].

We also compare with Algorithm *NoIndex*, which executes the workload under a baseline system configuration containing only primary and foreign key indexes that are created automatically. This baseline is chosen in order to gauge the maximum benefit of a particular index tuning algorithm.

[1] https://github.com/hyrise/index_selection_evaluation.

[2] https://github.com/Debabrota-Basu/rCOREIL-Learning-to-Tune-Databases.

[3] https://github.com/malingaperera/DBABandits.

Because of lack of space, results of empirical comparison with earlier and non-learning algorithms [2,6,9] will appear in the full version of the paper.

Workloads. We use the decision support benchmark TPC-H[4] with a scale factor 1, and the on-line transaction processing benchmark TPC-C[5] with five warehouses. Table 1 summarizes the properties of the benchmark datasets used in experiments. We also ran experiments on two other benchmarks. As the trends on these two benchmarks are similar, we omitted the details to conserve space.

Table 1. Key statistics of the benchmark datasets used in the experiment.

Benchmark	# tables	# records	# indexable columns	# query templates	size (MB)
TPC-H	8	8661245	43	22	1475
TPC-C	9	2592861	70	16	538

To compare different aspects of the tuning process, we design three types of workloads on TPC-H:

- *Shifting.* We divide query templates in the benchmark into 4 equal-sized groups. A template group is invoked for a total of 20 rounds, each round with different query instances of the templates in the group, after which the workload shifts to a new group of unseen queries, making a total 80 rounds in a shifting workload.
- *Noisy.* We consider *noise* to be any query that does not reflect the dominant traits of the current query distribution. We generate a noisy workload by injecting noises to a shifting workload. When a workload shifts, we inject four rounds of noises, making a total 96 rounds in a noisy workload. The number of noises is 10% of the total workload. This experiment evaluates a tuning algorithm's ability to resist unwanted transition oscillation.
- *Random.* We generate multiple instances of query templates in the benchmark and choose query instances randomly to make a random workload. The query instances are then divided into 25 equal-sized rounds. Random workloads are used to model more complex and dynamic settings and test the ability of a tuning algorithm to balance between shift and careful adaptation under returning workloads.

Parameter Setting. In *Lite*, the default value for smoothing coefficient λ, transition threshold β, and what-if call ratio α are set to 0.65, 1.1, and 0.5, respectively. We empirically set K, the maximum number of indexes in a configuration, to 10, and restricted the index width to 1 across all the reported experiments.

[4] https://www.tpc.org/tpch.
[5] https://www.tpc.org/tpcc.

In *Anytime*, the maximum number of indexes in a configuration is constrained to 10, and the maximum index width is set to its default value 3. *MAB* does not restrict the width of recommended indexes. The memory budget in *MAB* is set to 2G for the *shifting* workload, and 2.5G for both the *noisy* and *random* workloads under the TPC-H benchmark dataset.

Metrics. We report total end-to-end workload time for all rounds of different algorithms. To demonstrate the convergence of these algorithms, we compare also their total workload time per round. We present also the total workload time broken down by recommendation time, transition time, and workload execution time. We stop the execution of a query if it does not complete within 30 s, so that the experiments can finish in a reasonable amount of time. We record the elapsed time of these stopped queries as 30 s.

All experiments are performed on a server equipped with 2 Intel(R) Xeon(R) Gold 6240 CPUs with 18 cores and 756 GB RAM running Ubuntu 18.04.6. Unless otherwise specified, algorithms are evaluated over the PostgreSQL 12.0.

4.2 Parameter Impact Analysis

In this experiment, we investigate the impact of the three parameters of *Lite*, including smoothing coefficient λ, the what-if call ratio α, and the transition threshold β, on its index tuning performance. We generate parameter configurations for the study as follows: we varied λ from 0 to 1 in increments of 0.15. For each of the λ values, we varied β within $\{0.8, 0.9, 1.0, 1.1, 1.2, 1.5\}$, and α from 0.25 to 1 in increments of 0.15. Hence we generated 240 parameter configurations for each of the three workloads *shifting, noisy*, and *random* on TPC-H.

Impact of Parameter λ. Parameter λ controls the relative importance of the benefit of an index for the current query and its cumulated benefit for the past queries. A smaller λ discounts older data faster thereby placing greater relevance on the more current data, and vice versa. The impact of λ on the workload execution time of the TPC-H *shifting, noisy*, and *random* workloads is reported in Fig. 1(a), (b), and (c) respectively. The box chart for each λ summarizes the minimum, first quartile, median, third quartile and maximum workload execution time, as well as the mean and outliers, over different values of α and β.

As can be seen in Fig. 1, for all three workloads, after λ reaches 0.45, the variance of the workload execution time becomes smaller; it increases slightly when λ becomes larger than 0.75. The mean of the workload execution time converges once λ reaches 0.55. This demonstrates that when estimating the benefit of an index for future queries, it is more effective to balance the consideration of its benefit for past queries and for the current queries.

Impact of Parameter α. Parameter α determines the number of what-if optimization calls for computing the benefit of candidate indexes relevant to the incoming query (or queries). A larger α implies more accurate benefit estimation for a larger number of indexes relevant to the queries on hand, but on the other hand it also increases the computational overhead, i.e., the recommendation cost. Also, a larger α tends to encourage the configuration transition, henceforth increasing the transition time. This is confirmed by Fig. 2, which shows, for a fixed λ, the average workload execution, recommendation, transition, and total running time of *Lite* on the TPC-H *shifting*, *noisy*, and *random* workloads varying α, over different values of β. As we can see, the recommendation time grows linearly when α increase from 0.25 to 1. The transition time increases, albeit very slowly as α increases.

Impact of Parameter β. The transition threshold β determines whether the current configuration should be changed to the newly recommended one. A larger β makes it more difficult to change a configuration. In contrast, a smaller β encourages the configuration transition and makes it more likely that future queries will benefit when the workload shifts; however, it leads to a larger transition cost as a penalty.

In Fig. 3, we show the average workload execution, transition, and total running time of *Lite* on the TPC-H *shifting*, *noisy*, and *random* workloads varying β, when λ is set to 0.65, over different α values. Similar to α, we did not see a direct influence of β on the workload execution time, but we can see that the transition time decreases when β increases gradually.

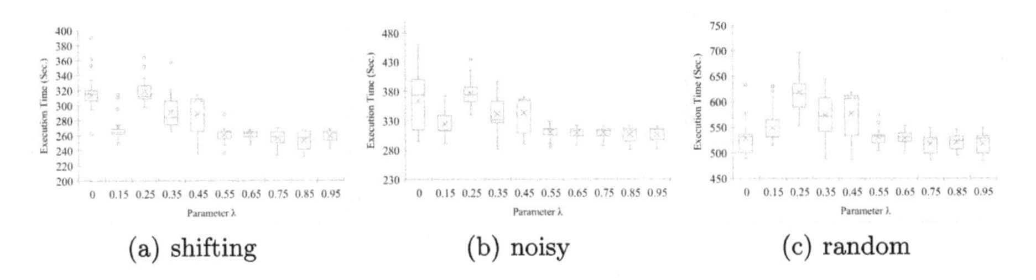

Fig. 1. Box plots of the execution time of the three types of TPC-H workloads on configurations recommended by *Lite* varying λ. Each box plot summarizes the distribution of workload execution time varying α and β for a fixed λ.

4.3 Index Tuning Performance Comparison

We report on the performance comparison of *Lite* with *Anytime* and *MAB* on the *shifting*, *noisy*, and *random* workloads under the TPC-H benchmark. The number of queries per round on these workloads are 5, 5, and 21, respectively. For each phase of the *shifting* and *noisy* workloads, *Anytime* is invoked once after the first round of queries, with those queries given as the representative of future rounds of the phase. On a random workload, *Anytime* is invoked every 4 rounds with queries from the last 4 rounds as the representative workload. This setting gives advantages to *Anytime* in the experiments. An online algorithm does

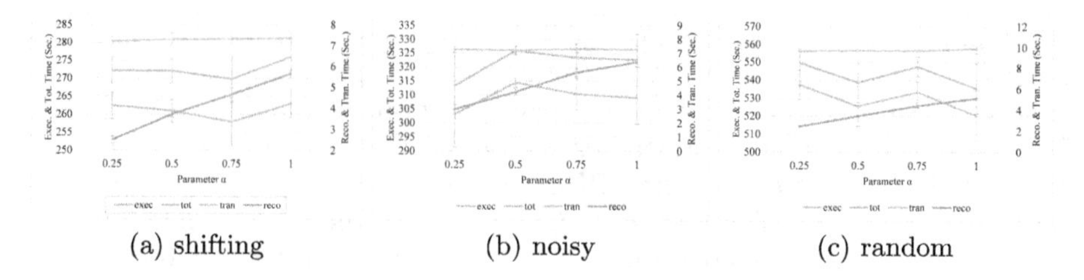

(a) shifting (b) noisy (c) random

Fig. 2. Average workload execution, recommendation, transition, and total running time of *Lite* on the TPC-H *shifting*, *noisy*, and *random* workloads varying α, for $\lambda = 0.65$ and β ranging in [0.8, 1.5].

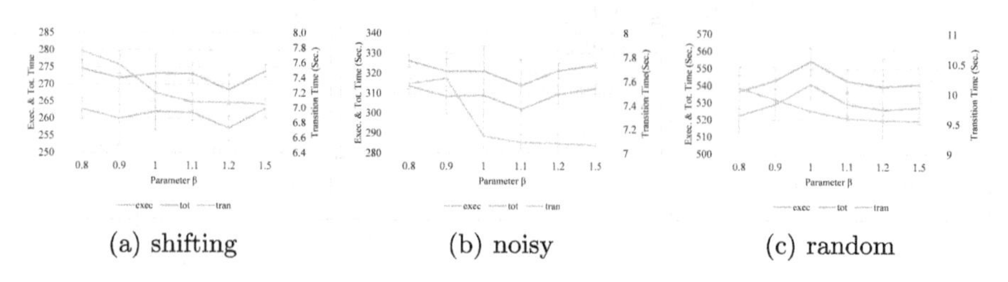

(a) shifting (b) noisy (c) random

Fig. 3. Average workload execution, transition, and total running time of *Lite* on the TPC-H *shifting*, *noisy*, and *random* workloads varying β, for $\lambda = 0.65$ and α ranging in [0.25, 1].

not use any workload information ahead of time, but instead observes workload sequence and reacts accordingly.

Tuning Performance on the *Shifting* Workload. A *shifting* workload consists of 4 phases, where the region of query interest shifts over time from one phrase to another. Each phase has 20 rounds of different query instances of a group of query templates. This experiment examines the ability of an online tuning algorithm to handle shifting workloads by automatically adapting the index configuration during different workload phases.

Lite has the best tuning performance among the algorithms in comparison. As shown in Fig. 5, the total workload execution time and the overall running time of *Lite* on the *shifting* workload are both substantially lower than *MAB*, providing about 3× and 3.4× speed-up, respectively. In particular, the total transition time of *MAB* is 21× larger than *Lite*. *Lite* outperforms *Anytime* by about 20% both in the workload execution time and the overall running time.

To obtain further insight from these data, we study the evolution of the efficiency (total time per round) of the algorithms from the beginning (i.e., round 1) of the *shifting* workload, as shown in Fig. 4(a). Both *Lite* and *Anytime* have large spikes in the first round of the workload. This is due to recommendation and creation of indexes for both algorithms. No other spikes can be seen in the *Lite* line. Further spikes can be observed for *Anytime* in rounds 41 and 61 where the workload shifts, mainly due to the invocation of the algorithm and the configuration transition. Other than those rounds, *Lite* and *Anytime* give similar performance. This demonstrates that *Anytime* has good tuning performance under a uniform, static setting where the future can be perfectly represented by a pre-determined workload. Unlike *Anytime*, *Lite* has no pre-knowledge of the workload but it automatically detects workload shifts and quickly adapts its configurations when the workload changes.

Different from *Lite*, *MAB* learns the benefits of indexes by actually creating indexes and observing the actual query execution time. The total workload time of *MAB* at some rounds is even higher than *NoIndex*, due to suboptimal indexes created (round 3) or because of a high index creation cost (rounds 22–24). In particular, *MAB* misses an index on *lineitem:l_partkey* for rounds 3 and 41–44, and another index on *lineitem:l_suppkey* from round 61 continuously till the end. This is mainly due to the error of the learned model. All these indexes are correctly detected and materialized by *Lite*. Missing those indexes leads to timeout for *MAB* on queries with templates 8, 15 and 18. The experiments illustrate the limits of the learning approach of *MAB* for online index tuning.

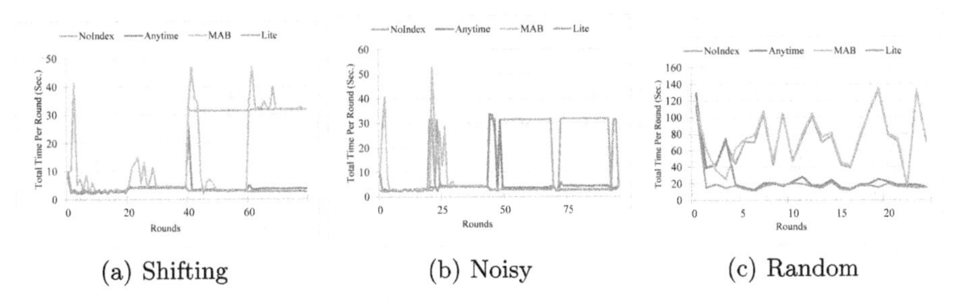

(a) Shifting (b) Noisy (c) Random

Fig. 4. Convergence comparison of *Lite* with *Anytime* on shifting workloads

Tuning Performance on the *Noisy* Workload. We generate a noisy workload by injecting four rounds of noise queries to a shifting workload when the workload shifts from one phase to the next one. Noise queries occur in concentrated bursts. Depending on the length of a burst, an on-line tuning algorithm may mistake it for a shift in the workload and change the configuration. The purpose of this experiment is to evaluate an algorithm's ability to resist unwanted transition oscillation.

As with the *shifting* workload, *Lite* has the best tuning performance with the *noisy* workload. As shown in Fig. 5, the total workload execution time and the overall running time of *Lite* on the *noisy* workload are both substantially smaller than *MAB*, providing about 26% and 79% speed-up respectively. In Fig. 4(b), we can see that *Anytime* has more large spikes on the *noisy* workload than on the *shifting* one due to injected noise (rounds 45–47 and 49). In contrast, *Lite* is resilient to short bursts of noise. *MAB* overacts to the noise resulting in large adaptive overheads (the index creation/dropping cost).

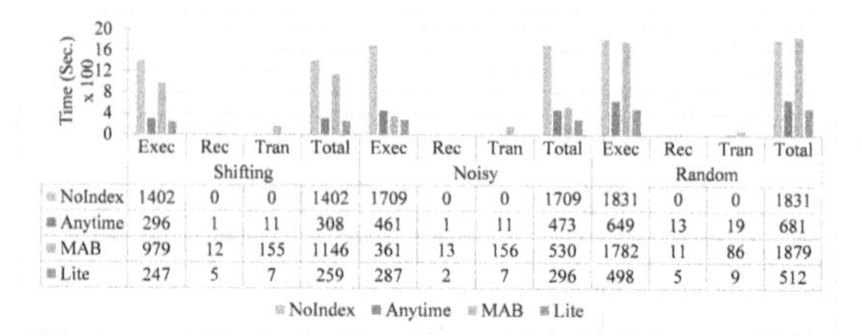

	Exec	Rec	Tran	Total	Exec	Rec	Tran	Total	Exec	Rec	Tran	Total
	Shifting				Noisy				Random			
NoIndex	1402	0	0	1402	1709	0	0	1709	1831	0	0	1831
Anytime	296	1	11	308	461	1	11	473	649	13	19	681
MAB	979	12	155	1146	361	13	156	530	1782	11	86	1879
Lite	247	5	7	259	287	2	7	296	498	5	9	512

Fig. 5. Total running time breakdown on TPC-H (Sec.)

Tuning Performance on the *Random* Workload. We simulate ad-hoc workloads using random workloads, which makes a worst scenario for an online tuning algorithm. The purpose of this experiment is to evaluate an online algorithm's ability to balance between shifts and careful adaptation under returning queries.

Lite again has the best tuning performance on the *random* workload. As we can see in Fig. 5, *Lite* greatly outperforms *MAB* by around 3.6× for both the total workload execution time and the total workload time. The workload time of *MAB* is even slightly higher than *NoIndex*. From Fig. 4(c), we see three major spikes for *Anytime* in rounds 2–4, where *Anytime* misses an index that boosts the performance of some queries. This is mainly due to the misestimates of the query optimizer. The performance of *MAB* is similar to *NoIndex* starting from round 5. This happens because *MAB* misses continuously an index needed for some queries. *MAB* is unable to detect that index due to the error of its learned cost model. In contrast, *Lite* correctly identifies and creates indexes needed for the workload.

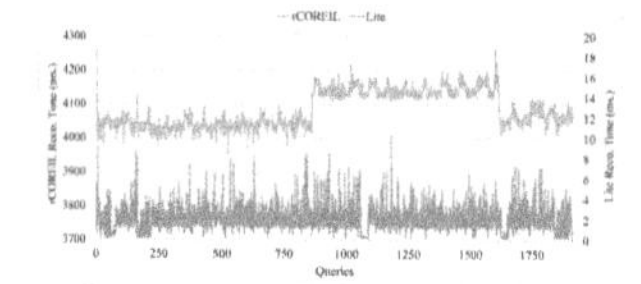

Fig. 6. Recom. time per query on TPC-C (ms.)

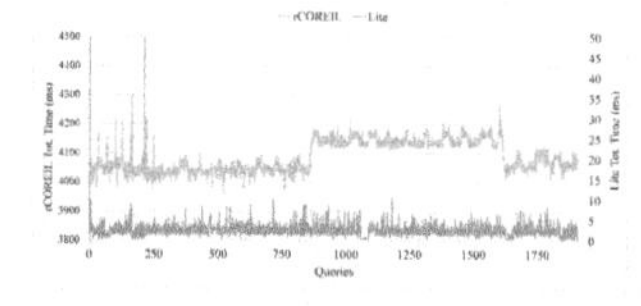

Fig. 7. Total time per query on TPC-C (ms.)

Time (Sec.)	rCOREIL	Lite
Execution	6.05	0.62
Recommendation	7801.57	4.2
Transition	1.42	0.39
Total	7809.04	5.21

Fig. 8. Total workload time breakdown on TPC-C (Sec.)

Tuning Performance on the TPC-C Workload. We compare *rCOREIL* with *Lite* on the TPC-C benchmark. *rCOREIL* runs on IBM's DB2 database (we use version 11.5 Enterprise Edition). *Lite* is evaluated over the PostgreSQL 12.0 as in the previous experiments. The workload was generated from the synthetic query set used in *rCOREIL* [1]. It consists of 1920 simple queries involving selection predicates on several attributes in a single table. We set the round size to one in this experiment as *rCOREIL* recommends configurations by query. *rCOREIL* does not constraint the configuration size, but sets the max index width to 2.

Figure 7 and Fig. 6 show respectively the evolution of the total workload time and recommendation time of the two algorithms from beginning of the workload. Figure 8 shows the total workload time breakdown of the algorithms. We can see that the recommendation of the tuning algorithm dominates the total cost. *Lite* greatly outperforms *rCOREIL* in all the performance metrics.

5 Related Work

We next give an overview of online index selection algorithms studied in the literature, while focusing on learning-based approaches proposed more recently.

Without the complete knowledge of the workload, non-learning online algorithms [2,6,9] commonly assume that queries in the near future will be similar to the queries seen from the recent past. Under this assumption, they design a heuristic policy which iteratively selects a configuration that is beneficial for the history workload. These heuristics are usually fixed, hence unlike learning-based approaches, their decision making is unable to learn from past experiences.

Two recent online algorithms *COREIL* [1] and *MAB* [7] apply machine learning techniques for index tuning in order to obviate the reliance on the query optimizer for query cost estimation. They approximate the cost function or index benefits using a linear model on features of configurations or indexes. We describe them in more detail below.

rCOREIL models the process of query execution and configuration transition as a Markov decision process. It develops an online policy iteration (PI) algorithm, which builds and updates overtime value/cost functions. To reduce the computational cost, it applies functional approximation techniques to estimate: (1) the cost of query execution and configuration transition, using linear projection to the feature space of the observed state and query; and (2) the cost-to-go function using Bellman Equations with a linear model on the feature vector of a configuration. The time complexity on building and updating value functions in *rCOREIL* is $O(d^3)$, where $d \in (0, 2^{|I|})$ and I denotes the set of all the possible indexes. This results in huge recommendation cost, as verified by our experimental results in Sect. 4.

MAB studies the online index tuning under the multi-armed bandit setting. It assumes that the benefit of each individual index is linearly dependent on a set of chosen features. It learns a linear regression model for index benefits sequentially by actually creating indexes and observing the execution time of the queries seen so far. A configuration is selected greedily based on estimated index benefits. The time complexity of the model learning is bounded by $O(d^3)$, where d is the number of candidate indexes at each stage and is bounded by $|I|$. The major problem with *MAB* is its large transition time (due to the actual index creation for learning index benefits) and its high variance (due to the instability of the learned cost model).

Unlike *rCOREIL* and *MAB*, *Lite* does not rely on sophisticated learning techniques but it exploits the simple exponential smoothing method to perform adaptive online index tuning. Our results in Sect. 4 verify its effectiveness.

6 Conclusion

We have designed an adaptive learning algorithm *LiteSelect* for online index selection framing the problem as a sequential decision problem under uncertainty. *LiteSelect* makes a decision based on the observations of index benefits for past queries and updates its estimates of index benefits for future queries overtime as new queries arrive. *LiteSelect* is lightweight, has controllable overheads, and can be tuned through a couple of parameters. An extensive experimental evaluation verifies the efficiency and effectiveness of *LiteSelect* as a promising solution for the online index selection problem and demonstrates its superiority over recent learning-based solutions using sophisticated learning techniques.

References

1. Basu, D., et al.: Regularized cost-model oblivious database tuning with reinforcement learning. Trans. Large Scale Data Knowl. Centered Syst. **28**, 96–132 (2016)
2. Bruno, N., Chaudhuri, S.: An online approach to physical design tuning. In: ICDE, pp. 826–835 (2007)
3. Chaudhuri, S., Narasayya, V.: Anytime algorithm of database tuning advisor for Microsoft SQL server (2020)
4. Chaudhuri, S., Narasayya, V.R.: Autoadmin 'what-if' index analysis utility. In: SIGMOD (1998)
5. Ding, B., Das, S., Marcus, R., Wu, W., Chaudhuri, S., Narasayya, V.R.: AI meets AI: leveraging query executions to improve index recommendations. In: SIGMOD, pp. 1241–1258 (2019)
6. Lühring, M., Sattler, K., Schmidt, K., Schallehn, E.: Autonomous management of soft indexes. In: ICDEW, pp. 450–458 (2007)
7. Perera, R.M., Oetomo, B., Rubinstein, B.I.P., Borovica-Gajic, R.: DBA bandits: self-driving index tuning under ad-hoc, analytical workloads with safety guarantees. In: ICDE, pp. 600–611 (2021)
8. Powell, W.B.: Reinforcement Learning and Stochastic Optimization: A Unified Framework for Sequential Decisions. Wiley, Hoboken (2022)
9. Schnaitter, K., Abiteboul, S., Milo, T., Polyzotis, N.: COLT: continuous on-line tuning. In: SIGMOD, pp. 793–795 (2006)
10. Schnaitter, K., Abiteboul, S., Milo, T., Polyzotis, N.: On-line index selection for shifting workloads. In: ICDEW, pp. 459–468 (2007)
11. Shi, J., Cong, G., Li, X.: Learned index benefits: machine learning based index performance estimation. Proc. VLDB Endow. **15**(13), 3950–3962 (2022)
12. Zhou, X., et al.: AutoIndex: an incremental index management system for dynamic workloads. In: ICDE, pp. 2196–2208 (2022)

IDAGEmb: An Incremental Data Alignment Based on Graph Embedding

Oumaima El Haddadi[1,2](✉), Max Chevalier[1], Bernard Dousset[1], Ahmad El Allaoui[2], Anass El Haddadi[2], and Olivier Teste[1]

[1] IRIT, SIG, Toulouse University, CNRS, Toulouse, France
{oumaima.elhaddadi,max.chevalier,bernard.dousset,olivier.teste}@irit.fr
[2] LSA, SDIC, Abdelmalek Essaadi University, Tetouan, Morocco
{elallaoui.ahmad,a.elhaddadi}@uae.ac.ma

Abstract. In the evolving digital environments, information systems are faced with a myriad of challenges such as data heterogeneity, the dynamic nature of data and integration complexities. These challenges impact on decision-making and data integration processes. We define data alignment as the process of aligning columns from different tabular sources using their schema and instances. Data alignment is emerging as an essential solution, ensuring data consistency between different sources and enabling effective integration and decision-making. However, existing solutions fail to take into account the dynamic nature of data in an incremental way. This study presents an incremental methodology that uses dynamic graph embedding techniques to progressively refine data alignments. Although the use of graph embedding techniques for data alignment is well established, their integration into incremental processing approaches remains less explored. This research attempts to fill this gap by evaluating the potential of incremental graph embedding techniques for data alignment. The adoption of this incremental technique has significantly improved the management of heterogeneous data in dynamic environments, while optimizing resource usage. Likewise, this study brings a new perspective to the field of data alignment at it aims to highlight the usefulness of dynamic embedding techniques for the exploration of dynamic datasets.

Keywords: Heterogeneous Data · Incremental Data Alignment · Dynamic Environment · Graph Embedding

1 Introduction

The evolution of data (schema and instances) within information systems requires advanced strategies for data alignment, which is vital for the integration and interoperability of heterogeneous datasets. Data alignment is no longer a static process but one that must evolve to reflect the ongoing changes within data environments. To address this complexity, our research delves into incremental data alignment, thus leveraging the potential of dynamic embedding

R. Wrembel et al. (Eds.): DaWaK 2024, LNCS 14912, pp. 19–33, 2024.
https://doi.org/10.1007/978-3-031-68323-7_2

techniques to facilitate continuous alignment adjustments. Specifically, we detail the selection of source and target tables to ensure a complete understanding of the matching process. The matching status of each source column (1:1 match, partial match or no match) is examined in detail to highlight the complexities involved.

We define data alignment as the process of aligning columns from different tabular sources using their schema and instances. Moreover, current embedding-based methods for data alignment (i.e. schema matching), such as those proposed by Cappuzzo et al. [1] and others [2–4] have shown promising results. However, the integration of an incremental perspective in these methods has not been explored. Our study aims to address this issue by evaluating how incremental embedding techniques can be applied to the incremental data alignment, thereby enhancing their adaptability and relevance.

The ability of embedding techniques to incorporate incremental changes is well recognised in some domains such as representation learning, as shown by models like Online Node2Vec and StreamNode2Vec [5,6]. Nevertheless, their application to data alignment has not been explored. Our research aims to fill this gap by investigating the adaptation of dynamic embedding methods to the requirements of evolving data alignment.

In this context, our study will focuses on several key questions that aim at clarifying the capabilities and performance of our incremental alignment method. These questions are essential to ensure that the approach is not only theoretically sound, but also practical for managing the dynamism of today's data ecosystems:

- RQ1: How does the proposed incremental alignment method compared with traditional techniques in terms of precision, recall and other key measures?
- RQ2: Can the proposed incremental approach based on graph embedding significantly reduce resource usage compared with static data alignment methods?
- RQ3: Given the potential variations in data models, how does the method guarantee consistent and reliable data alignment?

By addressing these issues, our research can extend the theoretical foundations and practical implementations of incremental data alignment. With the increasing growth of data and the dynamic nature of data, the need for incremental alignment methods becomes more crucial for data management.

This paper is organised as follows: Sect. 2 provides a background on data alignment and representation learning, highlighting the shortcomings of static alignment approaches and identifying the gaps that our research seeks to fill. In Sect. 3, we detail the methodology of the incremental approach. To do so, we attempt to explain the concept and the incremental embedding method that we adapted along with the processes steps. Section 4 presents the experimental setup, results, and discussion that follows, designed to address the research questions mentioned above. It includes a description of the datasets, metrics and measures that will be used to evaluate the performance of the incremental approach. Finally, Sect. 5 concludes with a summary of our contributions and suggestions for future research.

2 Background

The landscape of data alignment (i.e. schema matching) methodologies is rich and varied. It addressess the critical need for effective management of data and schema heterogeneity. This section explores a range of existing approaches, while highlighting the absence of incremental data alignment approaches in current literature, as outlined in Sect. 2.1 and Sect. 2.2 introduces dynamic graph embedding techniques, which we consider potential candidates for the development of incremental data alignment strategies. Finally, Sect. 2.3 provides a discussion of these topics.

2.1 Existing Data Alignment Approaches

Data alignment is a process of matching different data element, schema and instances that address the challenge of data heterogeneity. This process has traditionally been met with non-incremental alignment methods, broadly classified into schema-based, instance-based, and hybrid approaches [7]. Bernstein et al. [8] provide a comprehensive classification of these methods, highlighting their application in various contexts, from requirements-focused storage solutions like data integration and schema mapping [9] to broader applications such as data lakes [10] and ontology matching [11]. A particularly promising avenue in this field is the use of embeddings for alignment, which offers a robust framework for representing and comparing data from disparate sources [1,2]. Through the use of graphs, these methods do well in building complex connections between different data parts, greatly improving the data alignment process.

The advent of machine learning and natural language processing technologies has brought about significant advancements in data alignment methodologies. Incorporating representation-based learning, especially embedding, these modern approaches have redefined alignment strategies. Embeddings, essentially numerical vector representations of schema attributes or instances, leverage distance-based methods to compute alignments. Their applications extend beyond traditional databases to graph representations, offering a nuanced approach to data alignment [4,12].

2.2 Graph Embedding in Representation Learning

Exploring the domain of graph embedding has revealed several methods tailored for incremental embedding in temporal graphs. Each method addresses different aspects of dynamic graph analysis [13]. For instance, Online-Node2Vec [5] innovatively updates dynamic network representations in real-time. Liu's approach [6] involves generating embeddings for new nodes and revising the embeddings of nodes influenced by these additions. The FLDNE Framework [14] focuses on evolving networks, employing a combination function and alignment mechanism to adapt standard embedding techniques across various time steps.

These methods typically begin with an initialization phase using conventional embedding techniques. One of the main challenges they face is updating dynamic node representations (e.g. when new nodes are added). Most of these methods give priority to node additions, leaving a significant gap in the ability to handle node deletions and modifications in temporal networks. This gap highlights the need for more comprehensive solutions capable of handling fully dynamic sources (adding, deleting or modifying a schema element or instance in the source).

2.3 Discussion

Despite the absence of incremental data alignment to manage data evolution, dynamic embedding methods have shown potential. Even though they were not initially designed for data alignment, their ability to adapt to dynamic environments makes them promising candidates for application in this field. Furthermore, incremental graph embedding and representation learning techniques have demonstrated their efficacy in capturing the evolution of relationships and structural changes within complex data constructs [5, 13].

Adapting these methodologies for data alignment offers an interesting way of developing solutions that can accommodate real-time schema modifications, thereby improving the flexibility and efficiency of data alignment processes in dynamic environment.

3 Methodology

This section describes the methodology of the incremental approach, IDAGEmb. Section 3.1 introduces the concept, while Sect. 3.2 details the preliminaries and the algorithms implemented.

3.1 Research Design

Incremental data alignment is emerging as an essential solution for managing heterogeneous and dynamic data, as it avoids the need to recompute alignments from scratch. This approach is designed to manage data changes, ensuring that the alignment process remains both efficient and adaptive.

The changes taken into account fall into three main categories: the addition, modification, and deletion of 1) schema elements, 2) the entire schema itself, and 3) the instances within the data sources. For example, modifications have been planned to cover both minor changes (e.g. the removal of vowels from attribute names) and major changes (e.g. the complete encoding of attribute names). By considering these diverse evolutions, we aim to ensure that the proposed approach adequately addresses the dynamic nature of data sources and captures any changes that may impact the alignments. It is essential to account for these evolutions to maintain the efficiency and effectiveness of the alignment process over time. Moreover, such incremental dimension aims at reducing the computational cost of identifying the matching in data alignment (only matching concerned by data evolution should be updated).

Building on advances in representation learning, in particular studies of dynamic graph embedding, we have developed an incremental data alignment framework inspired by the [6] method, focusing on both additions of new nodes and modifications to existing nodes. Unlike previous approaches that focused solely on adding data, our goal is broader, targeting a full range of data source evolution to ensure up-to-date alignments. This strategy is essential for maintaining accurate data alignments in changing environments, which is crucial for sectors requiring real-time data analysis, such as healthcare IT and dynamic database management, improving system efficiency and reducing redundant computations.

3.2 Preliminaries

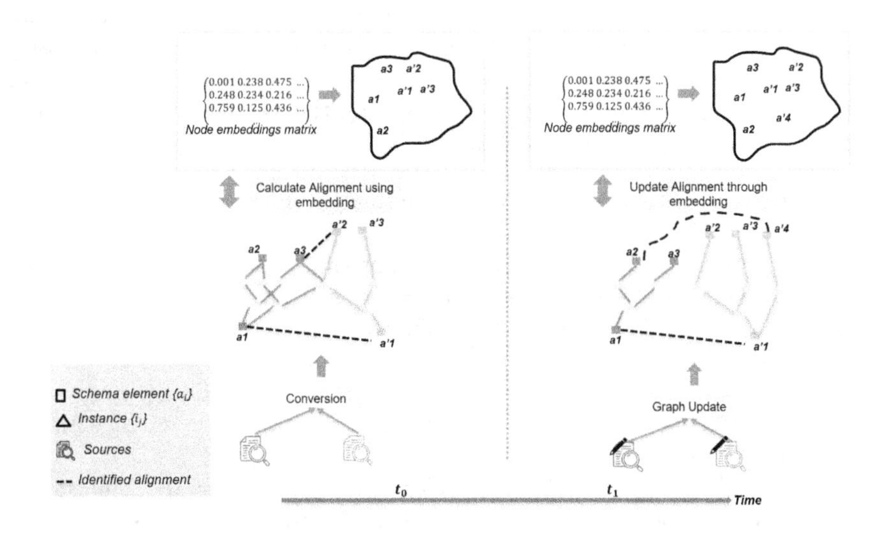

Fig. 1. Outline for the principle of Incremental Data Alignment.

To provide an overview of the process shown in Fig. 1, we begin by transforming the two data sources into a graphical representation. This allows the schema elements as well as the instances to establish a basic structure for later analysis. The process progresses with the generation of embedding for each node via graph embedding techniques, a critical step in capturing the nuanced characteristics of each node in a high-dimensional space.

Next, the similarity between nodes is determined using a vector distance measure, such as cosine similarity. This process facilitates the identification of data source columns in different graphs by quantifying the similarity of their embeddings. When changes are detected in the data sources, the strategy switches from recomputing the entire alignment to a more efficient approach. Specifically, we update the embeddings of modified nodes according to the algorithm proposed by [6], thus avoiding the process of recomputing from scratch. The foundational equation used within this algorithm is presented as follows:

$$Z_t^{(i+1)} = Z_t^{(i)} + \alpha_i \Gamma(i)[I_{k\times k} + \alpha_i^2 (\Gamma(i))^T \Gamma(i)]^{-1/2} \quad (1)$$

where:

- $Z_t^{(i)}$ is the embedding matrix at iteration i.
- α_i is the step size at iteration i.
- $\Gamma(i)$ is the search direction in the tangent space of the Stiefel manifold at iteration i.
- $I_{k\times k}$ is the identity matrix of size $k \times k$.

Each matrix embedding, as described by the set of n nodes vectors, is given by: $Z^{t_i} = [(z_{v_1})^T, z_{v_2})^T, ..., z_{v_n})^T]$.

By adopting incremental embedding methods, we can reduce computational overhead and ensure efficient updates to the alignments without the necessity to recompute the entire set of alignments. This process is depicted in Fig. 1 and elaborated through Steps 1–6 below, highlighting the seamless transition from data transformation to alignment updates.

1. Transform each data source (DS_1, DS_2) into a graph representation, denoted as $G_{t_i} = (V, E)$.
   Where t_i represents the state of the graph at time i, $V = v_1, v_2, ..., v_n$ is the a set of n nodes (vertices) and $E = \{e_{ij} = (v_i, v_j)|(v_i, v_j) \in (VXV)\}$ is a set of edges e_{ij} connecting pairs of nodes. This process involves transforming the data available at t_i in DS_1 and DS_2 into a graph that represents different types of nodes (schema nodes and instance nodes) and adding prefixes for each type to define the source node. At t_{i+1}, the graph is updated with the new state of the data sources.
2. Generate embeddings for the jth nodes (v_j) using the graph embedding method StreamNode2Vec (SN2V) [15]. Let Z_{t_i} represent the set of nodes embeddings for G_{t_i}. Where $Z_{t_i} = \{z(v_j)\}$.
3. Compute the similarity between pair of nodes' embedding by the Cosine Similarity at t_i, selected for its dimentionality independence and ability to effectively identify semantic relationships [16]:

$$\text{Similarity}\,(z(v_i), z(v_j)) = \frac{z(v_i) \cdot z(v_j)}{\|z(v_i)\|\|z(v_j)\|} \quad (2)$$

4. Compute data alignment set A_{t_i}.
5. If there are changes ΔG_i in the data sources (DS_1, DS_2), update graph, then update the embedding matrix using Eq. (1).
6. Update data alignments set $A_{t_{i+1}}$.

3.3 Adopted Algorithm for IDAGEmb

The process of the algorithm proposed by Liu et al. [6] for real-time streaming graph embedding can be described as follows:

1. **Identify Influenced Vertices:** Identify the set of vertices V_{infl} that are most influenced by the arrival of new vertices, as detailed in Algorithm 1 in [6].
2. **Generate Embeddings for each New Vertex:** For a new vertex v, generate its embedding $z(v)$ based on the linear summation of original embeddings of other vertices based on Eq. 1.
3. **Adjust Embeddings of Influenced Vertices:** Update the embeddings of vertices in V_{infl} considering the influence of the new vertex.

Algorithm 1. TRANSFORM(DS_1, DS_2, t_0)

```
1: G_0 ← TRANSFORMTOGRAPH(DS_1, DS_2)                    ▷ Step 1
2: Initialise Z_{t_0} as an empty set          ▷ Initialise embedding matrix
3: for all v ∈ V(G_0) do                    ▷ V(G_0) is the set of nodes
4:     z^{t_0}(v) ← EMBEDDING(v)                         ▷ Step 2
5:     Z_{t_0} ← Z_{t_0} ∪ [(z^{t_0}(v))^T]^T
6: end for
7: A_{t_0} ← CALCULATESIMILARITY(z^{t_i}(v), z^{t_i}(v'))   ▷ Eq. (2)
8: return A_{t_0}, Z_{t_0}, G_0
```

Inspired by this algorithm and the implementation presented in [15], our development incorporates mechanisms for creating and updating graphs, as well as their respective alignments. Algorithm 1 encompasses the first three steps of our methodology, while Algorithm 2 focuses on steps 4 to 6.

4 Experiments and Results

To evaluate the approach, we designed an experimental protocol that addresses the three research questions described in the introduction. This involves detailing the set-up implemented and the dataset used. For each research question, we develop the main objective as well as the results and discussion that follows. Before discussing the procedure and results associated with each research question, we incorporate an experiment (Experiment #1) devoted to optimizing the algorithm's hyperparameters. Section 4.1 describes the set-up and data sets used, followed by Sect. 4.2 to Sect. 4.4 which outline the objectives and present the relevant results for each experiment.

4.1 Experiment Configuration

For the three experiments we will follow the process described in Fig. 2, using the materials, datasets, and metrics/measures described below:

Algorithm 2. UPDATE(DS_1, DS_2, t, A_{t_0}, Z_{t_0}, G_0)

Initialise the set of alignment A_t
for $i = 1$ **to** $|t| - 1$ **do** ▷ Iterate over subsequent timestamps
  $G_i \leftarrow$ UPDATEGRAPH($G_{i-1}, \Delta G_i$) ▷ Step 5
  $Inf_{t_i}(u_j) \leftarrow$ DETECTINFLUENCEDNODES(G_i) ▷ Algorithm 1 [6]
  Compute embedding for new nodes
  $z^{(t_i)}(u_j) \leftarrow \frac{1}{|Inf^{(t_i)}(u_j)|} \sum_{v \in Inf^{(t_i)}(u_j)} z^{t_i-1}(v)$ ▷ Generated from Eq. (2)
  Update $z(u_j)$ based on $Inf_{t_i}(u_j)$
  $Z_{t_i} \leftarrow Z_{ti-1}$ ▷ Initialise $Z_{t[i]}$ with previous embeddings
  for all $v_j \in$ Influenced(V) **do**
    $z(v_j) \leftarrow$ UPDATEEMBEDDING($z(v_j), Inf_{t_i}(u_j)$) ▷ Algorithm 2 [6]
    $Z_{t_i} \leftarrow Z_{t_i} + z(v_j)$ ▷ Update Z_{t_i} with new embeddings
  end for
  $A_t \leftarrow$ UPDATEALIGNMENT(A_{t_0}, Z_{t_i}) ▷ Step 6
end for
return A_{t_i} ▷ Return the alignments for the last timestamp

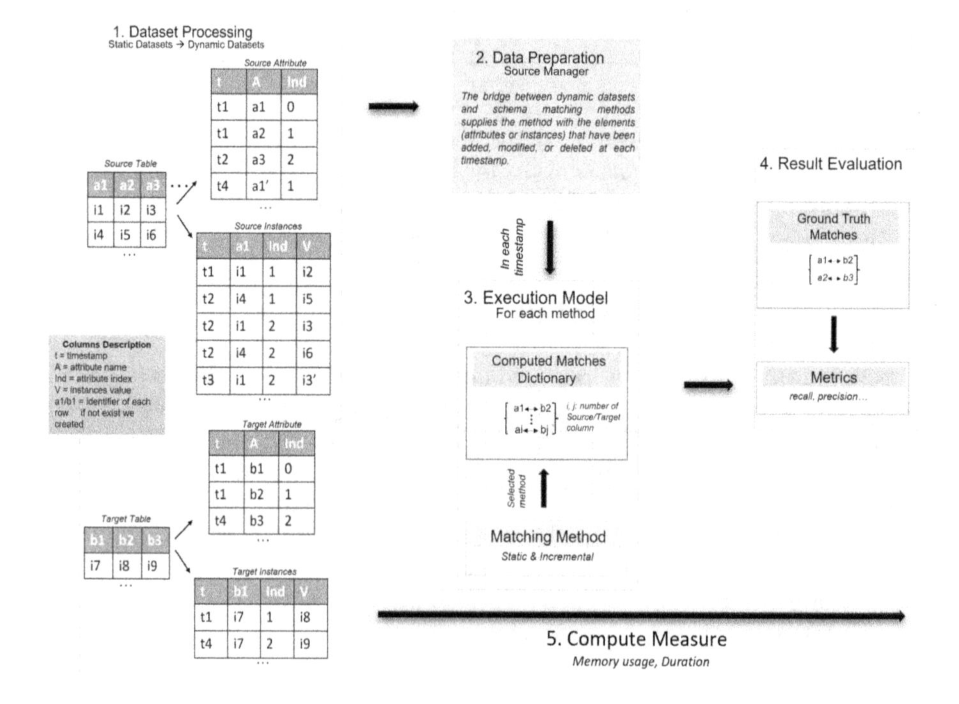

Fig. 2. Experiment Process.

Materials. The experiments will be conducted on a high-performance computing node to ensure efficient processing of complex data manipulations and model training. The node specifications are as follows: - Processor: AMD EPYC 7402 2.8 GHz dual processor. - Architecture: 48 processors. - Memory: 512 GB RAM. - Operating System: Linux Centos7.

Datasets. We have selected a set of datasets referenced in the literature [1,17] as described in Table 1. Three datasets, TPC-DI, Open Data and ChEMBL, were prepared according to the methodologies described by [17]. For each dataset, we identified four pairs of source and target tables. These datasets were selected because of their complexity and heterogeneity, characterised by differences when specific attributes in the source tables differ textually from their counterparts in the target tables. In addition, we encountered cases where attributes in the source tables matched textually those in the target tables, but differed semantically. Moreover, in some cases, only one column in each table could be accurately mapped. These tables range from 11 to 43 columns and from 7491 to 23254 rows.

Additionally, we use two raw datasets from [17] - Magellan and Wikidata. Magellan comprises 7 pairs, while Wikidata contains 4 pairs, with variations ranging from 331 to 10845 rows and 4 to 20 columns. The IMDB-Movielens dataset, sourced from [1], features a source table with 4529 rows and 11 columns, and a target table with 45346 rows and 13 columns. The later three dataset present simple heterogeneity with a significant number of rows. These tabular datasets comprise numerical, textual and noisy data.

Table 1. Datasets Characteristics.

Dataset	#Pairs	#Columns	#Rows
TPC-DI	4	11 to 22	7491 to 14982
Open Data	4	26 to 43	11627 to 23254
ChEMBL	4	14 to 20	7500
Magellan	7	4 to 9	331 to 64263
Wikidata	4	13 to 20	5422 to 10845
IMDB-Movielens	1	11 to 12	4529 to 45346

All datasets include files representing a list of true matches, as determined by researchers, which we use to compare the matches obtained from our model.

Given that these datasets are ‘static’, we processed them to simulate dynamic data. This was achieved through a splitting method that applied timestamps to distinguish schema (i.e. column attributes) from instances (i.e. rows) in each table, in order to simulate the addition, modification and deletion of data. For example, the addition, modification or deletion of schema was simulated at three splits, while the addition, modification or deletion of instances occurred at four splits. As a result, this process produced seven timestamps, assuming that the changes were not simultaneous; if they were, the number of timestamps could be lower. For each split, we selected part of the dataset. It was found that for large datasets, the starting point had to be 70% of the initial data, whereas for small datasets, it was sufficient to start with 30%. The purpose of this step (step 1 in Fig. 2) is to facilitate the detection of changes in both instances and schema.

Metrics and Measure. To evaluate the performance of the embedding methods and the overall model, the following metrics and measure will be used, focusing in particular on the F1 Score:

- For effectiveness:
  - F1 Score: To understand the balance between precision and recall in the model's performance.
  - Precision: Assessing the model's ability to correctly identify relevant matches.
  - Recall: Evaluate the model's capability to find all relevant instances.
- For efficiency:
  - Resource Usage: Evaluates the computational resources required by each method, including consumption and transformation duration and memory usage. This metric could compare the resource used in the static and incremental method.

Baseline Methods. We used static schema matching methods as a reference to compare the obtained alignments. Specifically, we selected the six methods used by [17], which fall into three categories:

- **Schema-based matching**
  - Similarity Flooding: A method that relies on graphical similarity flooding algorithms to find matches between schema elements
  - Cupid: Uses linguistic and structural analysis of schemas to match elements, taking into account both names and data types.
  - Coma_SHM: This technique, part of the COMA (Combined Approach) suite, combines several matching tools to improve the accuracy of schema matching through structural analysis.
- **Instance-based matching**
  - Coma_INS: Another variant of the COMA suite, which focuses on instances to determine matches between schemas.
  - Jaccard Levenstein: Uses a combination of Jaccard similarity and Levenshtein distance to match schema elements based on their instances
- **Hybrid matching**
  - EmbDi: An approach that integrates both schema and instance information, using embedding techniques to generate a complete match.

Our method, IDAGEmb, falls into the category of hybrid matching and is the only one that incorporates an incremental aspect.

4.2 Experiment #1: Embedding Method Selection

Objective. The main objective of this study is to identify the most efficient graph embedding technique among three distinct methodologies and to choose the best hyperparameters to generate node embedding at the first iteration. The first method considered is the node2vec (N2V) algorithm [18], which is readily available in the Python library. The second is the StreamNode2Vec (SN2V) method detailed by [15], and the third is EMBDI, a graph embedding method presented in [1]. The evaluation critically examines various hyperparameters: embedding dimension, walk length, window size, number of walks and min count. This evaluation is essential to achieve an optimal balance between the consumption of computing resources and the quality of the embedding results.

Description and Results. For this experiment, we focused on the following five hyperparameters for the three embedding graph methods:

- **Embedding dimension** = [64, 100, 128, 300]: Specifies the size of the vector space for embedding nodes, which is crucial for encapsulating the essential features of the graph.
- **Walk length** = [20, 40, 60, 80, 200]: Specifies the number of walks taken in each random walk, a parameter that influences the scope of local neighborhood exploration.
- **Window size** = [3, 5, 7, 10]: Defines the contextual window for the inclusion of neighboring nodes in the embedding process, influencing the amount of graph contextual information included in the embedding of each node.
- **Min count** = [0, 1, 2]: Defines the minimum frequency a node must have to be included in the embedding process, allowing us to filter out nodes with few occurrences.
- **Number of walk** = [10, 20, 30, 40, 100]: Indicates the number of random walks launched from each node, which plays an important role in the completeness of graph exploration.

Initially, we explored all possible combinations of hyperparameters for the three embedding methods, focusing on a relatively small, non-heterogeneous dataset to simplify the analysis. However, the EMBDI method proved particularly time-consuming, especially with higher dimensions. Therefore, we excluded the EMBDI method from further evaluations.

Subsequently, we conducted full experiments with the two remaining methods, N2V and SN2V, on three datasets of varying complexity: a simple dataset, an intermediate dataset characterized by its larger size, and a complex dataset with more heterogeneous columns. This strategic approach allowed us to evaluate the performance of both methods under diverse conditions, providing valuable insights into their applicability and effectiveness across datasets with varying characteristics and complexities. As a result, the F1 Score of SN2V is generally higher in most configurations than the F1 Score of N2V for the different datasets.

Based on these analyses, we determined that the optimal set of hyperparameters for different datasets is as follows: embedding dimension = 128, walk length = 40, window size = 10, minimum count = 0, and number of walks = 20.

4.3 Experiment #2: Comparison with Static Methods (effectiveness and Efficiency)

Objective. The main objective of this experiment is to evaluate the performance of the incremental alignment method, IDAGEmb, compared to traditional static methods. This evaluation, conducted on dynamic datasets, is designed to answer the two research questions, RQ1 and RQ2, by determining the effectiveness and efficiency of the embedding method in dynamical data environments, respectively.

Results and Evaluation. In this experiment, we aim to identify columns that are similar in different dynamic data sources. The data processing step was structured in three main phases. Firstly, we exclusively added, modified and deleted schema element (i.e. attributes column). Secondly, we simulated the addition, modification and deletion of instances (i.e. value cells). Finally, we simulated modifications encompassing both schema element and instances (e.g. modifying the numerical ages to a categorical value, modifying schema element values).

In all stages, the baseline methods relying on schema-based failed to match the correct columns when the attribute value or the instances values are modified, despite their faster performance. On the other hand, methods based on instance-based matching and those that match textually also failed to detect the correct columns when the instances were modified. However, the method that match semantically can match the columns correctly after the re-execution from scratch. Similarly, the baseline based on hybrid method, Embdi, was able to correctly match columns after re-execution of the whole process.

Effectiveness: The average F1 Score for all matchers across all datasets, as plotted in Fig. 3(a, b), reveals that our method, IDAGEmb, achieved the highest average F1 Score of 0.513, with a standard deviation of 0.283. In comparison, the Similarity Flooding method, with the second-highest mean, achieved an average F1 Score of 0.507, with a standard deviation of 0.352, while the EmbdI method recorded mean F1 Score of 0.416 with a standard deviation of 0.337. These results highlight the variability in performance between the matching techniques. Although the average effectiveness of Similarity Flooding is slightly lower than that of IDAGEmb, it shows considerable variability in the results, as suggested by its higher standard deviation. In contrast, IDAGEmb shows more consistent performance across the different scenarios, as indicated by its comparatively lower standard deviation. This consistency indicates that IDAGEmb may provide more stable performance under varying data conditions, offering a reliable, if slightly less accurate, matching solution compared to static methods. The high values of standard deviation and the variability observed in the results can be attributed to the different sizes, heterogeneity, and complexity of the datasets used. Improving and optimizing IDAGEmb could increase its accuracy and make it a competitive choice for data alignment in environments characterised by dynamic data.

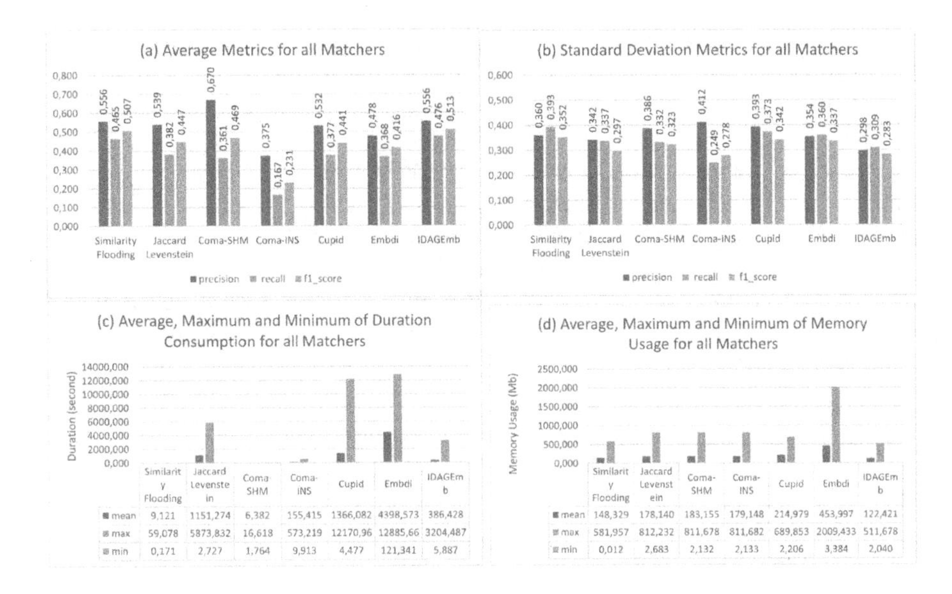

Fig. 3. Standard deviation and Mean of performance metrics (a, b) and resource usage (c, d).

Efficiency: The analysis of the consumption and transformation duration for all the matching methods, as shown in Fig. 3(c, d), shows that our method, IDAGEmb, requires $\approx$386 s on average, significantly less than the $\approx$4399 s required by the EmbDI method. EmbDI, a hybrid approach, also relies on graph embedding techniques similar to those used by IDAGEmb. In addition, the memory usage of IDAGEmb is significantly lower than that of EmbDI. These results indicate that IDAGEmb outperforms EmbDI in terms of processing speed and memory efficiency. In summary, IDAGEmb is more efficient than instance-based and graph-based methods such as EmbDI and Cupid. As a result, IDAGEmb appears to be a more appropriate option for data alignment tasks, particularly in dynamic data environments where resource optimization and fast processing are paramount.

4.4 Experiment #3: Model Sensitivity to Data Order Variation

Objective. The objective is to study the impact of changing the order of data entries. Specifically, we examine whether the order in which certain matched columns are presented at the beginning, middle or end of the alignment process affects model performance. The aim of this test is to understand the sensitivity of the model to the order in which the data is entered and to check whether the model's performance is consistent regardless of the order of the data.

Results and Evaluation. By analysing the results presented in the Fig. 4, we observed that the final F1 Scores at time #5, which represent the result of the alignment after data additions and modifications, retain a notable consistency. This consistency persists even if the datasets are randomly changed (e.g., an attribute that appears at the initial timestamp in one version of the dataset preparation may appear at the final timestamp in another). For the 'musician' dataset, the standard deviation of the final F1 Score is 0.02, implying very low variability and indicating the robustness of the alignment method. In contrast, the 'assays' dataset has a slightly higher standard deviation of 0.04. Although this indicates greater variability than the musician dataset, it still denotes a relatively stable final alignment accuracy across different dataset preparations. These observations suggest that the timing of data arrival has a negligible impact on the final alignment result.

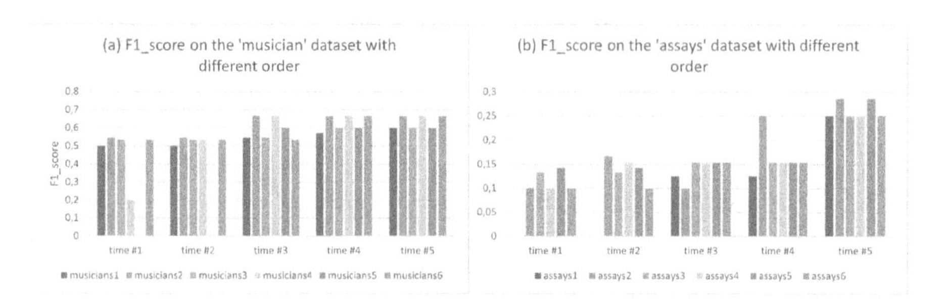

Fig. 4. F1 Score on two dataset with different order.

5 Conclusion and Outlook

This research tackles data alignment in evolving information systems, characterised by data heterogeneity, the dynamic data and complex integration processes. By introducing an incremental methodology based on dynamic graph embedding techniques, this study aims to improve data alignment progressively. It addresses the gap in applying incremental graph embedding techniques to incremental data alignement by evaluating their effectiveness (RQ1), efficiency (RQ2) and sensitivity (RQ3). The refined technique enhance the management of heterogeneous data in dynamic environments and optimizes resource consumption, offering a new perspective on data alignment through the integration of dynamicity for exploring constantly evolving data.

Outlook. In future work, we plan to enhance the process to align more complex data by incorporating external dictionaries and leveraging models that are pre-trained on other datasets, such as BERT found in Large Language Models (LLMs). In addition, we wish to evaluate the integration of incremental data alignment into broader data management processes, in order to improve the efficiency and consistency of data processing. This exploration will contribute to more effective management of complex and dynamic data environments.

References

1. Cappuzzo, R., Papotti, P., Thirumuruganathan, S.: Local embeddings for relational data integration. In: Proceedings of the 2020 ACM SIGMOD International Conference on Management of Data, pp. 1335–1349 (2020). https://doi.org/10.1145/3318464.3389742
2. Koutras, C., Fragkoulis, M., Katsifodimos, A., Lofi, C.: REMA: graph embedding-based relational schema matching. In: EDBT/ICDT Workshops (2020)
3. Rodrigues, D., da Silva, A.: A study on machine learning techniques for the schema matching network problem. J. Braz. Comput. Soc. **27**(1), 14 (2021). https://doi.org/10.1186/s13173-021-00119-5
4. Hättasch, B., Truong-Ngoc, M., Schmidt, A., Binnig, C.: It's AI match: a two-step approach for schema matching using embeddings. In: 2nd International Workshop on Applied AI for Database Systems and Applications (AIDB 2020) (2020)
5. Béres, F., Kelen, D.M., Pálovics, R., Benczúr, A.A.: Node embeddings in dynamic graphs. Appl. Netw. Sci. **2**(4), 64 (2019). https://doi.org/10.1007/s41109-019-0169-5
6. Liu, X., Hsieh, P.-C., Duffield, N., Chen, R., Xie, M., Wen, X.: Real-time streaming graph embedding through local actions. In: Companion Proceedings of The 2019 World Wide Web Conference, San Francisco USA, pp. 285–293. ACM (2019). https://doi.org/10.1145/3308560.3316585
7. Sutanta, E., Wardoyo, R., Mustofa, K., Winarko, E.: Survey: models and prototypes of schema matching. IJECE **6**(3), 1011 (2016)
8. Bernstein, P., Jayant, M., Rahm, E.: Generic schema matching, ten years later. PVLDB **4**, 695–701 (2011). https://doi.org/10.14778/3402707
9. Miller, R.J., Haas, L.M., Hernandez, M.A.: Schema mapping as query discovery. In: Very Large DataBase conference (VLDB), pp. 77–88 (2000)
10. Alserafi, A., Abelló, A., Romero, O., Calders, T.: Keeping the data lake in form: proximity mining for pre-filtering schema matching. ACM Trans. Inf. Syst. **2**(38), 3 (2020)
11. Aumueller, D., Do, H.-H., Massmann, S., Rahm, E.: Schema and ontology matching with COMA++. In: ACM International Conference on Management of Data (SIGMOD 2005), pp. 906–908. Association for Computing Machinery, New York (2005)
12. Zhao, Z., Castro Fernandez, R.: Leva: boosting machine learning performance with relational embedding data augmentation. In: Proceedings of the 2022 International Conference on Management of Data, pp. 1504–1517. ACM, Philadelphia (2022)
13. Barros, C.D., Mendonça, M.R., Vieira, A.B., Ziviani, A.: A survey on embedding dynamic graphs. ACM Comput. Surv. (CSUR) **55**, 1–37 (2021)
14. Bielak, P., Tagowski, K., Falkiewicz, M., Kajdanowicz, T., Chawla, N.V.: FILDNE: a framework for incremental learning of dynamic networks embeddings. Knowl.-Based Syst. **236** (2020). https://doi.org/10.1016/j.knosys.2021.107453
15. StreamNode2Vec. https://github.com/husterzxh/StreamNode2Vec
16. Wang, L., Luo, J., Deng, S., Guo, X.: RoCS: knowledge graph embedding based on joint cosine similarity. Electronics **13**, 147 (2024)
17. Koutras, C., Siachamis, G., Ionescu, A., Psarakis, K., et al.: Valentine: evaluating matching techniques for dataset discovery. In: IEEE 37th International Conference on Data Engineering (ICDE) (2021). https://doi.org/10.1109/ICDE51399.2021.00047
18. Node2vec python. https://github.com/eliorc/node2vec. Accessed 10 Mar 2024

Learning Paradigms and Modelling Methodologies for Digital Twins in Process Industry

Michael Mayr[1(✉)], Georgios C. Chasparis[1], and Josef Küng[2]

[1] Software Competence Center Hagenberg , Softwarepark 32a, 4232 Hagenberg, Austria
{michael.mayr,georgios.chasparis}@scch.at
[2] Institute for Application-Oriented Knowledge Processing, Johannes Kepler University, Altenberger Str. 69, 4040 Linz, Austria
jkueng@faw.jku.at

Abstract. Central to the digital transformation of the process industry are Digital Twins (DTs), virtual replicas of physical manufacturing systems that combine sensor data with sophisticated data-based or physics-based models, or a combination thereof, to tackle a variety of industrial-relevant tasks like process monitoring, predictive control or decision support. The backbone of a DT, i.e. the concrete modelling methodologies and architectural frameworks supporting these models, are complex, diverse and evolve fast, necessitating a thorough understanding of the latest state-of-the-art methods and trends to stay on top of a highly competitive market. From a research perspective, despite the high research interest in reviewing various aspects of DTs, structured literature reports specifically focusing on unravelling the utilized learning paradigms (e.g. self-supervised learning) for DT-creation in the process industry are a novel contribution in this field. This study aims to address these gaps by (1) systematically analyzing the modelling methodologies (e.g. Convolutional Neural Network, Encoder-Decoder, Hidden Markov Model) and paradigms (e.g. data-driven, physics-based, hybrid) used for DT-creation; (2) assessing the utilized learning strategies (e.g. supervised, unsupervised, self-supervised); (3) analyzing the type of modelling task (e.g. regression, classification, clustering); and (4) identifying the challenges and research gaps, as well as, discuss potential resolutions provided.

Keywords: Digital Twin · Review · Process Industry · Modelling Methods · Learning Paradigm · Self-Supervised · Transfer-Learning

1 Introduction and Motivation

The number of sensors and the corresponding data produced in the process industry is continuously increasing. This uptrend is part of the Industry 4.0 revolution, enabling a rich source of sensor data [20], which presents an unprecedented

R. Wrembel et al. (Eds.): DaWaK 2024, LNCS 14912, pp. 34–47, 2024.
https://doi.org/10.1007/978-3-031-68323-7_3

opportunity to harness complex sensor data for enhancing operational efficiency. The rapidly evolving market demands and the urgency for swift decision-making introduce considerable challenges to industrial manufacturers [37]. Furthermore, manufacturers within the European Union are obligated to comply with energy efficiency standards [1], aiming to neutralize the carbon footprint of the production facilities till 2050. An intelligently managed and regulated production is essential to remain competitive and meet efficiency goals.

The strategy of employing Digital Twins is recognized as a key enabler for this digital shift, aiming to increase competitiveness, productivity, and efficiency [18]. Rasheed et al. [28] define DT as a "virtual representation of a physical asset or process enabled through data and simulators for real-time prediction, optimization, monitoring, controlling, and improved decision-making". Recently, the evolution of DTs embraced cognitive capabilities, introducing Adaptive, Intelligent, and Cognitive DTs (e.g. [2]), showcasing their progression towards autonomy and intelligence.

Tracking concrete modelling methodologies, learning paradigms, and architectural designs is especially important for the latest concepts, i.e., cognitive or intelligent DTs, to help researchers and industrial practitioners adopt them more efficiently. Cognitive DTs aim to achieve elements of cognition, e.g. perception (i.e. abstracting meaningful data representations for subsequent processing), attention (i.e. focusing by intent or driven by signals on specific tasks, goals or data, e.g. focusing on certain aspects of the multi-dimensional and multi-modal high-volume and high-velocity data produced by Industrial Internet of Things (IIoT) devices), memory (i.e. working memory, episodic memory and semantic memory), learning (i.e. transforming insights from the physical Twin into generalizable knowledge for unseen scenarios), or reasoning [23]. Among the learning paradigms of machine learning models, transfer learning and self-supervised learning stand out as promising research directions for enabling cognitive DTs in the process industry. Transfer learning (TL) leverages knowledge, i.e. general representations, learned from pre-training on large-scale datasets and applies it to a target task with limited labelled data. Self-supervised learning does not rely on labelled data but instead learns the general representations from pre-text tasks like masked imputation, commonly using encoder-decoder-related structures (see [35]). Both learning paradigms have shown tremendous success in the natural language processing domain, where most of the success is traced back to the paper of Vaswani et al. [33]. The used learning paradigm and modelling methodology are promising research directions for DTs in general, but specifically for the process industry since the data accumulated in such industries is commonly of high volume, variety, variability and veracity, making manual labelling for specific use cases (e.g. detecting anomalies, predicting key performance indicators, classifying process states, simulating process behaviour based on different control inputs, etc.) a time-consuming task for experts and researchers, or even impossible if the modelled process does not allow for exploring the behaviour out-of-domain, i.e. experiments outside of the "normal", business-critical operation.

1.1 Research Questions (RQs)

Guided by the preliminary findings of this review, we propose several research questions aimed at furthering the understanding of modelling methodologies and learning paradigms of DTs in the industrial sector:

- **RQ 1:** What are the state-of-the-art modelling methodologies, and how has their usage evolved? Are Encoder-Decoder architectures on the rise, possibly signalling the adoption of transformer-like architectures to the industrial DT domain?
- **RQ 2:** What are the commonly utilized learning paradigms (i.e. unsupervised, supervised, self-supervised, transfer-learning, etc.)? What is the distribution of data-driven and hybrid modelling approaches? Is the self-supervised learning paradigm for generating general and transferable knowledge already explored in the industrial DT domain?
- **RQ 3:** Are DT research and application studies more focused on the evaluation of classification, clustering tasks (e.g. anomaly detection) or regression tasks (e.g. forecasting, imputation, etc.)?

1.2 Structure of Review

This review is organized into the following sections to systematically address the mentioned research questions, starting with the search strategy description, followed by the review report and detailed analysis of the selected primary studies. By structuring the review in this manner, we aim to provide a comprehensive overview of the current state of digital twin modelling methodologies and learning paradigms in the process industry, highlighting innovative practices and future opportunities for research and application.

- **Section 2 - Search Strategy:** Details the methodology used to select and analyze relevant literature, including data sources, selection criteria, and the procedure for selecting primary studies.
- **Section 3 - Reporting the Review:** Presents an overview of the studies included in the review, featuring publication trends and further synthesis modelling methodologies and learning paradigms based on selected primary studies.
- **Section 4 - Evaluating RQs on Primary Studies**: Offers a detailed analysis of the primary studies, focusing on their contributions to modelling methodologies, learning paradigms, and the tackled tasks in the context of the process industry. This section also discusses the challenges identified and potential solutions.
- **Section 5 - Discussion and Conclusion**: Synthesizes the review findings, discusses the implications for industry and academia, identifies current literature gaps, and proposes future research directions.

2 Literature Search Strategy

This review systematically searches academic and industry literature across several well-known databases, including IEEE Xplore[1], Scopus (Elsevier)[2], and ScienceDirect[3]. The search captures a broad spectrum of research focusing on DTs in the process industry w.r.t learning paradigms and modelling methodologies. The search string (see Table 1) captures all relevant papers related to DTs that mention modelling paradigms in the abstract, keywords or publication title. The selection criteria (see Table 2) ensure the selection of studies that are directly relevant to the objectives of this review.

Table 1. Screening criteria for literature search

Database	Scopus, IEEE Xplore, ScienceDirect
Search string	("digital twin*" OR "cognitive twin*" OR "cyber twin*" OR "adaptive twin*" OR "intelligent twin*") AND ("unsupervised" OR "supervised" OR "semi-supervised" OR "self-supervised")
Document type	Journal and conference papers

Table 2. Studies selection criteria

Inclusion Criteria	
I-1	Publications published as full research in relevant conferences and journals.
I-2	Scholarly literature, including books, book chapters, and technical reports.
I-3	Research focused on digital twins (DT) within the process industry.
I-4	Studies using machine learning, deep learning, other data-driven techniques, physics-based techniques, or a combination in the context of digital twins.
I-5	Contributions significantly advancing theory, methodology, or application of digital twins in the process industry.
Exclusion Criteria	
E-1	Non-full-length research articles, including abstracts, essays.
E-2	Publications lacking an accessible abstract or missing metadata.
E-3	Documents not written in English.
E-4	Studies on digital twins that omit explicit discussion of the modelling methodology and learning paradigm as well as tackled tasks.
E-5	Research does not specifically address the application of DTs in process industry.

[1] https://ieeexplore.ieee.org/.
[2] https://scopus.com/.
[3] https://sciencedirect.com/.

2.1 Quality Assessment Checks

Each research left after filtering based on the selection criteria (see Table 2) based on the abstract, keywords and title, gets checked on specific aspects of quality measurements in the full-text of the research, which are outlined in Table 3. **A score of 2** is awarded if the quality item is fully met, demonstrating comprehensive adherence to the criteria. **A score of 1** is given if the item is partially met, indicating that while there is some adherence to the criteria, certain aspects are lacking or not fully developed.**A score of 0** is assigned if the criterion is not met or the information is unavailable, reflecting a complete absence of the required information or a failure to meet the specified quality standard.

Table 3. Quality Assessment List

Quality Item	Description
Problem	Clearly defines the research problem and objectives
Context	Describes the research's industrial or practical context
Methodology	Details the research methodology, including concrete modelling methodologies
Learning	Details the learning paradigm and tasks
Architecture	Explains the architectural components
Evaluation	Evaluates the methodology on real-world or synthetic data
Limitation	Discusses study limitations and future research directions

2.2 Selection of Primary Studies

Each research left after filtering based on the selection criteria (see Table 2) is included as a primary study. All full texts, if available, of primary studies are scanned and evaluated based on the quality assessment list (see Table 3). This process ensures the fine-grained inclusion of research items that provide significant insights into modelling, learning, and architectural-related aspects of DT-creation for the process industry.

2.3 Data Synthesis and Analysis Approach

Data synthesis constitutes gathering, summarizing, and interpreting the data extracted from the primary studies. It is a crucial phase where qualitative and quantitative statistics are drawn and further analyzed. The approach focuses on synthesizing data specifically to address the pre-defined research questions (see Sect. 1.1. We will extract each primary study's data items specified in Table 4. This extracted data will be collectively analyzed to synthesize a comprehensive overview addressing each research question. For instance, we will gather and analyze all specific modelling methodologies mentioned in the studies (RQ 1) and subsequently visualize the findings. A similar approach will be applied to the other research questions, enabling us to identify patterns, trends, and areas lacking in the current literature.

Table 4. Data Items Extracted from Primary Studies

Data Item	Description	Relevant RQ
DOI	Unique identifier for research	Documentation
Title	Title of the study	Documentation
Keywords	Indexed Keywords	Demographics
Year	Year of publication	Demographics
Modelling Method	Modelling methodology used, e.g., CNN, AE, HMM, etc.	RQ 1
Learning Paradigm	Learning paradigm used, e.g., supervised, semi-supervised, unsupervised, self-supervised, etc.	RQ 2
Modelling Type	Modelling type used, e.g., data-based, physics-based, hybrid, etc.	RQ 2
Modelling Task	Model is evaluated or developed as a, e.g., regression-related model, classification-related model, clustering-related model	RQ 3
Architecture	Technology stacks, e.g., frameworks, concepts or stacks proposed to host model(s)	Documentation
Use-Case	Applied use-case, e.g. anomaly detection	Documentation

3 Reporting the Review

In this section, we report the intermediate steps of the selection process, give an overview of all the studies selected by the selection criteria, and give an overview of the primary studies after scanning the full texts (Table 5).

Table 5. Study Selection Process.

Selection Stage	Number of Studies
Studies matching the search query	326
Studies after deduplication and language filter	250
Studies after filtering (see Table 2)	40
Studies after reference snowballing	43
Studies after full-text review (see Table 3)	31

3.1 Overview of All Studies

This section presents an aggregate analysis of the studies identified through the search strategy. Critical trends over the years are highlighted in Fig. 1, illustrating the growing interest and evolution of DT-related modelling in the industrial sector.

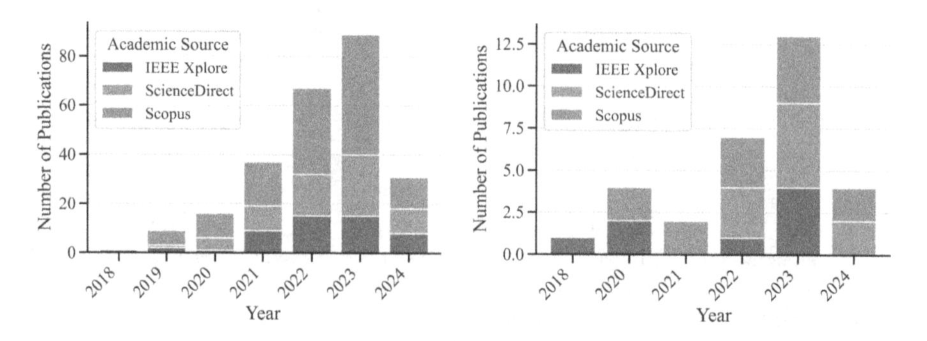

Fig. 1. (Left) The number of publications matching the search query over time. (Right) The number of publications matching the search query and passing the selection criteria over time.

3.2 Overview of All Primary Studies

The primary studies, selected based on their comprehensive coverage of digital twin technologies and their impact on industrial manufacturing, are examined in greater detail. This includes a deeper dive into the modelling methodologies, learning paradigms and architectural components discussed across these studies. The number of selected primary studies for each year divided into the different data sources is denoted in Fig. 1 on the right hand side. In addition, the proportions of the modelling task, the modelling type of the DT, and the utilized learning paradigm are visualized in Fig. 2. The concrete modelling methodologies for DT-creation are denoted in Fig. 3 as a matrix-like scatter plot over time. The size of the scatter point is based on the published research papers utilizing the corresponding methodology.

In addition, the individual extracted research items, including information on types, modelling task (MT), modelling methodology (MM), learning paradigm (LP), architecture (Arch) and experiments (Exp), are denoted in Table 6. For ensuring a compact table representation, the abbreviated terminology is used, e.g. data-based (D), hybrid (H), semi-supervised (SS), supervised (S), unsupervised (U), self-supervised (SFS), transfer learning (TL) and reinforcement learning (RL). The modelling methodologies are abbreviated as Autoencoder (AE), Long-Short-Term Memory (LSTM), Convolutional Neural Network (CNN), Physics-Informed Neural Network (PINN), Finite Element Analysis (FEA), Heat Transfer Equations (HTE), Decision Tree (DTR), Neural Network (NN), Attention Mechanism (AM), Bayesian Network (BN), Graph Neural Network (GNN), Gaussian Mixture Models (GMM), Support Vector Machine (SVM), Multi-Layer Perceptron (MLP), and Echo State Networks (ESN).

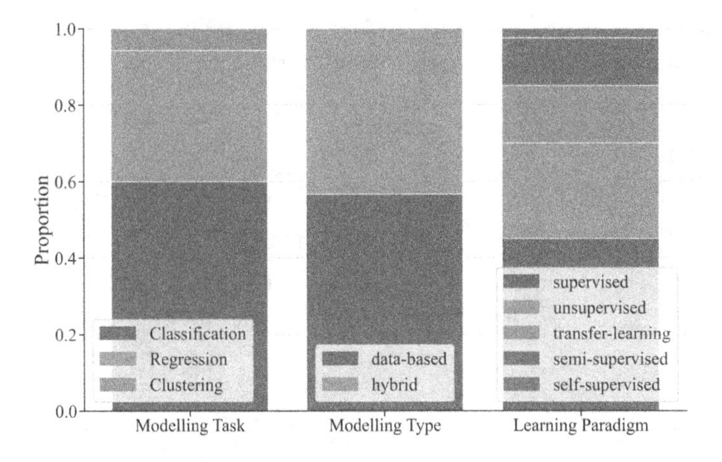

Fig. 2. Proportions of tasks, types and learning paradigms for the analyzed primary studies.

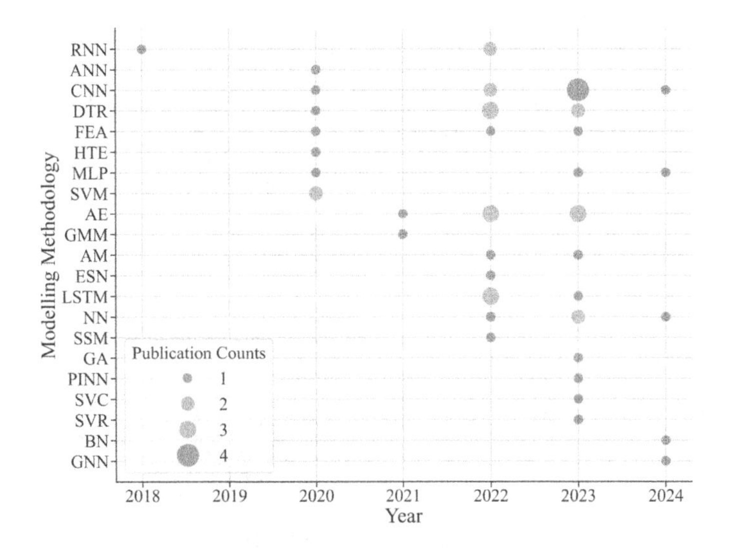

Fig. 3. The evolution of different used modelling methodologies for DTs.

Table 6. Summary of Primary Studies. Modelling Task (MT), Modelling Methodology (MM), Learning Paradigm (LP), Arch (Architecture), Exp (Experiments).

Type	MT	MM	LP	Arch	Exp	Author
D	Classification	AE, LSTM	SS	yes	yes	Lu et al. [22]
D	Classification	NN, GA	SS	no	yes	Orošnjak et al. [25]
H	Regression, Classification	RNN	S, RL	yes	no	Jaensch et al. [16]
D	Classification	CNN, AM	S	partly	yes	Li et al. [21]
D	Regression	AE	SFS, TL	no	yes	Yan et al. [35]
H	Classification	CNN	S, U	yes	yes	Yu et al. [36]
H	Regression	DTR, MLP, SVM, HTE, FEA	S, U, TL	partly	yes	Valdés et al. [32]
H	Regression	PINN	U	no	yes	Hosseini et al. [14]
H	Classification	FEA, DTR	S	partly	yes	Gawade et al. [11]
D	Classification	CNN	U, TL		yes	Sun et al. [30]
H	Classification	CNN	S, TL	partly	yes	Jauhari et al. [17]
H	Regression	SSM, RNN, ESN	S	partly	partly	Dettori et al. [9]
D	Regression, Classification	DTR, SVC, NN	S	no	yes	Chen et al. [8]
D	Clustering	AE	U	partly	yes	Cancemi et al. [6]
D	Classification	CNN, AE	S	partly	partly	Bauer et al. [4]
D	Classification	BN	SS	partly	yes	Qi et al. [27]
D	Classification	LSTM, AE	S, U	yes	yes	Hu et al. [15]
H	Regression	GNN	S	no	yes	Hernandez et al. [13]
H	-	CNN, ANN	S, TL	yes	partly	Alexopoulos et al. [3]
D	Classification, Regression	DTR, AM, LSTM	U	yes	yes	Zhang, Rui et al. [38]
D	Classification	AE	SS	partly	yes	Castellani et al. [7]
D	Classification, Clustering	DTR, NN	U, S	partly	yes	Naser [24]
H	Classification	SVM	S	partly	yes	Gaikwad et al. [10]
D	Classification	GMM	U	yes	yes	Ladj et al. [19]
H	Clustering	FEA	SS, TL	yes	yes	Xia et al. [34]
D	Classification	CNN	S	partly	yes	Tang et al. [31]
H	Regression	RNN, CNN, LSTM, AE	U	partly	yes	Gupta et al. [12]
H	Classification	CNN	S	no	yes	Parola et al. [26]
D	Regression	MLP, SVR, DTR	S	yes	yes	Boukredera et al. [5]
D	Regression	MLP, NN	S	partly	yes	Schroer et al. [29]
H	-	-	-	yes	no	Abburu et al. [2]*

4 Evaluating the Research Questions

This critical analysis of the primary studies aims to address the research questions outlined earlier. Each question is explored through the lens of the findings from these studies. This section summarizes the findings and synthesizes the evidence to provide a coherent understanding of the current landscape and future directions in digital twin technologies for industrial manufacturing.

Research Question 1: *What are the state-of-the-art modelling methodologies, and how has their usage evolved? Are Encoder-Decoder architectures and attention mechanisms on the rise, possibly signalling the adoption of transformer-like architectures to the industrial DT domain?*

The used modelling methodologies as the basis for DT-creation in the context of process industry are Convolutional Neural Network (CNN) e.g. [4,21,26,30], followed by Autoencoder (AE), e.g. [6,7,22,35]. Lue et al. [22] uses LSTMs in combination with Autoencoder for condition monitoring of magnesium furnace processes. The work combines data on electrical currents with segmented image features of the furnace flames. Li et al. [21] uses a customized CNN with deconvolution and convolution layers in combination with an attention layer on top to detect anomalies in the SWaT and WADI datasets. We have yet to see the adoption of state-of-the-art modelling methodologies of Large Language Models for the industrial DT-creation context. A notable exception in this context is Yan et al. [35], who uses a masked autoencoder that pre-trains in a self-supervised way using three novel pre-text tasks (inverse forecasting, coarser forecasting and anomaly forecasting). Results indicate robust performance and "[...] may pave the way for leveraging pre-training approaches for multivariate time-series forecasting in the context of digital twins", however, it is crucial to understand the behaviour of such pre-training strategies and the influence on various DT-related tasks [35].

Research Question 2: *What are the commonly utilized learning paradigms (i.e. unsupervised, supervised, semi-supervised, self-supervised or transfer-learning)? What is the distribution of data-driven and hybrid DT-modelling approaches? Is the self-supervised learning paradigm for generating general and transferable knowledge for various DT-related tasks already explored in the industrial DT domain?*
Most identified research items in this study use a supervised learning paradigm (e.g. [3,10,13]) to model the DT's use-case (see. Fig. 2). Unsupervised learning paradigms are commonly used in anomaly or fault detection ([6,19]). Ladj et al. use a Gaussian Mixture Model to determine the classification thresholds and also integrate a "data-knowledge closed loop system which combines a data-driven approach with a knowledge-driven approach, where expert rules are extracted and inferred in order to interpret and augment the results of data processing" [19]. The self-supervised learning paradigm, in combination with transferring knowledge to other DT-related tasks (e.g. prediction, simulation, anomaly detection), is not well-explored in the domain of DTs (notable exception is [35]). We think there is substantial potential in such learning paradigms, due to the vast availability of multi-modal, high-dimensional and high-volume IIoT-data, the abundance of labelled data, and the diverse expectations and use-case scenarios of DTs in industrial manufacturing. Around 40% of the analyzed primary studies employ a hybrid modelling approach, combining data-driven and physics-based modelling to tackle problems like small amounts of (labelled) data, or accuracy. Notable hybrid-based models are Hosseini et al. [14], who use physics-informed neural networks (PINNs) as a simulator for temperature profiles of laser powder bed fusion (LPBF) processes w.r.t. the different input process parameters and material thermal properties.

Research Question 3: *Are DT research and application studies more focused on evaluation on classification, clustering tasks (e.g. anomaly detection) or regression tasks (e.g. forecasting or imputation)?*
Given the conducted review results, it is evident that most research in the selected primary studies targets classification or clustering tasks, which are common for DT-related tasks like anomaly detection or failure identification. etc. ([6,19,21,36]). Less well represented are scientific publications in the context of DT in the process industry for regression tasks like forecasting or imputation ([5,13,29]). Hernandez et al. propose a method based on graph neural networks and encoder-decoder structures to predict the time evolution of an arbitrary dynamical system using both geometric and thermodynamic inductive biases; however, this work is limited to in-silico experiments, i.e. DT application of this methodology for industrial processes is still missing [13].

5 Discussion and Conclusion

In this study, we conduct a structured literature review to unravel the modelling methodologies, learning paradigms, and task-related evaluation aspects of Digital Twins in the context of the process industry. We first state the research questions, then formulate the search query and perform the search across various literature databases. We subsequently filter the resulting list based on includance and excludance rules and evaluate the full texts of the reduced list based on defined quality criteria to capture the relevancy of the analyzed papers for answering the specific research questions. The selected 31 primary papers apply DT-related modelling methodologies to various use cases in the industrial manufacturing domain. We synthesize the findings visually and textually to answer the research questions. We observe abundant research for self-supervised and transfer-learning paradigms (e.g. transfer knowledge to other DT-related tasks like prediction, simulation or anomaly detection). We have identified a research gap and think there is substantial potential in such learning paradigms due to the vast availability of multi-modal, high-dimensional and high-volume IIoT-data, the abundance of labelled data, and the diverse expectations and use-case scenarios of DTs in the industrial manufacturing domain.

Acknowledgments. This work received funding as part of the Trineflex project (trineflex.eu), which has received funding from the European Union's Horizon Europe research and innovation programme under Grant Agreement No 101058174. Funded by the European Union. Views and opinions expressed are however those of the author(s) only and do not necessarily reflect those of the European Union, the Austrian federal government or the federal state of Upper Austria. Neither the European Union nor the granting authority can be held responsible for them.

Disclosure of Interests. The authors have no competing interests to declare that are relevant to the content of this article.

References

1. Making the eu climate-neutral by 2050. https://ec.europa.eu/commission/presscorner/detail/en/ip_20_335
2. Abburu, S., Berre, A.J., Jacoby, M., Roman, D., Stojanovic, L., Stojanovic, N.: COGNITWIN - hybrid and cognitive digital twins for the process industry. In: 2020 IEEE International Conference on Engineering, Technology and Innovation (ICE/ITMC), pp. 1–8 (2020). https://doi.org/10.1109/ICE/ITMC49519.2020.9198403, https://ieeexplore.ieee.org/document/9198403
3. Alexopoulos, K., Nikolakis, N., Chryssolouris, G.: Digital twin-driven supervised machine learning for the development of artificial intelligence applications in manufacturing. Int. J. Comput. Integr. Manuf. **33**(5), 429–439 (2020). https://doi.org/10.1080/0951192X.2020.1747642
4. Bauer, M., Augenstein, C., Schäfer, M., Theile, O.: Artificial intelligence in laser powder bed fusion procedures - neural networks for live-detection and forecasting of printing failures. Procedia CIRP **107**, 1367–1372 (2022). https://doi.org/10.1016/j.procir.2022.05.159, https://www.sciencedirect.com/science/article/pii/S2212827122004437
5. Boukredera, F.S., Youcefi, M.R., Hadjadj, A., Ezenkwu, C.P., Vaziri, V., Aphale, S.S.: Enhancing the drilling efficiency through the application of machine learning and optimization algorithm. Eng. Appl. Artif. Intell. **126**, 107035 (2023). https://doi.org/10.1016/j.engappai.2023.107035
6. Cancemi, S.A., Lo Frano, R., Santus, C., Inoue, T.: Unsupervised anomaly detection in pressurized water reactor digital twins using autoencoder neural networks. Nucl. Eng. Des. **413**, 112502 (2023). https://doi.org/10.1016/j.nucengdes.2023.112502
7. Castellani, A., Schmitt, S., Squartini, S.: Real-world anomaly detection by using digital twin systems and weakly supervised learning. IEEE Trans. Ind. Inf. **17**(7), 4733–4742 (2021). https://doi.org/10.1109/TII.2020.3019788
8. Chen, L., et al.: Multisensor fusion-based digital twin for localized quality prediction in robotic laser-directed energy deposition. Robot. Comput. Integr. Manuf. **84**, 102581 (2023). https://doi.org/10.1016/j.rcim.2023.102581
9. Dettori, S., et al.: Optimizing integrated steelworks process off-gas distribution through economic hybrid model predictive control and echo state networks. IFAC-PapersOnLine **55**(40), 301–306 (2022). https://doi.org/10.1016/j.ifacol.2023.01.089
10. Gaikwad, A., Yavari, R., Montazeri, M., Cole, K., Bian, L., Rao, P.: Toward the digital twin of additive manufacturing: integrating thermal simulations, sensing, and analytics to detect process faults. IISE Trans. **52**(11), 1204–1217 (2020). https://doi.org/10.1080/24725854.2019.1701753, https://www.tandfonline.com/doi/full/10.1080/24725854.2019.1701753
11. Gawade, V., Singh, V., Guo, W.G.: Leveraging simulated and empirical data-driven insight to supervised-learning for porosity prediction in laser metal deposition. J. Manuf. Syst. **62**, 875–885 (2022). https://doi.org/10.1016/j.jmsy.2021.07.013, https://www.sciencedirect.com/science/article/pii/S0278612521001503
12. Gupta, R., Jaiman, R.: Three-dimensional deep learning-based reduced order model for unsteady flow dynamics with variable Reynolds number. Phys. Fluids **34**(3), 033612 (2022). https://doi.org/10.1063/5.0082741, http://arxiv.org/abs/2112.09302, arXiv:2112.09302 [physics]

13. Hernández, Q., Badías, A., Chinesta, F., Cueto, E.: Thermodynamics-informed graph neural networks. IEEE Trans. Artif. Intell. **5**(3), 967–976 (2024). https://doi.org/10.1109/TAI.2022.3179681
14. Hosseini, E., Scheel, P., Müller, O., Molinaro, R., Mishra, S.: Single-track thermal analysis of laser powder bed fusion process: parametric solution through physics-informed neural networks. Comput. Methods Appl. Mech. Eng. **410**, 116019 (2023). https://doi.org/10.1016/j.cma.2023.116019
15. Hu, W., Wang, X., Tan, K., Cai, Y.: Digital twin-enhanced predictive maintenance for indoor climate: a parallel LSTM-autoencoder failure prediction approach. Energy Build. **301**, 113738 (2023). https://doi.org/10.1016/j.enbuild.2023.113738
16. Jaensch, F., Csiszar, A., Scheifele, C., Verl, A.: Digital twins of manufacturing systems as a base for machine learning. In: 2018 25th International Conference on Mechatronics and Machine Vision in Practice (M2VIP), pp. 1–6. IEEE, Stuttgart (2018).https://doi.org/10.1109/M2VIP.2018.8600844, https://ieeexplore.ieee.org/document/8600844/
17. Jauhari, K., et al.: Modeling of deep learning applications for chatter detection in the milling process. Procedia CIRP **118**, 471–476 (2023). https://doi.org/10.1016/j.procir.2023.06.081, https://www.sciencedirect.com/science/article/pii/S2212827123003050
18. Kritzinger, W., Karner, M., Traar, G., Henjes, J., Sihn, W.: Digital twin in manufacturing: a categorical literature review and classification. IFAC-PapersOnLine **51**(11), 1016–1022 (2018). https://doi.org/10.1016/j.ifacol.2018.08.474
19. Ladj, A., Wang, Z., Meski, O., Belkadi, F., Ritou, M., Da Cunha, C.: A knowledge-based Digital Shadow for machining industry in a Digital Twin perspective. J. Manuf. Syst. **58**, 168–179 (2021). https://doi.org/10.1016/j.jmsy.2020.07.018, https://www.sciencedirect.com/science/article/pii/S027861252030128X
20. Lasi, H., Fettke, P., Kemper, H.G., Feld, T., Hoffmann, M.: Industry 4.0. Bus. Inf. Syst. Eng. **6**(4), 239–242 (2014). https://doi.org/10.1007/s12599-014-0334-4
21. Li, Z., Duan, M., Xiao, B., Yang, S.: A novel anomaly detection method for digital twin data using deconvolution operation with attention mechanism. IEEE Trans. Ind. Inf. **19**(5), 7278–7286 (2023). https://doi.org/10.1109/TII.2022.3231923
22. Lu, S., Wen, Y.: Semi-supervised condition monitoring and visualization of fused magnesium furnace. IEEE Trans. Autom. Sci. Eng. **19**(4), 3471–3482 (2022).https://doi.org/10.1109/TASE.2021.3124015, https://ieeexplore.ieee.org/abstract/document/9610130, conference Name: IEEE Transactions on Automation Science and Engineering
23. Mortlock, T., Muthirayan, D., Yu, S.Y., Khargonekar, P.P., Abdullah Al Faruque, M.: Graph learning for cognitive digital twins in manufacturing systems. IEEE Trans. Emerg. Topics Comput. **10**(1), 34–45 (2022). https://doi.org/10.1109/TETC.2021.3132251, https://ieeexplore.ieee.org/document/9642429/
24. Naser, M.Z.: Digital twin for next gen concretes: on-demand tuning of vulnerable mixtures through explainable and anomalous machine learning. Cement Concr. Compos. **132**, 104640 (2022). https://doi.org/10.1016/j.cemconcomp.2022.104640
25. Orošnjak, M., Jocanović, M., Penčić, M., Oros, D.: From signals to insights: uncovering latent degradation with deep learning as a stepping stone towards digital twins of failures (DTFs). In: 2023 7th International Conference on System Reliability and Safety (ICSRS), pp. 371–375. IEEE, Bologna, Italy (2023). https://doi.org/10.1109/ICSRS59833.2023.10381144, https://ieeexplore.ieee.org/document/10381144/

26. Parola, M., Galatolo, F.A., Torzoni, M., Cimino, M.G.C.A.: Convolutional neural networks for structural damage localization on digital twins. In: Fred, A., Sansone, C., Gusikhin, O., Madani, K. (eds.) Deep Learning Theory and Applications. CCIS, vol. 1858, pp. 78–97. Springer, Cham (2023). https://doi.org/10.1007/978-3-031-37317-6_5
27. Qi, Y., Hossain, M.S.: Semi-supervised Federated Learning for Digital Twin 6G-enabled IIoT: a Bayesian estimated approach. J. Adv. Res. (2024). https://doi.org/10.1016/j.jare.2024.02.012, https://www.sciencedirect.com/science/article/pii/S2090123224000742
28. Rasheed, A., San, O., Kvamsdal, T.: Digital twin: values, challenges and enablers from a modeling perspective. IEEE Access **8**, 21980–22012 (2020). https://doi.org/10.1109/ACCESS.2020.2970143, https://ieeexplore.ieee.org/document/8972429. conference Name: IEEE Access
29. Schroer, H.W., Just, C.L.: Feature engineering and supervised machine learning to forecast biogas production during municipal anaerobic co-digestion. ACS EST Eng. **4**(3), 660–672 (2024). https://doi.org/10.1021/acsestengg.3c00435
30. Sun, Z., Wang, Y., Chen, Z.: Fault diagnosis method for proton exchange membrane fuel cell system based on digital twin and unsupervised domain adaptive learning. Int. J. Hydrogen Energy **50**, 1207–1219 (2024). https://doi.org/10.1016/j.ijhydene.2023.10.148, https://www.sciencedirect.com/science/article/pii/S0360319923052564
31. Tang, K., et al.: Particle classification of iron ore sinter green bed mixtures by 3D X-ray microcomputed tomography and machine learning. Powder Technol. **415**, 118151 (2023). https://doi.org/10.1016/j.powtec.2022.118151
32. Valdés, J.J., Tchagang, A.B.: Deterministic numeric simulation and surrogate models with white and black machine learning methods: a case study on direct mappings. In: 2020 IEEE Symposium Series on Computational Intelligence (SSCI), pp. 2485–2494 (2020). https://doi.org/10.1109/SSCI47803.2020.9308370, https://ieeexplore.ieee.org/document/9308370?signout=success
33. Vaswani, A., et al.: Attention is all you need. In: Advances in Neural Information Processing Systems. vol. 30. Curran Associates, Inc. (2017). https://proceedings.neurips.cc/paper_files/paper/2017/hash/3f5ee243547dee91fbd053c1c4a845aa-Abstract.html
34. Xia, P., Huang, Y., Tao, Z., Liu, C., Liu, J.: A digital twin-enhanced semi-supervised framework for motor fault diagnosis based on phase-contrastive current dot pattern. Reliab. Eng. Syst. Saf. **235**, 109256 (2023). https://doi.org/10.1016/j.ress.2023.109256, https://www.sciencedirect.com/science/article/pii/S0951832023001710
35. Yan, L., Zhang, Z., Wang, X., Zhang, Y., Gu, Y.: Toward digital twin: leveraging pre-training approaches for multivariate time series forecasting, pp. 1106–1111 (2023). https://doi.org/10.1109/ICEICT57916.2023.10245025
36. Yu, X., Yang, Y., Du, M., He, Q., Peng, Z.: Dynamic model-embedded intelligent machine fault diagnosis without fault data. IEEE Trans. Ind. Inform. **19**, 1–10 (2023). https://doi.org/10.1109/TII.2023.3245677
37. Zeng, Y., Yin, Y.: Virtual and physical systems intra-referenced modelling for smart factory. Procedia CIRP **63**, 378–383 (2017). https://doi.org/10.1016/j.procir.2017.03.105, https://www.sciencedirect.com/science/article/pii/S2212827117302512
38. Zhang, R., Zeng, Z., Li, Y., Liu, J., Wang, Z.: Research on remaining useful life prediction method of rolling bearing based on digital twin. Entropy **24**(11), 1578 (2022). https://doi.org/10.3390/e24111578

Entity Matching and Similarity

MultiMatch: Low-Resource Generalized Entity Matching Using Task-Conditioned Hyperadapters in Multitask Learning

John Bosco Mugeni[1,2(✉)], Steven Lynden[2], Toshiyuki Amagasa[1], and Akiyoshi Matono[2]

[1] University of Tsukuba, Tsukuba, Japan
bosco@kde.tsukuba.ac.jp
[2] National Institute of Advanced Industrial Science and Technology (AIST), Tokyo, Japan

Abstract. Generalized Entity Matching (GEM) is a variant of entity matching that identifies whether entity descriptions from diverse data sources with heterogeneous data formats refer to the same real-world entity. State-of-the-art single-task fine-tuning approaches have shown limitations in handling scenarios with entity distribution shifts, particularly in low-resource settings, and can also require significant amounts of computationally expensive fine-tuning when applied to the GEM problem. This paper addresses these challenges by deploying task-conditioned adapters for low-resource GEM. We present MultiMatch, which explores the benefits of sharing knowledge across related tasks while improving the efficiency and accuracy of models used for GEM. Furthermore, we propose a loss composition strategy that leverages the heteroscedastic uncertainty of individual tasks to adjust the loss terms for each task before computing the overall loss. Empirically, we observe regulatory effects on the model's variance. Lastly, we analyze the carbon impact of fine-tuning different systems. Results are promising: our approach generalizes over eight GEM benchmarking tasks while reducing CO_2 emissions by 85.0%.

Keywords: Generalized Entity Matching · Data Integration · Databases

1 Introduction

Entity Matching (EM) is a task that identifies and classifies the relationship between entity pairs from diverse data sources such as product feeds, customer databases, product catalogues, etc. [4]. The main objective is to establish whether or not correspondence exists between entity pairs, categorizing them as a match or non-match [12]. This process plays a crucial role in data integration as it can be used to establish links between database entities from different data sources, enabling organizations to perform data cleaning, wrangling and other tasks to

R. Wrembel et al. (Eds.): DaWaK 2024, LNCS 14912, pp. 51–65, 2024.
https://doi.org/10.1007/978-3-031-68323-7_4

uncover previously unseen patterns and insights that would otherwise remain concealed within individual data sources.

Typically, the size and complexity of performing EM is challenging. In scenarios such as this, the complete EM workflow is decomposed into two steps (i) blocking and (ii) matching. Blocking [14] narrows down the comparison space, which in worst case scenarios would be $O(n^2)$, by grouping similar instances together, enabling efficient execution of matching. Matching is the process of analyzing pairs of instances to establish their correspondence as a binary (match/non-match) classification problem. The primary focus of the research outlined in this paper is limited to matching, yet it is noteworthy that the approach delineated herein is adaptable to a diverse array of blocking techniques that could be effectively deployed within comprehensive, real-world entity matching scenarios.

The database research community has proposed various approaches towards constructing accurate and effective matching solutions [4,10,13] and these efforts have resulted in some degree of benchmark saturation, a phenomenon observed in performance evaluation where the performance metrics or scores of different EM approaches [19,27,29] reach a plateau or peak level, especially since the emergence of powerful models such as the Transformer [25]. Consequently, EM-related research has moved on to more challenging problems, exemplified by the development of significantly more challenging datasets, such as the Machamp [27] Generalized Entity Matching (GEM) dataset and the recently revamped WDC product benchmark [19]. GEM focuses on the much more challenging problem of introducing different data formats, e.g. semi-structured data such as JSON, XML etc. and the problem of matching again different data formats such as unstructured text etc. Nevertheless, we note that most works tested against these new benchmarks still follow a single-task fine-tuning paradigm, with little or no effort in the direction of Multi-Task Learning (MTL). Current evaluations [18] indicate that single-task objective models still struggle with entity distribution shifts e.g., when the model must identify that two entities represented as 'project-based, with variable hours ranging from 10–20 hours per week' and 'casual or freelance' co-refer which can be seen as a temporal shift brought on by different naming conventions. These challenges are especially pronounced in low-resource settings (i.e., limited training data) accentuated by class imbalance. In such scenarios, we anticipate that single-task models will face considerable difficulty in robustly capturing the nuanced underlying patterns.

Additional challenges are compounded by the fact that data integration encompasses tasks from several sub-domains, such as bibliographic citations, products, and geographic references. When a Pre-trained Language Model (PrLM) must be sequentially trained for each target task, it is parameter inefficient and concerns rise about environmental sustainability within the context of Green AI initiatives [26]. Addressing the computational and resource-intensive nature of this process becomes imperative, especially considering the attention the database community has devoted to such sustainability matters [5]. This study is aimed at developing a solution that is efficient and scalable while maximizing performance in low-resource settings.

Specifically, our questions are:

- How does the carbon cost compare to the achieved gains in terms of F1 score when training a model across various settings (i.e., single and multitask fine-tuning)?
- How do we identify challenging tasks and combine multiple losses in a principled way while capturing the nuances among various *target* domains?

We investigate these research questions by incorporating hyperadapters in a multi-objective model, which has several benefits, including alleviating catastrophic forgetting and negative transfer.

Additionally, while previous works examine sample-level weighting such as entropy [21] or temperature-based [9] which may ignore inherent task complexity, we propose to weight tasks by their *heteroscedastic* uncertainty as a determinant of task complexity. This process assigns higher weights to tasks with higher uncertainty, ensuring the optimization focuses on resolving the challenges associated with those tasks. To assess the benefits of this, we carry out experiments to examine the impact of task order and the effect of ablating tasks in MTL settings as well as omitting the aforementioned weighting strategy. The key contributions of our work are as follows:

1. We propose a novel framework called MultiMatch[1], which is used to explore the application of MTL to the GEM problem. To the best of our understanding, our study is the first to investigate MTL using hyperadapters for entity matching.
2. We conduct various task ablation studies. Specifically, we investigate the impact of task position and task omission on MTL performance.
3. We analyze and report the CO_2 footprint emissions from fine-tuning different matching systems.
4. We propose a scheme to weight individual task losses using heteroscedastic uncertainty which is shown to reduce model variance in MTL settings.

2 Background

2.1 Problem Formulation

Traditional approaches tackle linkage between collections that share the same format and have aligned schemas. However, in real-world applications, this is rarely the case as data exists in various formats, such as semi-structured files including JSON and XML, in addition to unstructured data such as natural text. Furthermore, the complexity is exacerbated when different schemas are involved. Additionally, in classic EM settings, entities are identical, whereas, in real-world settings, it may be preferable to establish a general or relevant relationship. This problem of matching entities with different formats and schemas is called Generalized Entity Matching (GEM) [27].

[1] https://github.com/boscoj2008/MultiMatch.

Our problem setup aligns with the adoption of PrLMs to develop a classification model and is closely related to adapter-based GEM solutions [15]. Nevertheless, our work distinctively focuses on joint learning of several GEM tasks for simultaneous data integration while dealing with the challenging landscape of loss balancing e.g., when the loss of one or more tasks dominates the training process— a non-trival problem in MTL. Moreover, we use a hyperadapter to further achieve a compact model by sharing adapter parameters as opposed to utilizing several adapters in our model which would lead to overparametization. Finally, our problem setup investigates other boundaries such as perturbing the task order and ablating tasks from the combined task pool. These dimensions, which serve as a way of examining sensitivity/robustness of the underlying MTL model, have received less focus in previous works involving hyperadapters [8,9,24] and imperative for MTL studies building upon hyperadapters.

2.2 Entity Matching with Single-task Objective Models

The single-task objective fine-tuning paradigm has been extensively explored for building EM models, exploiting the power of fine-tuning a pre-trained model. Typically, the model is initialized with a set of pre-trained weights, Φ, and adapted to several downstream tasks sequentially. In this paper, we categorize the single-task objectives into two groups: (i) fully fine-tuning methods and (ii) parameter-efficient fine-tuning methods.

2.3 Fully Fine-tuning Methods

The workflow of fine-tuning Transformers for EM involves organizing the input as sentence pairs preceded by the [CLS] token to mark the start of a sentence and delimited/ended by a [SEP] token. Later, the [CLS] token encodes contextual information to be fed to a linear layer to learn class weights. Thus, by training the entire Transformer model including the pre-trained layers on task-specific data using this input format, the model can learn to capture complex matching patterns and achieve strong performance on EM tasks. Several works explore this paradigm such as Ditto [12], JointBERT [16], R-SupCon [17], and DADER [23]. However, a discrepancy exists between the objective forms of pre-training and fine-tuning, leading to catastrophic forgetting when the models are sequentially fine-tuned. This effect may be more pronounced in low-resource settings due to insufficient data to tune the model parameters. In addition, a new version of the model has to be stored for each task catalyzing considerable storage resource costs and makes model sharing costly.

2.4 Parameter-Efficient Fine-tuning Methods

Parameter-efficient fine-tuning (PEFT) methods such as prompt-tuning [11] and adapter-tuning [20] have emerged as promising alternatives to the expensive fully fine-tuning approach by adding fewer trainable parameters ϕ while reusing the bulk of the pre-trained weights for various tasks. This efficiency reduces the discrepancy that exists between pre-training and fine-tuning. PEFT Prompt-tuning and adapter-tuning can be applied to EM as described below.

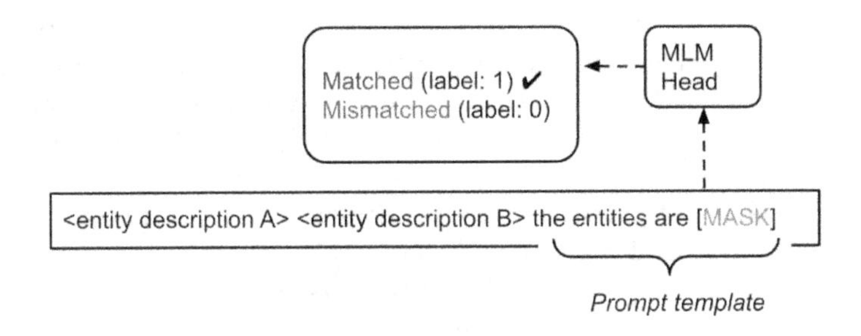

Fig. 1. Prompt-tuning method used for GEM (PEFT)

Prompt-Tuning: Prompt-tuning has been applied to EM by PromptEM [28] which re-purposes the *upstream* pre-trained parameters of an NLP model for various tasks and utilizes entity sequence pairs with additional tokens inserted in a prompt template $T(\cdot)$. During fine-tuning, the model is conditioned (Fig. 1) using the template $T(\cdot)$ for the correct match status, i.e., the correct class label Y, $P_{\Phi}(Y|T(x))$. The requirements of a template mean that prompt engineering is non-trivial, demanding domain knowledge. Furthermore, additional tokens in $T(\cdot)$ increase the input sequence length, resulting in elevated computational requirements for the attention mechanisms, which are governed by $(O(n^2 \cdot d))$ where n is the input sequence length and d is the language models dimension, e.g., 768 for RoBERTa base [30].

Adapter-Tuning: In constract to prompt-tuning, adapter-tuning involves adding new parameters ϕ between the layers of a language model while keeping the remaining parts of the model Φ fixed, where $\phi \ll \Phi$. As a result, adapters are said to be *modular*. In this context, we focus on AdapterEM [15] which examines the single-task adapters and explores various combinations of deploying them. These combinations include fine-tuning a blank task adapter and integrating it with pre-trained adapters. Given that the PrLM is parameterized by ϕ, the blank task adapter acquires domain knowledge during downstream fine-tuning on the current task. Although single-task adapters can offer efficiency and modularity, they may miss out on the potential benefits of cross-task knowledge transfer. To mitigate this, we investigate a solution that involves sharing knowledge across adapters using shared hypernetworks. This approach is explained in the next subsection.

2.5 Entity Matching with Parameter-Efficient Multi-task Models

A hypernetwork (hypernet) is a small network designed to generate weights for specific layers of a larger network [6]. Recently, researchers have explored the use of hypernets to facilitate weight generation for adapters layers [8,9,24]. In such approaches, the main network is fused with adapters and conditioned on sequence pairs for feature extraction, while the hypernet is conditioned on a task embedding, which is a noise vector generated from a normal distribution to generate task feature maps which initialize the adapter layers simultaneously. This approach allows for efficiency and weight sharing across adapter layers, enabling effective MTL. Given the promising performance of this approach, we propose a *canonical* hard-parameter MTL system with distinct heads while emphasizing modularity. This contrasts with context-dependent variants of MTL achieved in encoder-decoder architectures [8,9]. Additionally, how to mitigate competition of tasks in the loss landscape is rarely studied for such a hypernet-based system which becomes a critical objective of this work. We explain the details in Sect. 3. Next, we highlight the details of the canonical system, including the source hypernet and task-conditioned adapters.

MTL Architectures: The goal of Multi-Task Learning (MTL) is to learn all tasks simultaneously by sharing knowledge across multiple related tasks thus improving the generalization capability of the model. Modular MTL architectures add task-specific layers into the shared bulk of a Pre-trained Language Model (PrLM) followed by multiple task-specific output layers. In a modular hard-sharing architecture (Fig. 2), tasks benefit from the pre-trained layers as shared modules. The task-conditioned adapters (initialized by a source hypernet) and task-specific output layers (which tailor the model predictions to the required output of each specific task) act as task-specific modules that specialize in a particular task.

Source Hypernet and Task-Conditioned Adapters: The source hypernet is a simple two-layer network that generates the adapter parameters W_U, W_D, b_U and b_D. Let the task embedding generated from a standard normal distribution $Z \sim \mathcal{N}(0, 1)$ be ξ and φ be the parameter vector of the hypernetwork. Given ξ, the following adapter parameters are generated:

$$h = ReLU(W_0\xi + b_0) \tag{1}$$

$$W_U = W_1 h + b_1 \qquad W_D = W_2 h + b_2 \tag{2}$$

$$b_U = W_3 h + b_3 \qquad b_D = W_4 h + b_4 \tag{3}$$

These adapter parameters from the hypernet are then used to initialize the adapter layers that follow the Houlsby architecture [7], namely feedforward-up and feedforward-down (refer to Fig. 2). Given an adapter A_i at position i, the feedforward-down operation, via a hyperparameter called the reduction factor, reduces the dimensionality of the output from a Transformer block, applies a non-linearity, and passes it to the feedforward-up operation, which expands the vector back to the original dimension of the Transformer governed by Eq. 4 below:

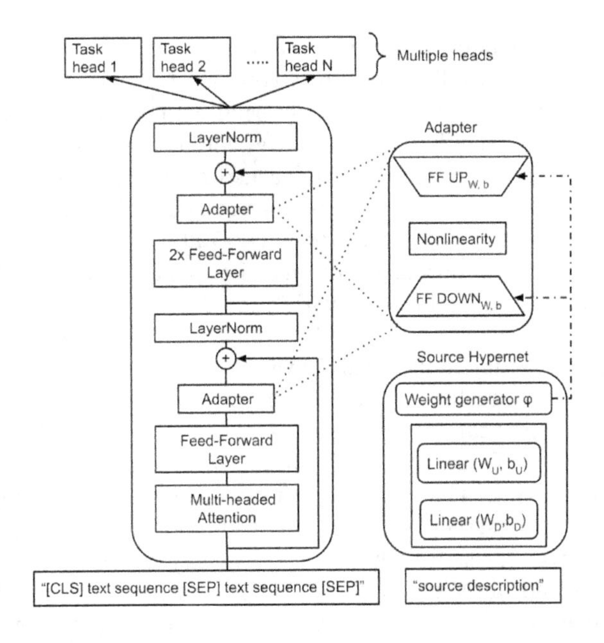

Fig. 2. Canonical Hard Parameter Sharing Architecture

$$A_i(x) = W_{U_i} \cdot ReLU(W_{D_i}x + b_{D_i}) + b_{U_i} \tag{4}$$

where W and b are the weight and bias terms generated by the hypernet and x denotes the vector of a *serialized* sequence pair.

3 MultiMatch Training

This section explains the serialization process, uncertainty loss weighting approach and training procedure used in MultiMatch.

Attribute Value Serialization: As a Transformer encoder expects token sequences, the serialization process involves concatenating attribute values and names using unique tags: '[COL]' and '[VAL]'. This process maintains the structure of the input at the sequence level. Specifically, a sequence of tokens is serialized as '[COL] attribute_name [VAL] attribute_value.' This serialization approach follows the intuition used in several state-of-the-art (SoTA) EM/GEM systems discussed in Sect. 2.2. The goal is to enable the Transformer to effectively utilize additional information when distinguishing between attribute values and names during the encoding process.

Loss Weighting and MultiMatch Training: In MTL, the aim is to learn a shared latent representation that enhances performance on a subset of tasks.

However, there is a phenomenon called negative transfer, where the performance on specific tasks may actually deteriorate compared to single-task models. This negative transfer can be attributed to internal conflicts that arise when optimizing individual objective functions and the dominance of specific tasks during training that impacts the quality of the learned representations.

One approach to address this is to combine multiple objectives using aleatoric uncertainty, particularly *homoscedastic* uncertainty [3]. Homoscedastic uncertainty refers to a constant parameter or quantity, such as noise $\sigma \sim \mathcal{N}(0, 1)$, that remains the same regardless of input variations. Moreover, aleatoric uncertainty encompasses *heteroscedastic* uncertainty, which is distinct from homoscedastic uncertainty. Heteroscedastic uncertainty varies across model inputs, such as the entropy of different data batches, and is influenced by the observed data. We consequently contend that, challenging tasks can be identified based on their corresponding loss ℓ. To reap the benefit in this, we propose Task-dependent Entropy Loss Weighting, which leverages heteroscedastic uncertainty or individual task entropy to construct the overall loss function $\mathcal{L}$.

Specifically, given a set of tasks $\mathrm{T} = \{t_1, \ldots, t_n\}$ where $t_n = \{X, Y\}$ are the training samples and an MTL model $f(\cdot; \Phi, \phi, \varphi)$ where the model parameters Φ are fixed and ϕ and φ are the adapter and hypernet parameters. Let $\mathbf{z} := g(\cdot; \Phi, \phi, \varphi)$ be the logits s.t $\sigma(g(\cdot; \Phi, \phi, \varphi)) = f(\cdot; \Phi, \phi, \varphi)$ where σ is a softmax function. We compute the logits and task loss using $f(\cdot)$ and based on this, the task weights λ_n using Eq. 5:

$$\lambda_n = -1 \cdot \sum_n \sigma(\mathbf{z}) \cdot log(\sigma(\mathbf{z})) \tag{5}$$

$$\mathcal{L} = \sum_n \lambda_n \cdot \ell_n \tag{6}$$

Finally, this task weight λ_n, an entropy-based weight, is used in the final loss computation in Eq. 6.

Power and Energy Monitoring: In order to understand the computational impact of training our proposed approach, metrics to measure computation and resource utilization are required. Deep learning can yield impressive results in various tasks, but extensive training on specialized hardware accelerators contributes to an energy-intensive workload, to address this issue we utilized the CarbonTracker [1] tool which tracks and predicts the energy and carbon footprint of training deep learning models to foster responsible computing in machine learning and encourage research into energy-efficient deep neural networks. CarbonTracker monitors CPU and GPU usage to compute energy usage and carbon footprint (grams), which we report in our experimental results.

4 Experiments

Datasets: We conduct experiments on Machamp which builds upon prior benchmarks such as Magellan [10], DeepMatcher [13], and the WDC product benchmark [22] to address the issue of benchmark saturation on commonly used evaluation EM datasets. It aims to overcome this saturation by offering a diverse range of matching scenarios that encompass varying levels of task difficulty, formats, and dataset structures. There are a total of 7 tasks (summarized in Table 1), each targeting specific types of data matching. Namely, these are;

- R-HET: Matches entities in structured tables with different schemas.
- S-HOM: Links entities in similar semi-structured data.
- S-HET: Matches entities in dissimilar semi-structured data.
- S-REL: Finds relationships between semi-structured and structured tables.
- S-TEX: Matches semi-structured tables with unstructured text (computer and watch product data).
- R-TEX: Matches structured tables with unstructured text.
- G-HET: Matches structured tables from the geospatial domain [2].

Table 1. Dataset statistics †.

Dataset	Domain	#Tuples (L-R)	#Labeled	Train rate	Train set(P\N)
R-HET	restaurant	534 - 332	567	5%	3\28
S-HOM	citation	2,616 - 64,263	17,223	2.5%	80\350
S-HET	book	22,133 - 23,264	1,240	5%	24\38
S-REL	movie	29,180 - 32,823	1,309	5%	27\38
S-TEX-w	product	9,234 - 9,234	5,540	10%	64\490
S-TEX-c	product	20,897 - 20,897	12,538	10%	89\538
R-TEX	citation	2,616 - 2,295	7,417	10%	133\608
G-HET	geospatial	2,469 - 2,788	2,500	5%	36\89

† Note: The ratio of P\N may change depending on seed value. P denotes positive (matching) pairs and N denotes negative (non-matching) pairs. L and R denote the left and right tuples of respective datasets.

Baseline Methods: we use the following baseline categories (described in Sect. 2):

- **Full Fine-tuning**: using the Magellan SoTA Ditto [12] which incorporates a full parameter update. During training, the domain knowledge and data augmentation packages are turned off.
- **Prompt-tuning**: using the PromptEM [28] framework for GEM. Specifically, we use the default prompts given in the published source code.

- **Adapter-tuning**: In this category, we adopt AdapterEM [15] using a single-task adapter as the domain expert. Fine-tuning only updates the adapter using backpropagation. We replicate the experimental setup as published in their source code.

Implementation Details: We utilize an Ubuntu workstation with an Nvidia GPU graphics card (RTX A6000) with 48 GiB of memory. The Transformer methods[2] from the Huggingface library are utilized, and all programming is done in Python 3.9.7, CUDA 11.7, and PyTorch version 1.13.1. Our experimentation uses the SoTA RoBERTa-base model. We set the adapter size to 32, which corresponds to the reduction factor, and the learning rate is fixed to 2e−4 for the adapter-based models. We use a batch size of 32 or 64 according to the model requirements, and the number of training epochs is fixed at 40 for all scenarios. To evaluate the performance, we conduct 6 runs with different seeds for all experiments, including the baselines, and select the best model for evaluation on the unseen test set in each scenario. Finally, we report the average F1-score as the evaluation metric.

5 Analysis

5.1 Single Versus Multiple Objective Models

Table 2. Results of single-task vs multi-task objective models on Machamp

Methods (F1)	R-HET	S-HOM	S-HET	S-REL	S-TEX-c	S-TEX-w	R-TEX	G-HET	AVGF_1
Full-FT	$0.0_{\pm 0.0}$	$93.0_{\pm 3.0}$	$10.4_{\pm 17.0}$	$\mathbf{93.7}_{\pm 7.3}$	$79.0_{\pm 5.0}$	$42.2_{\pm 13.8}$	$46.9_{\pm 11.5}$	$64.3_{\pm 32.4}$	$53.7_{\pm 11.3}$
Prompt-tuning	$80.4_{\pm 26.6}$	$93.4_{\pm 0.2}$	$26.5_{\pm 2.0}$	$87.0_{\pm 3.2}$	$79.0_{\pm 1.3}$	$31.0_{\pm 5.5}$	$\mathbf{59.2}_{\pm 2.2}$	$77.0_{\pm 1.7}$	$66.7_{\pm 5.3}$
Adapter-tuning	$73.5_{\pm 3.9}$	$86.2_{\pm 1.0}$	$18.5_{\pm 32.0}$	$43.4_{\pm 3.0}$	$0.01_{\pm 0.02}$	$0.03_{\pm 0.05}$	$11.0_{\pm 20.2}$	$67.4_{\pm 2.0}$	$37.5_{\pm 7.8}$
MultiMatch†	$77.8_{\pm 40.4}$	$91.9_{\pm 3.8}$	$\mathbf{87.1}_{\pm 13.4}$	$91.8_{\pm 13.0}$	$80.7_{\pm 4.7}$	$57.7_{\pm 7.4}$	$47.7_{\pm 9.6}$	$\mathbf{82.6}_{\pm 8.4}$	$77.2_{\pm 12.6}$
MultiMatch‡	$\mathbf{94.5}_{\pm 14.0}$	$\mathbf{94.6}_{\pm 2.8}$	$77.2_{\pm 21.4}$	$91.6_{\pm 6.2}$	$\mathbf{81.1}_{\pm 7.2}$	$\mathbf{60.0}_{\pm 3.4}$	$58.9_{\pm 5.2}$	$81.6_{\pm 4.0}$	$\mathbf{80.0}_{\pm 3.4}$

† represents the unweighted loss function while ‡ represents the proposed weighted loss.

In the analysis comparing single-task and multi-task fine-tuning on the primary datasets, encompassing 7 tasks, including G-HET from the geospatial domain, we evaluated performance along two dimensions: the average F_1 score and the carbon footprint of the models. The findings presented in Table 2, show that adapter-tuning encounters challenges on 4 out of 8 tasks (S-TEX-c, S-TEX-w, R-TEX, S-HET). We attribute this to several factors, such as the inherent complexity of the tasks, limitations in the available training materials (5–10%), and the reduced number of trainable parameters in the PrLM, resulting in sub-optimal performance. In constrast, prompt-tuning outperforms full fine-tuning, achieving a 13-point improvement in average F_1 score. Notwithstanding, significant challenges persist in handling distribution shifts modelled in G-HET,

[2] The sequence length is set 512 tokens for all models.

S-HET and S-TEX-w. Interestingly, when utilizing MTL, there is noticeable improvement in the performance metrics for these tasks. Additionally, the utilization of the weighted loss approach enhances the average F_1 score by a margin of 13.3 points compared to prompt-tuning. Furthermore, we also observe a lower count of trainable parameters (Table 3) in the case of MTL compared to adapter-tuning. It might have been anticipated that fewer parameters would result in a less capable MTL model, aligning with the observed trend in adapter-tuning. However, this is counteracted by the effectiveness of MTL in leveraging shared representations when learning features from similar tasks. Finally, our proposed weighted loss proves effective in reducing model variance in MTL settings, with the exception of S-HET/S-TEX-c. In summary, MultiMatch‡ (weighted) attains the highest average F_1 score at 80 points.

Table 3 shows the carbon footprint of the fine-tuned models reported by the CarbonTracker tool. In the aspect of single objective models, full fine-tuning uses the least footprint (however, we suspect that this footprint would be significantly higher than other methods for sufficient-resource settings owing to overparametization) followed by adapter then prompt-tuning. MTL is comparable to full fine-tuning in terms of footprint emitted represented by a 21% margin.

Table 3. Energy consumption and trainable parameter size of each model

Methods	CO2eq (g) ↓	Trainable Parameters (M)
Full-FT	90.6	129
PromptEM	757.6	150
Adapter-tuning	366.0	14.1
MultiMatch†	113.3	5.20
MultiMatch‡	114.2	5.20

Note: ↓ emphasises that lower is better.

5.2 Task Ablation Experiments

Effect of Dropping Tasks: The heatmap in Fig. 3 shows the effect of dropping tasks from the MTL model in low-resource settings. The x-axis indicates the omitted task during training while the y-axis indicates the resulting change in F_1 score on the remaining tasks following the ablated task. The colour grid to the right reflects the intensity of the observed change: a lighter shade represents a positive transfer and a darker shade denotes a negative impact. Consequently, our analysis reveals that three tasks (R-TEX, S-TEX-w, and R-HET) are highly sensitive to changes in the feature space due to ablation. For instance, when S-HOM is ablated, R-TEX experiences a substantial negative loss of −58.9 F1 points. We conjecture that in such scenarios the underlying task interactions in the shared-layers did not adequately encompass ample information to effectively cover the remaining tasks, resulting in negative interference.

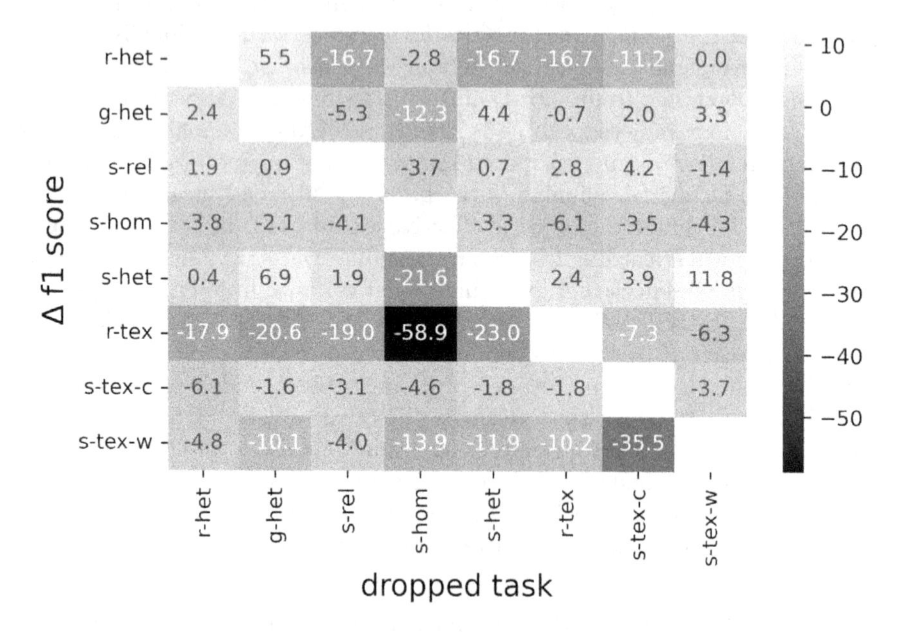

Fig. 3. Heatmap summarizing the effect of dropped tasks

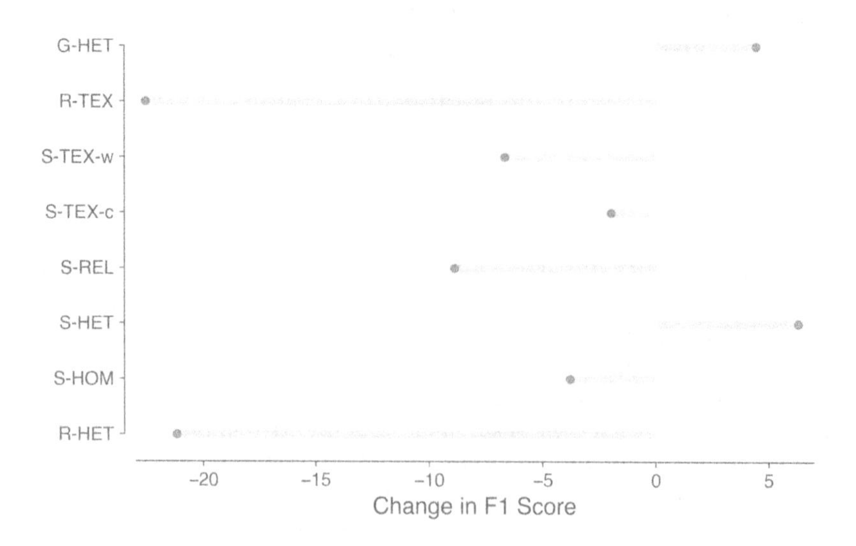

Fig. 4. Bar chart summarizing the effect of swapping task order during training

Next, we examine the impact of task order by swapping the position of the tasks in the main experiments of Table 2. Initially task order was as follows; R-HET, G-HET, S-REL, S-HOM, S-HET, R-TEX, S-TEX-c, and S-TEX-w respectively. This ordering is inspired by curriculum learning, where the tasks are gradually presented to the model in the order of increasing complexity. Figure 4 illustrates our observations. The $y-$axis denotes the task under observation while the $x-$axis the direction of the observed change in F_1, i.e., positive or negative. Our results indicate that task order is critical suggesting the existence of task dependencies where certain tasks rely on foundational features learned during initial training. Thus, changing the order of tasks disrupts these dependencies, leading to task interference. This interference is particularly pronounced in the case of S-REL, R-TEX, and R-HET. Interestingly, G-HET and S-HET experience positive task transfer.

6 Conclusions and Future Work

In this work, we present MultiMatch (MultiMatch[‡]), a system that incorporates learning in a single step for multiple data integration tasks. We demonstrate that MTL can enhance downstream performance on related tasks. An important and significant finding presented in this paper is that our system drastically reduces trainable model parameter size and the carbon footprint compared with existing approaches. This addresses the important challenge of constructing quality EM models while minimizing the fine-tuning footprint. Our heteroscedastic-based loss achieved improved performance on datasets with challenging distribution shifts, such as S-HET and S-TEX-w. In the future, we plan to explore the strengths and limitations of our loss-balancing method by incorporating challenging distribution shifts. Solidifying the position of our heteroscedastic-based loss is an open direction for future works that may prove useful in handling data distribution shifts in MTL systems.

Acknowledgement. This paper is based on results obtained from "Research and Development Project of the Enhanced Infrastructures for Post-5G Information and Communication Systems" (JPNP20017), commissioned by the New Energy and Industrial Technology Development Organization (NEDO) - JPNP20006, JST CREST Grant Number JPMJCR22M2, and JSPS KAKENHI Grant Number JP23K24949.

References

1. Anthony, L.F.W., Kanding, B., Selvan, R.: Carbontracker: Tracking and predicting the carbon footprint of training deep learning models. arXiv (2020)
2. Balsebre, P., Yao, D., Cong, G., Hai, Z.: Geospatial entity resolution. In: Proceedings of the ACM Web Conference (2022). https://doi.org/10.1145/3485447.3512026
3. Cipolla, R., Gal, Y., Kendall, A.: Multi-task learning using uncertainty to weigh losses for scene geometry and semantics. pp. 7482–7491 (2018). https://doi.org/10.1109/CVPR.2018.00781

4. Ebraheem, M., Thirumuruganathan, S., Joty, S., Ouzzani, M., Tang, N.: Distributed representations of tuples for entity resolution. Proc. VLDB Endow. (2018). https://doi.org/10.14778/3236187.3236198
5. Guo, B., Yu, J., Yang, D., Leng, H., Liao, B.: Energy-efficient database systems: a systematic survey. ACM Comput. Surv. **55**(6) (2022). https://doi.org/10.1145/3538225
6. Ha, D., Dai, A.M., Le, Q.V.: Hypernetworks. ArXiv **abs/1609.09106** (2016). https://api.semanticscholar.org/CorpusID:208981547
7. Houlsby, N., et al.: Parameter-efficient transfer learning for NLP. ArXiv **abs/1902.00751** (2019). https://api.semanticscholar.org/CorpusID:59599816
8. Ivison, H., Peters, M.: Hyperdecoders: instance-specific decoders for multi-task NLP, pp. 1715–1730. Association for Computational Linguistics (2022). https://api.semanticscholar.org/CorpusID:247476141
9. Karimi Mahabadi, R., Ruder, S., Dehghani, M., Henderson, J.: Parameter-efficient multi-task fine-tuning for transformers via shared hypernetworks, pp. 565–576. Association for Computational Linguistics (2021). https://api.semanticscholar.org/CorpusID:235309789
10. Konda, P., et al.: Magellan: toward building entity matching management systems. Proc. VLDB Endow. **9**(12), 1197–1208 (2016)
11. Lester, B., Al-Rfou, R., Constant, N.: The power of scale for parameter-efficient prompt tuning, pp. 3045–3059. ACL (2021). https://doi.org/10.18653/v1/2021.emnlp-main.243
12. Li, Y., Li, J., Suhara, Y., Doan, A., Tan, W.C.: Deep entity matching with pre-trained language models. Proc. VLDB Endow. **14**(1), 50–60 (2020). https://doi.org/10.14778/3421424.3421431
13. Mudgal, S., et al.: Deep learning for entity matching: a design space exploration, pp. 19–34. In: SIGMOD 2018, Association for Computing Machinery (2018)
14. Mugeni, J.B., Amagasa, T.: A graph-based blocking approach for entity matching using contrastively learned embeddings. SIGAPP Appl. Comput. Rev. **22**(4), 37–46 (2023). https://doi.org/10.1145/3584014.3584017
15. Mugeni, J.B., Lynden, S., Amagasa, T., Matono, A.: Adapterem: pre-trained language model adaptation for generalized entity matching using adapter-tuning, pp. 140–147. IDEAS 2023, Association for Computing Machinery (2023). https://doi.org/10.1145/3589462.3589498
16. Peeters, R., Bizer, C.: Dual-objective fine-tuning of BERT for entity matching. Proc. VLDB Endow. **14**(10), 1913–1921 (2021). https://doi.org/10.14778/3467861.3467878
17. Peeters, R., Bizer, C.: Supervised contrastive learning for product matching. In: Companion Proceedings of the Web Conference 2022. ACM (2022). https://doi.org/10.1145/3487553.3524254
18. Peeters, R., Bizer, C.: Entity matching using large language models. arXiv (2024)
19. Peeters, R., Der, R.C., Bizer, C.: WDC products: a multi-dimensional entity matching benchmark. arXiv (2023)
20. Pfeiffer, J., et al.: AdapterHub: a framework for adapting transformers, pp. 46–54. Association for Computational Linguistics (2020). https://doi.org/10.18653/v1/2020.emnlp-demos.7
21. Pilault, J., hattami, A.E., Pal, C.: Conditionally adaptive multi-task learning: Improving transfer learning in NLP using fewer parameters & less data. In: International Conference on Learning Representations (2021). https://openreview.net/forum?id=de11dbHzAMF

22. Primpeli, A., Peeters, R., Bizer, C.: The WDC training dataset and gold standard for large-scale product matching, pp. 381–386. Association for Computing Machinery (2019). https://doi.org/0.1145/3308560.3316609
23. Tu, J., et al.: Domain adaptation for deep entity resolution, p. 443–457. In: SIGMOD 2022, Association for Computing Machinery (2022). https://doi.org/10.1145/3514221.3517870
24. Üstün, A., Bisazza, A., Bouma, G., van Noord, G., Ruder, S.: Hyper-X: a unified hypernetwork for multi-task multilingual transfer. In: Goldberg, Y., Kozareva, Z., Zhang, Y. (eds.) Proceedings of the 2022 Conference on Empirical Methods in Natural Language Processing, pp. 7934–7949. Association for Computational Linguistics, Abu Dhabi, United Arab Emirates (2022). https://doi.org/10.18653/v1/2022.emnlp-main.541
25. Vaswani, A., et al.: Attention is all you need. In: Proceedings of the 31st International Conference on Neural Information Processing Systems, pp. 6000–6010 (2017). http://arxiv.org/abs/1706.03762
26. Verdecchia, R., Sallou, J., Cruz, L.: A systematic review of green AI. Wiley Interdisc. Rev. Data Min. Knowl. Disc. **13**(4), e1507 (2023)
27. Wang, J., Li, Y., Hirota, W.: Machamp: a generalized entity matching benchmark, pp. 4633–4642. In: CIKM 2021, Association for Computing Machinery (2021). https://doi.org/10.1145/3459637.3482008
28. Wang, P., et al.: PromptEM: prompt-tuning for low-resource generalized entity matching. Proc. VLDB Endow. **16**(2), 369–378 (2022). https://doi.org/10.14778/3565816.3565836
29. Wang, T., et al.: Bridging the gap between reality and ideality of entity matching: a revisiting and benchmark re-construction (2022 arxiv-2205.05889
30. Zhuang, L., Wayne, L., Ya, S., Jun, Z.: A robustly optimized BERT pre-training approach with post-training. In: Li, S., (eds.) et al Proceedings of the 20th Chinese National Conference on Computational Linguistics. pp. 1218–1227. Chinese Information Processing Society of China, Huhhot, China (2021). https://doi.org/10.1007/978-3-030-84186-7_31

Embedding-Based Data Matching for Disparate Data Sources

Nour Elhouda Kired[1,2(✉)], Franck Ravat[1], Jiefu Song[1], and Olivier Teste[2]

[1] Université Toulouse Capitole, IRIT, Toulouse, France
nour-elhouda.kired@irit.fr
[2] Université Toulouse II Jean Jaurès, IRIT, Toulouse, France

Abstract. Dealing with heterogeneous sources is an important challenge in the field of knowledge discovery and management. Schema matching methods are employed to solve this problem using three approaches: schema-based, instance-based, or a combination. This paper focuses on mapping between a schema-available (only) data source and a data source containing both schema and instance (both). Given the lack of suitable methods for aligning these two types of sources, we propose an approach using embedding models to provide vector modelling of sources and calculate similarities between data. Our solution consists in combining domain-specific embedding models and cross-domain embedding models to make data matching possible and efficient between the above-mentioned data sources. We have conducted several experiments using the Valentine datasets to evaluate our data matching method on several disparate tabular data. The result indicate effectiveness in terms of stability and ablation handling.

Keywords: Schema Matching · Disparate Data Source · Embeddings

1 Context and Main Issues

Data lakes store diverse data types, making schema-on-read essential for determining data schema during access. Aligning these disparate sources at the schema level is crucial for comprehensive data analysis, employing schema-based, instance-based, or hybrid methods. Integrating unstructured data like images and videos requires metadata, posing a challenge when matching sources with schemas and instances to those with only schemas. Existing methods, including rule-based techniques like Regular Expressions, often fall short due to disparate data structures and semantic complexities, despite improvements from incorporating syntactic and semantic relations [1]. This paper proposes a novel approach combining cross-domain and domain-specific embeddings to effectively match non-comparable data source structures, preserving the structure of the input sources.

R. Wrembel et al. (Eds.): DaWaK 2024, LNCS 14912, pp. 66–71, 2024.
https://doi.org/10.1007/978-3-031-68323-7_5

In schema matching, pre-trained embedding models, particularly cross-domain embedding models, have shown effectiveness in generalizing knowledge across different domains. Zhang et al. [2] and Dash et al. [3] use BERT-based embeddings to perform schema matching, leveraging deep domain knowledge. The BART model [4] excels in Natural Language Inference (NLI) for understanding complex semantic relationships, essential for schema matching [5]. Cappuzzo et al. [6] use domain-specific embeddings tailored to the characteristics of specific domains, employing graph embedding to uncover connections within datasets and enhance pattern matching accuracy. While cross-domain embeddings generalize knowledge across domains without considering data structure, domain-specific embeddings focus on relationships within structured data but struggle with disparate data sources. Recent research in Open Domain Question Answering (ODQA) and information retrieval uses a dual-model approach, fine-tuning BERT-based models for domain-specific data retrieval [7], but these methods are not fully adapted for schema matching with varied data structures.

This paper contributes by developing an approach to align schemas from disparate data source structures, focusing on mapping schema-based data with both schema and instance-level data (e.g., CSVs, Excels) in Data Lakes. Our method combines cross-domain embeddings, capturing broad semantic relationships, with domain-specific embeddings, using graph embeddings for deeper dataset insights. This approach efficiently aligns disparate data structures by leveraging both broad and deep domain knowledge.

2 Proposed Framework

2.1 Problem Statement

We present an approach using the BART model for cross-domain embeddings and EMBDI for domain-specific embeddings to map two disparate data sources: $S1$ (schema $Sc1$) and $S2$ (schema $Sc2$ and instances $I2$). The goal is to align $Sc1$ and $Sc2$, producing matches $(a_i, b_j, score)$ that indicate the degree of similarity between elements from the two schemas, with scores ranging from 0 to 1.

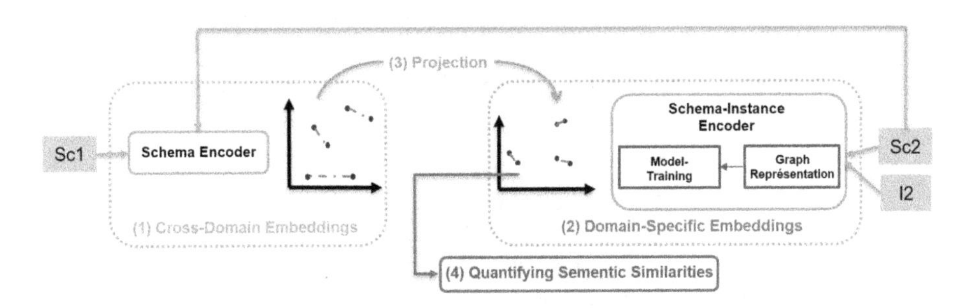

Fig. 1. Framework Overview

2.2 Overview

Our schema-matching approach with four key components significantly contributing to achieving our goal. Figure 1 illustrates our approach with these components:

(1) Cross-Domain Embedding. We employ the BART model's Natural Language Inference (NLI) capabilities to compare the semantic and syntactic relationships between two schemas, Sc_1 and Sc_2. Each attribute in Sc_2 is treated as a hypothesis against the attributes in Sc_1. The BART model outputs logits for these attribute pairs, which are then transformed into probabilities to form a similarity matrix textitSimMatrix. Let: A = $\{a_1, a_2, \ldots, a_m\}$ be the set of attributes in schema Sc_1, B = $\{b_1, b_2, \ldots, b_n\}$ be the set of attributes in schema Sc_2. For each pair (a_i, b_j), the BART model computes logits l_{ij}, which are converted to probabilities p_{ij} representing the likelihood of entailment: $p_{ij} = \text{softmax}(l_{ij})$

The similarity matrix *SimMatrix* is then defined as:

$$\textbf{SimMatrix}_{ij} = p_{ij} \quad \text{for} \quad i = 1, \ldots, m \quad \text{and} \quad j = 1, \ldots, n$$

(2) Domain-Specific Embedding. The EmbDI algorithm constructs domain-specific embeddings by representing relational data as a heterogeneous graph $G = (V, E)$. Nodes V include token nodes (T), Record ID nodes (RID), and Column ID nodes (CID). Edges E represent relationships between these nodes. Random walks on G generate sequences of nodes, forming sentences for the embedding training corpus. The sentences are used to train embeddings with algorithms like word2vec. Let $\mathbf{E}_G(v)$ be the embedding vector for node $v \in V$.

(3) Projection of Sc_1 Attributes onto Domain-Specific Embeddings. Attributes from Sc_1 are projected onto the domain-specific embeddings using the similarity matrix *SimMatrix*.

For each attribute $a_i \in A_1$, the embedding vector $\mathbf{E}_{Sc_1}(a_i)$ is computed as:

$$\mathbf{E}_{Sc_1}(a_i) = \sum_{j=1}^{n} \textbf{SimMatrix}_{ij} \cdot \mathbf{E}_{Sc_2}(b_j)$$

where : $\mathbf{E}_{Sc_1}(a_i)$ is the embedding vector of attribute $a_i \in Sc_1$ and $\mathbf{E}_{Sc_2}(b_j)$ is the embedding vector of attribute $b_j \in Sc_2$.

(4) Quantifying Semantic Similarities. To quantify the semantic similarities between attributes in Sc_1 and Sc_2, we use cosine similarity. For each pair (a_i, b_j):

$$\text{similarity}(a_i, b_j) = \frac{\mathbf{E}_{Sc_1}(a_i) \cdot \mathbf{E}_{Sc_2}(b_j)}{\|\mathbf{E}_{Sc_1}(a_i)\| \|\mathbf{E}_{Sc_2}(b_j)\|}$$

This results in a list of matches (a_i, b_j, score) where $a_i \in Sc_1$ and $b_j \in Sc_2$, with each match assigned a similarity score.

3 Experiments

In this section, we describe experiments to evaluate the effectiveness of our schema matching approach, focusing on two main research questions:

- **RQ1(Effectiveness & stability)** How significantly does the proposed schema mapping approach enhance the F1 score when aligning data schemas of disparate structures? And how consistent is our method when the configuration and dataset are held constant?
- **RQ2(Ablation)** How does the removal of a specific phase of the proposed schema matching approach affect its overall performance?

We address these questions through in-depth experiments, exploring various configurations and parameters to refine our approach to schema matching. In our experiments, we used four dataset categories from the "Valentine" collection [8]. The experimental setup utilized PyTorch with NVIDIA CUDA on 12 Dell servers with Intel Xeon processors. Comprehensive results and additional experiments (e.g., handling noise) are available on the companion website.[1].

Table 1. Performance Metrics Across Dataset Categories and Relation Types

Dataset categories	Relations (Datasets)	F1 Score		Recall		Precision	
		Mean	Std	Mean	Std	Mean	Std
ChEMBL (180 datasets)	Joinable (48)	0.92	0.08	0.95	0.09	0.91	0.11
	Semantically-Joinable (48)	0.95	0.07	0.98	0.06	0.93	0.11
	Unionable (36)	0.98	0.02	0.95	0.05	1.00	0.00
	View-Unionable (48)	0.96	0.05	0.98	0.04	0.94	0.08
TPC-DI (180 datasets)	Joinable (48)	0.76	0.22	0.97	0.06	0.67	0.25
	Semantically-Joinable (48)	0.74	0.21	0.96	0.06	0.65	0.26
	Unionable (36)	0.98	0.02	0.97	0.03	1.00	0.00
	View-Unionable (48)	0.75	0.21	0.95	0.06	0.67	0.25
Wikidata (4 datasets)	Joinable (1)	0.67	0.01	0.67	0.02	0.67	0.00
	Semantically-Joinable (1)	0.93	0.04	0.87	0.08	1.00	0.00
	Unionable (1)	0.87	0.04	0.77	0.06	1.00	0.00
	View-Unionable (1)	0.66	0.02	0.67	0.00	0.66	0.04
Magellan	Unionable (7)	0.93	0.06	0.88	0.11	1.00	0.00

[1] https://github.com/user28060/Embedding-based-data-matching-for-disparate-data-sources.git.

3.1 RQ1. Effectiveness and Stability

Discussion: The results in Table 1, using 96 configurations, demonstrate the effectiveness and reliability of our schema matching algorithm with an average accuracy of 0.85 across all datasets and relations. Low standard deviations in ChEMBL (0.055), Wikidata (0.022), and Magellan (0.06) indicate robustness. However, variability in TPC-DI (0.75–0.98) and some Wikidata instances (0.66 for joinable and view-unionable) highlights challenges in matching semantically similar but differently labeled attributes, especially when these attributes have high similarity scores within the same domain.

A detailed stability analysis was conducted on datasets D1-D10, each tested with different hyperparameter configurations. We focused on three key hyperparameters affecting random walks, resulting in 12 configurations, each run 10 times. The results show high stability, with D1-D7 achieving mean F1, Recall, and Precision scores of 1.00. The lower F1 score for D10 (0.91) is due to difficulties distinguishing between 'givenNameLabel' and 'familyName.' Slight variations in standard deviation were observed for D8 (0.01), D9 (0.002), and D10 (0.018), confirming the algorithm's consistent performance across different configurations.

3.2 RQ2. Ablation

The ablation study evaluated two approaches: the combined approach and the cross-domain approach. A domain-specific model alone was unfeasible due to the incompatibility of heterogeneous dataset structures with the EMBDI algorithm. Experiments were conducted on various datasets with noise. The study used 96 configurations to compare the performance of the two methods, with results shown in Fig. 2.

Discussion: The combined approach outperforms the cross-domain approach with a higher median F1 score and tighter interquartile range, demonstrating better and more consistent performance. Figure 2.b indicates that the combined approach is, on average, 0.3 F1 score points better than the cross-domain app-

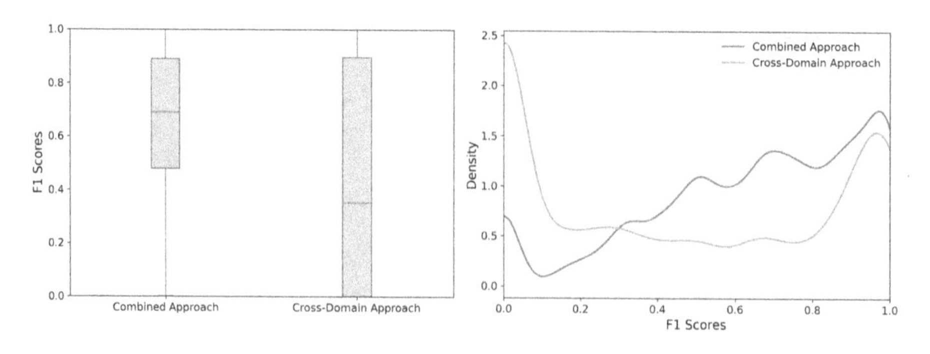

Fig. 2. Comparison of Cross-Domain Embeddings and Combined Approaches

roach, making it preferable for applications where performance consistency is crucial.

4 Conclusion

Mapping disparate data source structures is critical within data lakes. This paper introduces a novel schema matching method to align schema-based data sources with those containing both schemas and instances. By leveraging embedding models, we created vector representations to compute data similarities. Our technique, merging domain-specific and cross-domain models, was rigorously evaluated using Valentine dataset categories. Results highlight its effectiveness, stability, and proficiency in ablation, indicating its potential in data matching.

Future work will focus on enhancing our framework by incorporating entity resolution and developing diverse models for cross-domain and domain-specific contexts, aiming to improve both schema matching and entity resolution.

References

1. Christodoulou, K., Fernandes, A.A.A., Paton, N.W.: Combining syntactic and semantic evidence for improving matching over linked data sources. In: Wang, J., et al. (eds.) WISE 2015. LNCS, vol. 9418, pp. 200–215. Springer, Cham (2015). https://doi.org/10.1007/978-3-319-26190-4_14
2. Zhang, Y., et al.: Schema matching using pre-trained language models. In: 2023 IEEE 39th International Conference on Data Engineering (ICDE), pp. 1558–1571. IEEE, Anaheim, CA, USA (2023). https://doi.org/10.1109/ICDE55515.2023.00123
3. Dash, S., Bagchi, S., Mihindukulasooriya, N., Gliozzo, A.: Linking tabular columns to unseen ontologies. In: Payne, T.R., et al. (eds.) ISWC 2023. LNCS, vol. 14265, pp. 502–521. Springer, Cham (2023). https://doi.org/10.1007/978-3-031-47240-4_27
4. Lewis, M., et al.: BART: denoising sequence-to-sequence pre-training for natural language generation, translation, and comprehension (2019). http://arxiv.org/abs/1910.13461
5. Liu, H., Cui, L., Liu, J., Zhang, Y.: Natural language inference in context - investigating contextual reasoning over long texts. In: AAAI, vol. 35, pp. 13388–13396 (2021). https://doi.org/10.1609/aaai.v35i15.17580
6. Cappuzzo, R., Papotti, P., Thirumuruganathan, S.: Creating embeddings of heterogeneous relational datasets for data matching tasks. In: Proceedings of the 2020 ACM SIGMOD International Conference on Management of Data, pp. 1335–1349. ACM, Portland OR USA (2020). https://doi.org/10.1145/3318464.3389742
7. Bosch, N., Shalmashi, S., Yaghoubi, F., Holm, H., Gaim, F., Payberah, A.H.: Fine-tuning BERT-based language models for duplicate trouble report retrieval. In: 2022 IEEE International Conference on Big Data (Big Data), pp. 4737–4745. IEEE, Osaka, Japan (2022). https://doi.org/10.1109/BigData55660.2022.10020825
8. Koutras, C., et al.: Valentine: evaluating matching techniques for dataset discovery (2021). http://arxiv.org/abs/2010.07386

Subtree Similarity Search Based on Structure and Text

Takuya Mizokami[1(✉)], Savong Bou[2], and Toshiyuki Amagasa[2]

[1] Graduate School of Science and Technology, University of Tsukuba, Tsukuba, Ibaraki, Japan
mizokami@kde.cs.tsukuba.ac.jp

[2] Center for Computational Sciences, University of Tsukuba, Tsukuba, Ibaraki, Japan
{savong-hashimoto,amagasa}@cs.tsukuba.ac.jp

Abstract. Given a query tree, the subtree similarity search problem is finding all subtrees in a document tree that are similar to the query tree. The previous scan-based method extracts candidate subtrees based on the size difference, which only considers the structural differences and ignores the differences in the contents represented by the trees. For this reason, it suffers from the following two issues. First, for queries against a tree with a regular structure, it is difficult to differentiate subtrees in terms of structural similarity, yielding a large number of candidate results to verify. Second, the candidates are verified by computing the tree edit distance, which is cubic to the number of tree nodes. In this paper, we propose a solution for the subtree similarity search problem based on the structure and contents of the trees. We demonstrate through experiments that our proposed method outperforms the previous scan-based methods in terms of speed and is competitive with index-based methods.

Keywords: Approximate Matching · Similarity search · Tree edit distance

1 Introduction

We can naturally represent data with a hierarchical structure as a tree. One can store a value or a label in a node, and two nodes can be connected by an edge if they have a parent-child relationship. Specifically, we consider in this work rooted, ordered, and labeled trees where there exists a distinguished root node, and there is a toral order among the child nodes for a (parent) node. JSON, HTML, and XML are typical examples of such trees.

In this paper, we study the subtree similarity search problem, where, given a large target tree D and a query tree Q, find all subtrees in D that are similar to Q. For example, in Fig. 1, subtree D' in the target tree D is considered similar to the query tree, and thus is included in the search results, while D'' is deemed dissimilar and is excluded from the search results. Subtree similarity search is used in various applications, such as duplicate XML/JSON data detection [9,17], duplicate code fragment search [7,8], and similar RNA secondary structure search [1,10].

R. Wrembel et al. (Eds.): DaWaK 2024, LNCS 14912, pp. 72–87, 2024.
https://doi.org/10.1007/978-3-031-68323-7_6

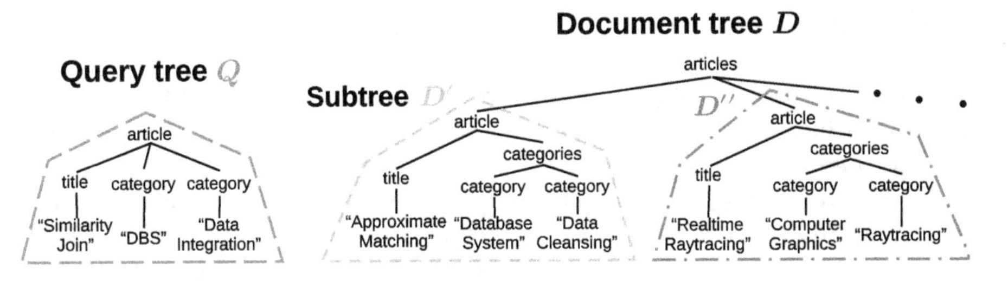

Fig. 1. An example of subtree similarity search. The tree edit distances between Q and D' and between Q and D'' are both 4 (rename 3 leaf nodes and insert a "categories" node). But the contents of D' are more similar to Q than those of D''.

Previous methods evaluate the similarity between two trees based on the tree edit distance (TED) [2,4,16]. TED is defined as the minimum number of node edit operations (insertion, deletion, and rename) required to transform one tree into another [26]. Methods for subtree similarity search can be classified into two categories: scan-based methods that scan the entire tree and index-based methods that use an index structure to search for similar subtrees. The state-of-the-art scan-based and index-based methods are TASM [2] and SlimCone [16], respectively. These methods adopt a filter-and-verification approach; i.e., we first filter out subtrees that are unlikely to be the final result from D using filter functions, and then we verify each candidate to remove false positives to get the final result by computing the TED between query Q and each candidate.

The previous scan-based method suffers from two issues. First, this method extracts candidate subtrees based on the difference in the number of nodes between the query tree and these subtrees. Therefore, for a document tree composed of many subtrees with structures similar to the query, many candidate subtrees are extracted. Second, the computational cost of computing the TED is cubic to the number of nodes in the tree, making it difficult to search large document trees efficiently.

In this paper, we propose a solution for the subtree similarity search problem based on both the structure and the contents of the trees. We evaluate the structural similarity based on the TED and the textual similarity based on the similarity of the text strings of the leaf nodes. We demonstrate through experiments that our proposed method outperforms previous scan-based methods in terms of speed and is competitive with index-based methods.

The rest of the paper is organized as follows. In Sect. 2, we define the subtree similarity search problem. In Sect. 3, we review related works. In Sect. 4, we provide preliminary knowledge. In Sect. 5, we describe our proposed method. In Sect. 6, we demonstrate the effectiveness of our proposed method through experiments. Finally, in Sect. 7, we conclude this paper.

2 Problem Definition

We address the problem of subtree similarity search. In contrast to the traditional definition, which measures similarity only by edit distance, this uses combined similarity metrics.

We assume rooted, ordered, and labeled trees. Let D be a document tree and Q be a query tree. The number of vertices is indicated by $|Q|$ and $|D|$, respectively. Let D_i be a subtree of D rooted at a vertex v_i, which includes v_i and all of its descendants. The similarity between Q and D_i is defined in Eq. 1, using the structural similarity $\mathrm{sim_s}(\cdot,\cdot)$, the textual similarity $\mathrm{sim_t}(\cdot,\cdot)$, a penalty based on the size difference $\beta(\cdot,\cdot)$, and a weight coefficient $\alpha \in (0,1]$.

$$\mathrm{sim}(Q, D_i) = \alpha \mathrm{sim_s}(Q, D_i) + (1-\alpha)\beta\left(|Q|, |D_i|\right)\mathrm{sim_t}(Q, D_i) \tag{1}$$

$$\left(\beta(|Q|, |D_i|) = \frac{\min\{|Q|, |D_i|\}}{\max\{|Q|, |D_i|\}}\right)$$

We will explain $\mathrm{sim_s}(\cdot,\cdot)$ and $\mathrm{sim_t}(\cdot,\cdot)$ in detail in Sect. 4.

We define the subtree similarity search problem in Eq. 2. Let $\theta \in (0,1]$ be the similarity threshold. The subtree similarity search problem is to find the subset $\mathcal{R}$ of all subtrees in D, such that each subtree in $\mathcal{R}$ has a similarity score of at least θ.

$$\mathcal{R} = \{D_i | D_i \in \mathit{SubTrees}(D) \ \wedge \theta \leq \mathrm{sim}(Q, D_i)\} \tag{2}$$

3 Related Works

This section provides the related works on tree edit distance and its upper/lower bounds, subtree similarity search methods, and related methods for joining similar trees.

3.1 Tree Edit Distance

The classical tree edit distance algorithms are surveyed by Bille [3] and one of them by Zhang and Shasha [26] runs in $O(n^4)$ time and $O(n^2)$ space for trees with n nodes. Demaine [5] further proposed the new algorithm, which reduces the runtime to $O(n^3)$. Subsequently, Pawlik and Augsten proposed RTED/APTED [18,19] an $O(n^3)$ algorithm that exhibits good practical performance. However, these algorithms are not suitable for computing the TED between two trees with large n.

To reduce the computational cost of TED, several algorithms to compute restricted TED have been proposed. These algorithm leverages the similarity of the input trees. These algorithms compute TED if it is less than a given upperbound τ, and returns ∞ otherwise. The algorithm proposed by Touzet [21] prunes the search space of potential solutions from Zhang and Shasha's dynamic programming, and runs in $O(n\tau^3)$ time and $(n\tau)$. Pawlik and Augsten [20] proposed TopDiff, the improvement of Touzet's algorithm, which exibits good practical performance. These algorithms, however, can't be applied directly to compute the TED without a given upper bound τ. In this paper, we first perform pruning based on lower and upper bounds, and then use TopDiff to compute the restricted TED more quickly. Same approach is also used in existing methods for joining similar trees [12,15].

3.2 Lower Bounds of Tree Edit Distance

Li [17] surveyed lower bounds of the tree edit distance and compared their performance. They compare three lower bounds: string-based [9], histogram-based [13] and set-based lower bounds [25]. They showed that the string-based lower bound is the tightest through experiments. Thomas [12] proposed a structural filter that further tightens the lower bound based on the difference of the label set. However, this structural filter is incorporated into the nested-loop tree join algorithm, and cannot be applied to our method for subtree similarity search. We use the thresholded string-based lower bound to prune candidate subtrees, which can be computed quickly and provides a tight bound.

3.3 Upper Bounds of Tree Edit Distance

Most upper bound algorithms of the tree edit distance have been reviewed by Kan et al. [14]. All the algorithms reviewed and proposed in [14] require $O(n^2)$ time and are not suitable for large trees. Among them, Wang and Zhang's constrained edit distance [23] is the only algorithm that require less than quadratic space, specifically $O(n \log n)$.

The most recent and fastest upper bound algorithm is the Label-Guided Mapping (LGM) proposed by Hütter et al. [12]. LGM computes the upper bound in $O(n\tau)$ time, where τ is the threshold. Specifically, LGM constructs edit mapping by mapping node pairs between two trees in greedy fashion. This algorithm reduces the computational cost by pruning any node pairs that would lead to a mapping with a cost higher than τ. We use the LGM algorithm, which can compute the upper bound in linear time for the number of nodes n.

3.4 Subtree Similarity Search

Several algorithms attacking the subtree similarity search problem measure similarity by the tree edit distance, such as TASM [2], StructureSearch [4], and SlimCone [16]. TASM is a scan-based method that scans the entire document tree to find similar subtrees, while StructureSearch and SlimCone are index-based methods that search using an index created from the document tree. SlimCone is the best in terms of execution time, index creation time, and index size. The existing scan-based method, TASM, extracts candidate subtrees that are close in size to the query tree, so it takes time to search for a document tree with many subtrees that have structure similar to the query tree. In this paper, we filter out candidate subtrees by textual similarity to improve the search efficiency.

3.5 Other Related Problems

There are several related but different problems, such as searching for similar trees from a set of tree records [11] and tree similarity join [12,15,22]. Our subtree similarity search problem is different from these in terms of the problem definition.

For search problems that finds the best k objects with the highest overall scores, the TA Algorithm is well known [6]. Our method is used to find objects with high scores that are the sum of two scores, but it is different from our method because our method is not top-k search problem.

4 Preliminaries

This section provides the definition of the tree edit distance, the structural similarity, and the textual similarity we're using in this paper.

Definition 1. *The* **tree edit distance** *[26] $TED(T, T')$ between two trees T and T' is the minimum number of edit operations required to transform T into T'. Here, the edit operations are assumed: (1)* Rename *the node's label, (2)* Delete *the node v and connect its descendants to the ancestor of v, and (3)* Insert *a new node v between the i-th and j-th nodes of the descendants of node u and connect $C = \{c_i, \cdots, c_j\}$ to v. In Delete and Insert operations, the sibling order is preserved.*

Example 1. In Fig. 1, to transform the left query tree Q into the subtree D' of the document tree, we need to insert the *categories* node between the *article* and *category* nodes and replace the labels of the three leaf nodes. Therefore, $TED(Q, D') = 4$. Similarly, $TED(Q, D'') = 4$ for the subtree D'' on the right side of D.

Definition 2. *The* **structural similarity** *between trees T and T' is defined in Eq. 3, using the normalized distance $TED(T, T')/(|T| + |T'|)$.*

$$\mathrm{sim_s}(Q, D_i) = 1 - \frac{TED(Q, D_i)}{|Q| + |D_i|} \tag{3}$$

Example 2. In Fig. 1, $TED(Q, D') = 4$, $TED(Q, D'') = 4$, $|Q| = 7$, and $|D'| = |D''| = 8$, so the structural similarity is $\mathrm{sim_s}(Q, D') = \mathrm{sim_s}(Q, D'') = 1 - 4/(7 + 8) = 0.73$.

The textual similarity is a similarity metric that considers the difference in the label data of the leaf nodes. We define the textual similarity using the Jaccard coefficient of the word sets of the leaf nodes.

Definition 3. *The* **textual similarity** $\mathrm{sim_t}(T, T')$ *between trees T and T' is defined as follows. Let $\mathcal{L}_T$ and $\mathcal{L}_{T'}$ be the sets of leaf nodes of T and T', respectively. Let $words(v)$ be the multiset of words contained in the label of a node v.*

Then, $\mathrm{sim_t}(T, T')$ *is defined in Eq. 4* ($A \Cap B$ *denotes the intersection between two multisets, A and B*)

$$\mathrm{sim_t}(T, T') = \frac{|\mathcal{W}_T \Cap \mathcal{W}_{T'}|}{|\mathcal{W}_T \uplus \mathcal{W}_{T'}|} = \frac{|\mathcal{W}_T \Cap \mathcal{W}_{T'}|}{|\mathcal{W}_T| + |\mathcal{W}_{T'}| - |\mathcal{W}_T \Cap \mathcal{W}_{T'}|} \tag{4}$$

where $\mathcal{W}_T = \biguplus_{v \in \mathcal{L}_T} words(v)$.

Example 3. In Fig. 1, the word sets of the leaf nodes of the query tree Q and the subtrees D' and D'' of the document tree are as shown in Table 1. Then, the textual similarity for the query tree is $\mathrm{sim_t}^J(Q, D') = 1/10 = 0.1$ and $\mathrm{sim_t}^J(Q, D'') = 0/10 = 0$.

Table 1. The word sets of the leaf nodes of the tree Q, D' and D'' in Fig. 1. Common words are shown in bold.

T	$\lvert\mathcal{W}_T\rvert$	$\mathcal{W}_T$
Q	5	{"Similarity", "Join", "DBS", "**Data**", "Integration"}
D'	6	{"Approximate", "Matching", "Database", "System", "**Data**", "Clensing"}
D''	5	{"Realtime", "Raytracing", "Computer", "Graphics", "Raytracing"}

5 Proposed Method

The approach of our proposed method is similar to the existing approach, TASM [2], which scans the entire document tree to find similar subtrees. Similar to TASM, we implement pruning based on the size of the subtrees and subsequently calculate the similarity for the resultant candidate subtrees.

However, calculating the computationally expensive structural similarity (tree edit distance) for all candidate subtrees is time-consuming. To address this, we first calculate the less expensive textual similarity and use the results to further prune the subtrees. This reduces the number of expensive structural similarity calculation.

Despite this, the computational cost of the structural similarity remains high at $O(|Q|^4)$, making the aforementioned pruning insufficient. To further improve efficiency, we leverage the similarity between the query tree and the candidate subtrees that have passed the initial pruning. By using an algorithm [9,12,20] that computes the tree edit distance in linear time relative to the number of nodes, we can achieve further improvement.

Similar to TASM [2], our algorithm represents the document tree as a postorder queue, as shown in Fig. 2. The postorder queue holds the tree nodes in postorder sequence, where each element is a pair of (*label*, *size*), with *label* representing the node label and *size* the size of the subtree rooted at the node. To construct all candidate subtrees within the upperbound size $\mathrm{maxD_i}$, which will be discussed

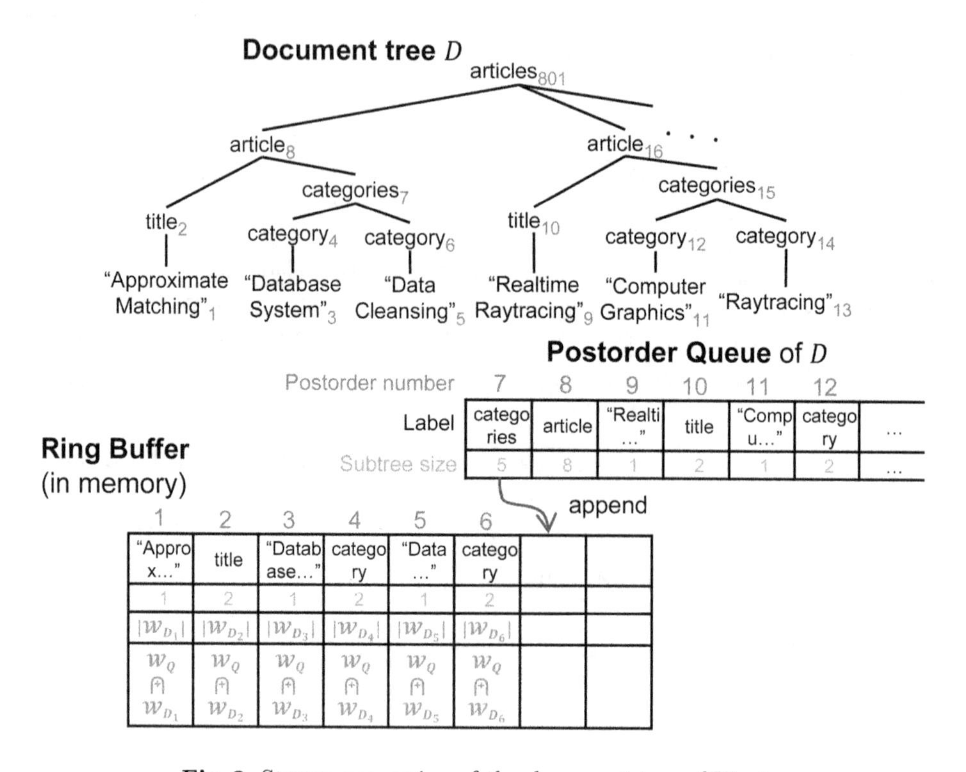

Fig. 2. Stream processing of the document tree of Fig. 1.

later, we use a ring buffer, similar to TASM (see Fig. 2). The postorder queue is constructed with low memory footprint because it dynamically adds nodes to the end as needed during the traversal of the document tree on disk. The ring buffer also contributes to low memory footprint as its size is independent of the document size.

We present our proposed algorithm in Algorithm 1. The detailed process of Algorithm 1 is as follows.

1) **Computing minimum and maximum size of candidate subtree.** At Line 2 of Algorithm 1, before scanning the document tree, We calculate the size range $[\mathrm{minD_i}, \mathrm{maxD_i}]$ of the candidate subtree used for pruning based on the size of subtrees. The condition that the candidate subtree D_i must satisfy is expressed in Eq. 5. The right-hand side of Eq. 5 represents an upper bound of the similarity, calculated using the lower bound of the tree edit distance $||Q| - |D_i||$ ($\leq TED(Q, D_i)$) [16] and the maximum value of the textual similarity, which is 1. Equation 5 can be expressed for the cases of $|Q| \geq |D_i|$ (cf. Eq. 6) and $|Q| < |D_i|$ (cf. Eq. 7).

$$\theta \leq \alpha \left(1 - \frac{||Q| - |D_i||}{|Q| + |D_i|}\right) + (1-\alpha)\frac{\min\{|Q|, |D_i|\}}{\max\{|Q|, |D_i|\}} \tag{5}$$

$$\theta \leq \alpha \frac{2|D_i|}{|Q| + |D_i|} + (1-\alpha)\frac{|D_i|}{|Q|} \quad (|Q| \geq |D_i|) \tag{6}$$

$$\theta \leq \alpha \frac{2|Q|}{|Q| + |D_i|} + (1-\alpha)\frac{|Q|}{|D_i|} \quad (|Q| < |D_i|) \tag{7}$$

For each of Eq. 6 and Eq. 7, we calculate the range of $|D_i| > 0$ that satisfies the equation, obtaining Eq. 8 and Eq. 9, respectively:

$$|D_i| \geq \begin{cases} \left\lceil \frac{-(1+\alpha-\theta)+\sqrt{(1+\alpha-\theta)^2+4\theta(1-\alpha)}}{2(1-\alpha)}|Q| \right\rceil & \text{if } \alpha \neq 1 \\ \left\lceil \frac{\theta}{2-\theta}|Q| \right\rceil & \text{if } \alpha = 1 \end{cases} \quad (|Q| \geq |D_i|) \tag{8}$$

$$|D_i| \leq \left\lfloor \frac{-(\theta - 1 - \alpha) + \sqrt{(\theta - 1 - \alpha)^2 + 4\theta(1-\alpha)}}{2\theta}|Q| \right\rfloor \quad (|Q| < |D_i|) \tag{9}$$

In the algorithm, the right-hand sides of Eq. 8 and Eq. 9 are calculated as $\mathrm{minD_i}$ and $\mathrm{maxD_i}$, respectively.

2) **Initializing ring buffer.** At Line 3 of Algorithm 1, we initialize a ring buffer R that holds up to $\mathrm{maxD_i}$ nodes of the document tree read from the postorder queue. The capacity of the ring buffer is $\mathrm{maxD_i} + 1$, including the terminal sentinel.
   By storing the continuous subsequence of the postorder array of D with length up to $\mathrm{maxD_i}$, we can extract all subtrees whose size is up to $\mathrm{maxD_i}$. All nodes of the candidate subtrees are located within the continuous subsequence of the postorder array of D with a length up to $\mathrm{maxD_i}$ (see e.g., [2]).
3) **Extracting candidate subtrees.** At Line 8 of Algorithm 1, We read nodes from the postorder queue and store them in the ring buffer to obtain the subtree D_i within the size range. This process is performed by the pull_candidate function, as shown in Algorithm 2.
   In the pull_candidate function, we pre-calculate the information required for the calculation of textual similarity. Specifically, for the subtree D_i rooted at the i-th node obtained from the postorder queue, we calculate the size of the bag of words $|\mathcal{W}_{D_i}|$ and the intersection of the set with the query tree $\mathcal{W}_Q \Cap \mathcal{W}_{D_i}$ (As shown in Lines 8 to 12 of Algorithm 2). Results are stored in the ring buffer. All calculations are performed in a bottom-up fashion, and the total time complexity is $O(|\mathcal{W}_Q||D| + |\mathcal{W}_D|)$.
4) **Computing textual similarity.** At Line 13 of Algorithm 1, we calculate the textual similarity Sim_t between the query tree Q and the candidate subtree D_i. This process utilizes $|\mathcal{W}_{D_i}|$ and $\mathcal{W}_Q \Cap \mathcal{W}_{D_i}$, which were pre-calculated in the pull_candidate function, enabling the similarity to be computed in $O(1)$ time.
5) **Filtering by textual similarity.** We perform pruning based on the calculated textual similarity value Sim_t. First, if $\beta(|Q|, |D_i|)Sim_t$ alone exceeds θ, we add the candidate subtree D_i to the output result and skip the subsequent processing (Lines 14 to 16 of Algorithm 1). Then, we calculate a tighter upper

bound $SimUB$ by replacing 1 (the upper bound of the textual similarity) in Eq. (5) with the actual calculated value Sim_t (At Line 17). if $SimUB$ is less than θ, D_i cannot be similar and omit the subsequent calculations.

6) **Computing structural similarity.** At Lines 21 to 26 of Algorithm 1, for the candidate subtree D_i that has passed the previous pruning steps. This calculation uses an algorithm [9,12,20] that computes the tree edit distance between similar trees in linear time relative to the number of nodes.
We use the precomputed textual similarity Sim_t to determine the threshold τ for the tree edit distance when the similarity condition $\mathrm{sim}(Q, D_i) \geq \theta$ is met. This threshold τ helps to limit the search space and speed up the computation of the tree edit distance. We calculate the tree edit distance as follows.
   (a) **Computing lower bound of TED.** At Line 22 of Algorithm 1, We calculate the lower bound of the tree edit distance by calculating the maximum value of the Levenshtein distance between the arrays arranged in preorder/postorder of the nodes of the tree [9], and prune the subtree that exceeds τ. The time complexity is $O(\max\{|Q|, |D_i|\}\tau)$, and the space complexity is $O(\tau)$.

Algorithm 1 Subtree-Similarity-Search(Q, pq, θ, α)

Input: Query tree Q, postorder queue pq of a document tree D, similarity threshold $\theta \in (0, 1]$, weight coefficient for structural similarity $\alpha \in [0, 1]$.
Output: $\mathcal{O} = \{D_i | sim(Q, D_i) \geq \theta\}$.

1: $\mathcal{O} \leftarrow \emptyset$
2: Calculate $\mathrm{minD_i}$ and $\mathrm{maxD_i}$ from $|Q|$, θ, α.
3: $R \leftarrow$ Ring buffer of size $\mathrm{maxD_i} + 1$.
4: $s \leftarrow 1,\ \ e \leftarrow 1$ // Start/End index of ring buffer R.
5: /* Ring buffer R stores s-th to $(e-1)$-th elements of pq. */
6: $Rsize \leftarrow \mathrm{maxD_i} + 1$ // Capacity of ring buffer R.
7: **while** true **do**
8: $D_i, s, e \leftarrow$ pull_candidate($pq, R, Rsize, s, e, \mathrm{minD_i}, \mathrm{maxD_i}$)
9: $Sim_t \leftarrow 0$
10: **if** $D_i = \emptyset$ **then**
11: break
12: **if** $\alpha < 1$ **then**
13: $Sim_t \leftarrow \mathrm{sim_t}(Q, D_i)$
14: **if** $(1-\alpha)\beta(|Q|, |D_i|)Sim_t \geq \theta$ **then**
15: $\mathcal{O} \leftarrow \mathcal{O} \cup \{D_i\}$
16: continue
17: $SimUB \leftarrow \alpha\left(1 - \frac{||Q|-|D_i||}{|Q|+|D_i|}\right) + (1-\alpha)\beta(|Q|, |D_i|)Sim_t$
18: **if** $SimUB < \theta$ **then**
19: continue
20: **else if** $0 < \alpha$ **then**
21: $\tau \leftarrow \left\lfloor \left(1 - \frac{\theta-(1-\alpha)\beta(|Q|,|D_i|)Sim_t}{\alpha}\right)(|D_i| + |Q|)\right\rfloor$
22: **if** LowerBoundED(Q, D_i, τ) $\leq \tau$ **then**
23: **if** UpperBoundED(Q, D_i, τ) $\leq \tau$ **then**
24: $\mathcal{O} \leftarrow \mathcal{O} \cup \{D_i\}$
25: **else if** TED(Q, D_i, τ) $\leq \tau$ **then**
26: $\mathcal{O} \leftarrow \mathcal{O} \cup \{D_i\}$

(b) **Computing upper bound of TED.** At Line 23 of Algorithm 1, We calculate the upper bound of the tree edit distance using the LGM proposed by Thomas et al. [12], output the subtree that is less than or equal to τ, and skip the subsequent processing. The time complexity is $O(\max\{|Q|, |D_i|\}\tau)$, and the space complexity is $O(\max\{|Q|, |D_i|\})$.

(c) **Computing TED.** At Line 25 of Algorithm 1, We calculate the thresholded tree edit distance using the TopDiff proposed by Mateusz et al. [20] for the remaining candidate subtrees. The time complexity is $O(nd\tau \min\{l, \tau\})$, and the space complexity is $O(n\tau)$, where $n = \max\{|Q|, |D_i|\}$, $d = \max\{depth(Q), depth(D_i)\}$, and $l = \max\{|\mathcal{L}_Q|, |\mathcal{L}_{D_i}|\}$.

Algorithm 2 pull_candidate(pq,R,$Rsize$,s,e,minD$_\mathrm{i}$,maxD$_\mathrm{i}$)

Input: Postorder queue of a document tree pq, ring buffer R, capacity of ring buffer $Rsize$, start/end index of ring buffer s and e, minimum/maximum size of candidate subtree minD$_\mathrm{i}$ and maxD$_\mathrm{i}$.
Output: Candidate subtree D_i, start/end index of ring buffer s and e.

```
1: D_i ← ∅
2: while pq ≠ ∅ do
3:     /* λ : label. size : size of subtree.                              */
4:     (pq, (λ, size)) ← dequeue(pq)
5:     R[e mod Rsize] ← (λ, size)                        // Push (λ, size) to R.
6:     i ← e
7:     e ← e + 1
8:     if size ≤ maxD_i then
9:         if size = 1 then
10:            compute |W_{D_i}| and W_Q ⋒ W_{D_i} from label λ.
11:        else
12:            compute |W_{D_i}| and W_Q ⋒ W_{D_i} from D_i's children.
13:    if minD_i ≤ size and size ≤ maxD_i then
14:        lmld ← i − size + 1     // The index of the leftmost leaf descendant of D_i.
15:        /* Extract the subtree rooted at i-th node.                    */
16:        D_i ← extract_subtree(R, lmld, i)
17:    if e − s ≥ maxD_i then
18:        s ← s + 1                          // Pop the oldest element from R.
19:    if D_i ≠ ∅ then
20:        break
```

6 Experiments

We conduct experiments to verify the effectiveness of the proposed method, evaluating both the search speed and accuracy in the context of duplicate detection. All methods are implemented in C++20 and run on an Intel Core i7-8700K 3.20 GHz and 16 GB RAM, Ubuntu 24.04 LTS. The compiler is g++ 13.2.0, with the optimization option -O3.

6.1 Dataset

We use the MEDLINE[1] database, which collects literature information mainly in the field of life sciences, provided in XML format. The collection date is January 19, 2024, and it contains about 34 million pieces of literature information.

In this experiment, we randomly select literature records from the MEDLINE database to construct a document tree. Specifically, we use one piece of literature from MEDLINE as the query tree. The document tree consists of several duplicate literature records that match the query tree and other non-relevant literatures as child subtrees of the root node. For the duplicate records, we add noise to the label of each leaf node of the query tree with an 80% probability by performing one of the following two operations: (1) Randomly delete one word from the label. (2) Randomly select one word from the label and replace it with a synonym or similar word with a similarity of 0.5 or higher using the MeSH (Medical Subject Headings)[2] dictionary of MEDLINE. The similarity between synonyms or similar words is calculated based on the method described in [24].

Table 2 shows the statistics of the generated query tree and document tree. The datasets `D25`, `D50`, and `D100` vary only in the size of the document tree (# of publication records in D). In contrast, the datasets `Q150`, `Q300`, and `Q600` vary only in the size of the query tree ($|Q|$).

Table 2. Data set characteristics.

Name	Size D [MB]	$\|Q\|$	# Pub. in D	# Dup. subtrees with Q in D	Size [Nodes]		# diff. labels	
					$\|D\|$	avg. $\|D_i\|$	Q	D
D25	**25.9**	150	3750	100	1.2×10^6	6.54	102	87.7×10^3
D50	**51.7**	150	7500	100	2.3×10^6	6.54	102	152×10^3
D100	**106.1**	150	15000	100	4.8×10^6	6.56	102	272×10^3
Q150	25.7	**150**	3750	100	1.2×10^6	6.54	102	86.3×10^3
Q300	26.0	**300**	3750	100	1.2×10^6	6.55	153	86.4×10^3
Q600	27.1	**600**	3750	100	1.2×10^6	6.56	241	86.6×10^3

6.2 Methods

We compare the proposed method TASMj with the two state-of-the-art methods: TASM (index-free) [2] and SlimCone (index-based) [16]. The implementation of SlimCone is available online[3]. We evaluate the search speed and search accuracy (Precision/Recall) while decreasing the similarity threshold θ from 1.0 to 0.85 by 0.05. Since the compared methods are top-k search, we perform the search while increasing the parameter k from ten by ten. Precision and Recall are calculated using Eq. 10 and 11, respectively. TP (True Positive) is the number of duplicate

[1] https://ftp.ncbi.nlm.nih.gov/pubmed/baseline/.
[2] https://nlmpubs.nlm.nih.gov/projects/mesh/MESH_FILES/xmlmesh/.
[3] https://frosch.cosy.sbg.ac.at/dkocher/kocher-sigmod2019-reproducibility.

trees correctly detected as duplicates. The denominator of Recall is 100, which is the number of duplicate trees in the dataset.

$$\text{Precision} = \frac{\text{TP}}{\#\ \text{subtrees in result}} \tag{10}$$

$$\text{Recall} = \frac{\text{TP}}{100}, \tag{11}$$

6.3 Effect of the Recall

We evaluate the effect of recall on the query time by varying the similarity threshold θ (or k) for the dataset Q300. For TASMj, the query time and the number of candidates are improved by orders of magnitude compared to TASM (cf. Figure 3). By utilizing LGM and TopDiff, TASMj remains competitive in terms of running time, even in high recall settings where many subtrees need to be processed. For TASMj with $\alpha = 0.5$, the number of candidates is comparable to the index-based SlimCone, due to pruning based on textual similarity with respect to the label information. This approach is similar to SlimCone, which initially extracts subtrees with the most unique labels common to the query tree by looking up an inverted list index.

6.4 Effect of the Document Size

Next, we evaluate the effect of the document size by varying the document size while keeping the query size fixed at $|Q| = 150$ and recall at approximately 0.8. The query time and the number of candidates are improved by orders of magnitude compared to the existing scan-based method TASM (cf. Fig. 4). TASMj reduces the number of candidates through effective pruning, as shown in Fig. 4, where the number is comparable to SlimCone. While TASM, a scan-based method, shows an increase of candidates proportional to the document tree size, TASMj with $\alpha = 0.5$ maintains a comparable number of candidates to SlimCone.

6.5 Effect of the Query Size

Finally, we evaluate the effect of the query size varying the query size while keeping the document size fixed at approximately 25 MB and recall at approximately 0.8. When the query tree size increases, TASMj outperforms TASM and SlimCone in terms of query time (cf. Fig. 5). As the size of the query tree increases, the computational cost of similarity verification rises, leading to an increase in query time. However, the increase in query time is more gradual compared to TASM and SlimCone due to the reduction in computational cost achieved by effective pruning during the structural similarity calculation.

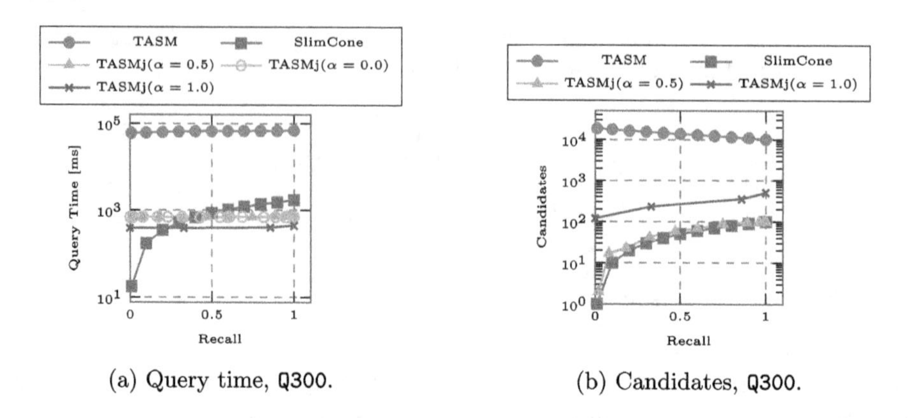

(a) Query time, Q300. (b) Candidates, Q300.

Fig. 3. State of the art vs. TASMj: Query time and number of candidates over varying recall, Document size ≃25 MB, $|Q| = 300$.

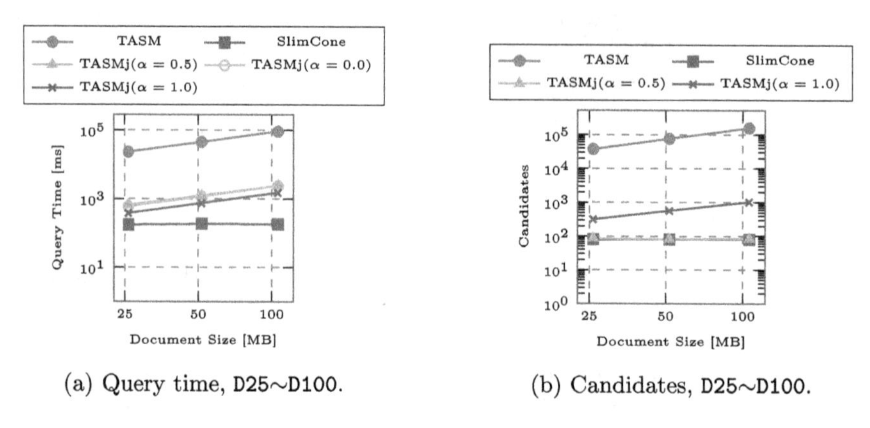

(a) Query time, D25∼D100. (b) Candidates, D25∼D100.

Fig. 4. State of the art vs. TASMj: Query time and number of candidates over document size, recall ≃0.8, $|Q| = 150$.

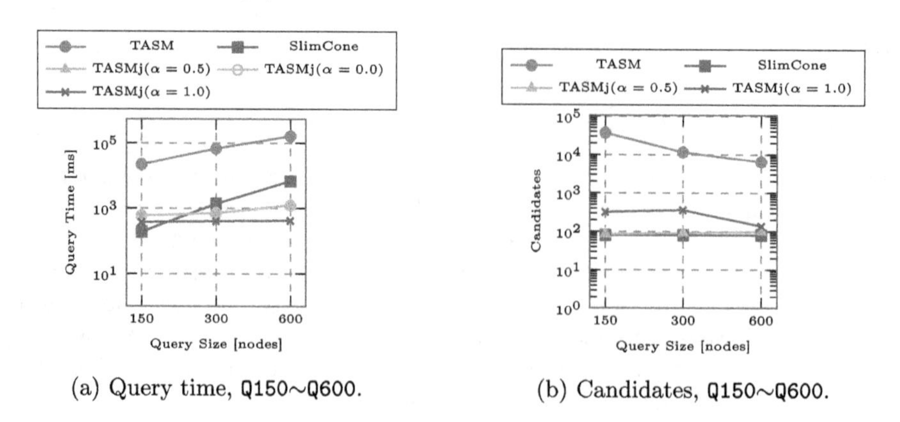

(a) Query time, Q150∼Q600. (b) Candidates, Q150∼Q600.

Fig. 5. State of the art vs. TASMj: Query time and number of candidates over query size, recall ≃0.8, Document size ≃25 MB.

6.6 Accuracy

All plotted points have a search accuracy (Precision) of 100%. This high precision is because, when generating duplicate records, only the labels of the query tree's leaf nodes are modified, while the structure and other labels of the inner nodes remain the same as the query tree. Consequently, all top-ranked similar subtrees correctly match the duplicate records.

7 Conclusion

In this paper, we proposed TASMj, an extension of the index-free method TASM, which combines structural similarity and textual similarity. TASMj calculates textual similarity using the label information of leaf nodes and reduces the number of structural similarity calculations through pruning based on textual similarity. Additionally, it further reduces the computational cost of structural similarity by leveraging the similarity between the query tree and candidate subtrees. Empirical evaluations show that the proposed method improves search speed by orders of magnitude compared to the existing scan-based method TASM and achieves competitive search speed compared to the index-based method SlimCone.

Acknowledgements. This paper is based on results obtained from JSPS KAKENHI Grant Numbers JP23K16888 and JP23K24949, JST CREST Grant Number JPMJCR22M2, and "Research and Development Project of the Enhanced infrastructures for Post-5G Information and Communication Systems" (JPNP20017), commissioned by the New Energy and Industrial Technology Development Organization (NEDO).

References

1. Akutsu, T.: Tree edit distance problems: algorithms and applications to bioinformatics. IEICE Trans. Inf. Syst. **E93.D**(2), 208–218 (2010). https://doi.org/10.1587/transinf.E93.D.208
2. Augsten, N., Barbosa, D., et al.: TASM: top-k approximate subtree matching. In: 2010 IEEE 26th International Conference on Data Engineering (ICDE 2010), pp. 353–364 (2010). https://doi.org/10.1109/ICDE.2010.5447905
3. Bille, P.: A survey on tree edit distance and related problems. Theor. Comput. Sci. **337**(1), 217–239 (2005). https://doi.org/10.1016/j.tcs.2004.12.030. https://www.sciencedirect.com/science/article/pii/S0304397505000174
4. Cohen, S.: Indexing for subtree similarity-search using edit distance. In: Proceedings of the 2013 ACM SIGMOD International Conference on Management of Data, SIGMOD 2013, pp. 49–60. Association for Computing Machinery, New York (2013). https://doi.org/10.1145/2463676.2463716
5. Demaine, E.D., Mozes, S., et al.: An optimal decomposition algorithm for tree edit distance. ACM Trans. Algorithms **6**(1) (2010). https://doi.org/10.1145/1644015.1644017

6. Fagin, R., Lotem, A., et al.: Optimal aggregation algorithms for middleware. In: Proceedings of the Twentieth ACM SIGMOD-SIGACT-SIGART Symposium on Principles of Database Systems, pp. 102–113 (2001)
7. Falleri, J.R., Morandat, F., et al.: Fine-grained and accurate source code differencing. In: Proceedings of the 29th ACM/IEEE International Conference on Automated Software Engineering, ASE 2014, pp. 313–324. Association for Computing Machinery, New York (2014). https://doi.org/10.1145/2642937.2642982
8. Fluri, B., Wursch, M., et al.: Change distilling: tree differencing for fine-grained source code change extraction. IEEE Trans. Softw. Eng. **33**(11), 725–743 (2007). https://doi.org/10.1109/TSE.2007.70731
9. Guha, S., Jagadish, H.V., et al.: Approximate XML joins. In: Proceedings of the 2002 ACM SIGMOD International Conference on Management of Data, SIGMOD 2002, pp. 287–298. Association for Computing Machinery, New York (2002). https://doi.org/10.1145/564691.564725
10. Herrbach, C., Denise, A., et al.: Average complexity of the Jiang-Wang-Zhang pairwise tree alignment algorithm and of a RNA secondary structure alignment algorithm. Theor. Comput. Sci. **411**(26), 2423–2432 (2010). https://doi.org/10.1016/j.tcs.2010.01.014
11. Hütter, T., Augsten, N., et al.: JEDI: these aren't the JSON documents you're looking for... In: Proceedings of the 2022 International Conference on Management of Data, SIGMOD 2022, pp. 1584–1597. Association for Computing Machinery, New York (2022). https://doi.org/10.1145/3514221.3517850
12. Hütter, T., Pawlik, M., et al.: Effective filters and linear time verification for tree similarity joins. In: 2019 IEEE 35th International Conference on Data Engineering (ICDE), pp. 854–865 (2019). https://doi.org/10.1109/ICDE.2019.00081
13. Kailing, K., Kriegel, H.-P., Schönauer, S., Seidl, T.: Efficient similarity search for hierarchical data in large databases. In: Bertino, E., et al. (eds.) EDBT 2004. LNCS, vol. 2992, pp. 676–693. Springer, Heidelberg (2004). https://doi.org/10.1007/978-3-540-24741-8_39
14. Kan, T., Higuchi, S., et al.: Segmental mapping and distance for rooted labeled ordered trees. Fundam. Informaticae **132**(4), 461–483 (2014). https://doi.org/10.3233/FI-2014-1054
15. Karpov, N., Zhang, Q.: Syncsignature: a simple, efficient, parallelizable framework for tree similarity joins. Proc. VLDB Endow. **16**(2), 330–342 (2022). https://doi.org/10.14778/3565816.3565833
16. Kocher, D., Augsten, N.: A scalable index for top-k subtree similarity queries. In: Proceedings of the 2019 International Conference on Management of Data, SIGMOD 2019, pp. 1624–1641. Association for Computing Machinery, New York (2019). https://doi.org/10.1145/3299869.3319892
17. Li, F., Wang, H., et al.: A survey on tree edit distance lower bound estimation techniques for similarity join on XML data. SIGMOD Rec. **42**(4), 29–39 (2014). https://doi.org/10.1145/2590989.2590994
18. Pawlik, M., Augsten, N.: RTED: a robust algorithm for the tree edit distance. Proc. VLDB Endow. **5**(4), 334–345 (2011). https://doi.org/10.14778/2095686.2095692
19. Pawlik, M., Augsten, N.: Tree edit distance: robust and memory-efficient. Inf. Syst. **56**, 157–173 (2016). https://doi.org/10.1016/j.is.2015.08.004
20. Pawlik, M., Augsten, N.: Minimal edit-based diffs for large trees. In: Proceedings of the 29th ACM International Conference on Information & Knowledge Management, CIKM 2020, pp. 1225–1234. Association for Computing Machinery, New York (2020). https://doi.org/10.1145/3340531.3412026

21. Touzet, H.: Comparing similar ordered trees in linear-time. J. Discret. Algorithms **5**(4), 696–705 (2007). https://doi.org/10.1016/j.jda.2006.07.002
22. Wang, J., Yang, J., Zhang, W.: Top-k tree similarity join. In: Proceedings of the 30th ACM International Conference on Information & Knowledge Management, CIKM 2021, pp. 1939–1948. Association for Computing Machinery, New York (2021). https://doi.org/10.1145/3459637.3482304
23. Wang, L., Zhang, K.: Space efficient algorithms for ordered tree comparison. In: Deng, X., Du, D.-Z. (eds.) ISAAC 2005. LNCS, vol. 3827, pp. 380–391. Springer, Heidelberg (2005). https://doi.org/10.1007/11602613_39
24. Xu, P., Lu, J.: Towards a unified framework for string similarity joins. Proc. VLDB Endow. **12**(11), 1289–1302 (2019). https://doi.org/10.14778/3342263.3342268
25. Yang, R., Kalnis, P., Tung, A.K.H.: Similarity evaluation on tree-structured data. In: Proceedings of the 2005 ACM SIGMOD International Conference on Management of Data, SIGMOD 2005, pp. 754–765. Association for Computing Machinery, New York (2005). https://doi.org/10.1145/1066157.1066243
26. Zhang, K., Shasha, D.: Simple fast algorithms for the editing distance between trees and related problems. SIAM J. Comput. **18**(6), 1245–1262 (1989). https://doi.org/10.1137/0218082

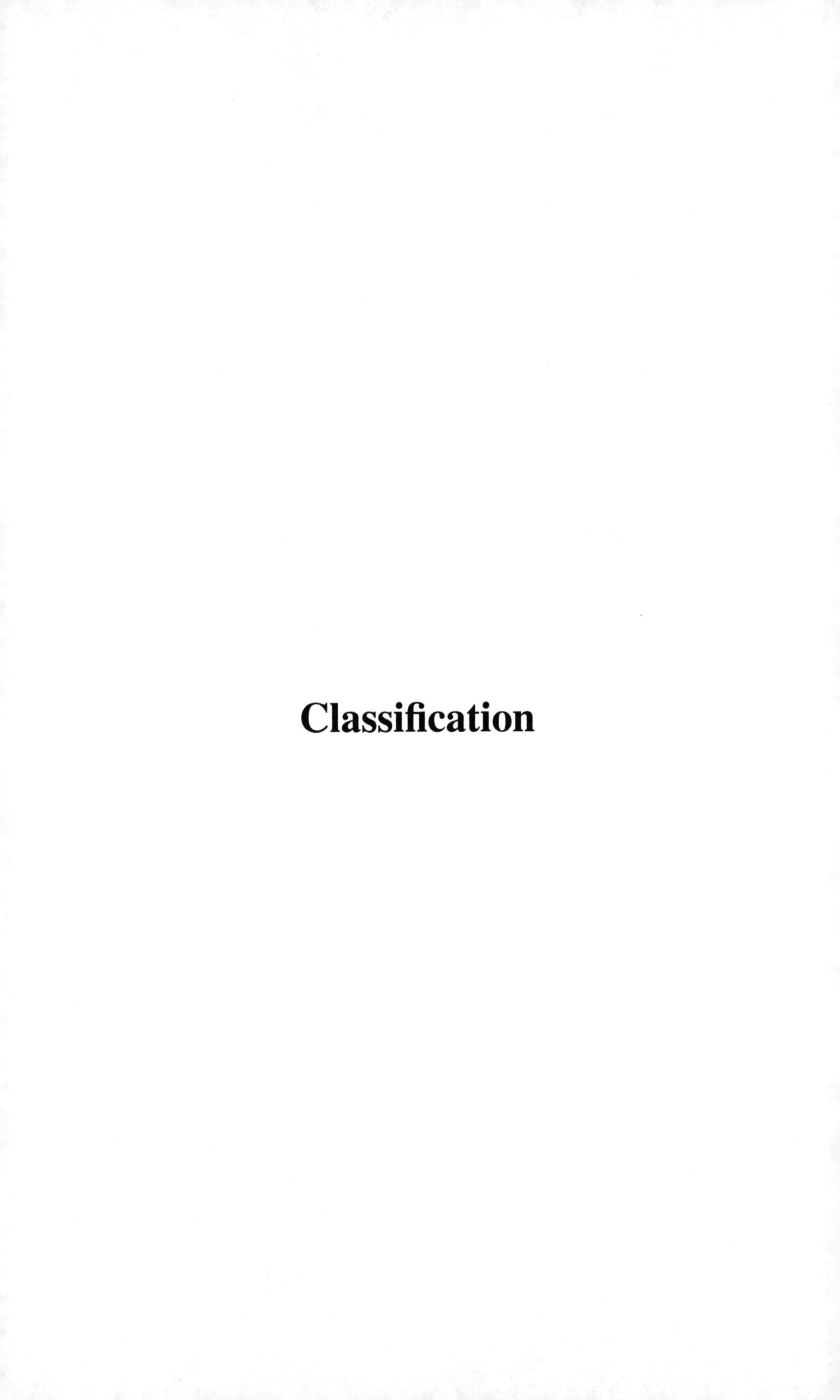

Classification

Towards Hybrid Embedded Feature Selection and Classification Approach with Slim-TSF

Anli Ji(✉), Chetraj Pandey, and Berkay Aydin

Georgia State University, Atlanta, USA
{aji1,cpandey1,baydin2}@gsu.edu

Abstract. Traditional solar flare forecasting approaches have mostly relied on physics-based or data-driven models using solar magnetograms, treating flare predictions as a point-in-time classification problem. This approach has limitations, particularly in capturing the evolving nature of solar activity. Recognizing the limitations of traditional flare forecasting approaches, our research aims to uncover hidden relationships and the evolutionary characteristics of solar flares and their source regions. Our previously proposed Sliding Window Multivariate Time Series Forest (Slim-TSF) has shown the feasibility of usage applied on multivariate time series data. A significant aspect of this study is the comparative analysis of our updated Slim-TSF framework against the original model outcomes. Preliminary findings indicate a notable improvement, with an average increase of 5% in both the True Skill Statistic (TSS) and Heidke Skill Score (HSS). This enhancement not only underscores the effectiveness of our refined methodology but also suggests that our systematic evaluation and feature selection approach can significantly advance the predictive accuracy of solar flare forecasting models.

Keywords: Multivariate Time Series Classification · Solar Flare Prediction · Interval-based Classification

1 Introduction

Solar weather events, encompassing phenomena like solar flares, coronal mass ejections (CMEs), solar wind variations, and geomagnetic storms, hold significant importance for Earth's environment and human technological systems. Among many solar phenomena, solar flares are one of the most intense localized explosions of electromagnetic energy emanating from the Sun's atmosphere. When such energy bursts out, it usually travels near the speed of light ranging from several minutes to hours. It often does not occur alone but alongside other events like coronal mass ejections (CMEs) or solar wind, which can trigger severe geomagnetic storms, extensive radio blackouts on Earth's daylight side, and interfere with delicate instruments onboard near-Earth space equipment. Recent studies have employed physics-based or data-driven models [20,28,31] to

R. Wrembel et al. (Eds.): DaWaK 2024, LNCS 14912, pp. 91–105, 2024.
https://doi.org/10.1007/978-3-031-68323-7_7

predict solar flares using data primarily sourced from solar magnetograms [33]. Many of these approaches tend to predict solar flares as a classification problem using point-in-time measurements (where a single time point is applied to represent a single event). Such methods often do not consider the intrinsic temporal evolution nature of data [10] by evaluating different observations as separate entities, meaning the dynamic essence of flares is usually overlooked.

The characteristics of solar flare evolution are important as they are intricately linked to the dynamic behavior of solar active regions, as delineated in prior research [6,24,25]. Analyzing these temporal characteristics of flares, it becomes possible to reveal potential implicit relationships and capture unidentified patterns between flares and their originating regions. In our prior study [18], we utilized ensembles of interval-based classification models on multivariate time series data for event prediction. However, this method presented a limitation in understanding which features were more pivotal in decision-making and the rationale behind these decisions. Traditional interval-based classifiers often do not support systematic evaluation through random sub-interval sampling, leading to a process where the identification of relevant features (or intervals from the time series) was arbitrarily generated, thereby missing out on extracting meaningful insights from the model. In our subsequent studies [15] and [17], we aimed to identify crucial interval features from multivariate time series data using multi-scale sliding windows with varying interval sizes and step sizes as well as an innovative feature ranking schema for identifying feature importance. This advancement seeks to introduce interpretability into previously opaque models, enhancing our understanding of the decision-making processes underlying model prediction.

In this study, we expand our previous work focusing more on the systematic evaluation of our Sliding Window Multivariate Time Series Forest (Slim-TSF) model. This involves strategically selecting relevant features to enhance our grasp of the temporal dynamics crucial for solar flare prediction. We've introduced an indexing function to improve the model selection process. This function enables us to identify optimal models using a concise set of parameters and features that have shown promise in prior research. Additionally, we employ a customized internal validation schema to cross-verify our findings, ensuring the robustness and reliability of our results. This approach has led to a noticeable improvement in our model's performance. Specifically, we've achieved an average increase of 5% in True Skill Statistics (TSS) and Heidke Skill Score (HSS) compared to our original Slim-TSF outcomes. This improvement underscores the value of a systematic feature selection and validation strategy in enhancing the accuracy of solar flare predictions.

The rest of the paper is organized as follows: Sect. 2 provides background information on existing time series classification models pertinent to flare prediction. In Sect. 3, we provide our problem formulation and introduce our multivariate time series classification model and feature ranking method used for extracting relevant feature intervals from provided time series data. Section 4 presents our experimental setup and evaluation framework. Finally, Sect. 5 provides conclusions from our study and discusses potential avenues for future research.

2 Related Work

From the proliferation of available time series datasets [32] and a wide spectrum of machine learning-based techniques proposed for time series classification, similarity-based and feature-based algorithms are two notable categories utilized for these predictive tasks. Similarity-based methods predict by measuring the similarity between training and testing instances, using metrics like Euclidean distance or Dynamic Time Warping (NN-DTW) [4,5,21,30]. In contrast, feature-based algorithms generate predictions by extracting temporal features from entire time series or subsequences within them. For solar flare prediction, both full-disk (e.g., [24,26,27]) and active region-based (e.g., [8,13,14,18]) approaches have shown significant impact by utilizing derived time series features.

Using feature-based algorithms that capture associations between target variables and time series instances through derived features, this distinction is particularly evident in tasks like solar flare prediction or other tasks (such as anomaly detection [12]). For example, [23] extracted basic statistical features like mean and standard deviation from global time series to feed a multi-layer perceptron network, though this method neglected localized informative characteristics. In contrast, [11] enhanced model interpretability by considering local attributes through piecewise constant modeling and pattern extraction, though it often resulted in simplistic features during selection. Furthermore, [?] incorporated an extensive range of features such as wavelets and Chebyshev coefficients, but this method faced high computational costs and lacked inherent interpretability in high-dimensional data spaces.

It is a challenging task for many feature-based classification methods when dealing with multivariate time series data because they require additional intricate information across features. Such discriminating features are usually hard to generate in high-dimensional space due to the unknown interrelations among input parameters, adding complexity to model construction. To address this problem, various techniques have been attempted to ensemble univariate models from individual feature spaces instead of considering the global correlations between them. These methods focus on extracting relevant features in univariate aspects and then applying traditional machine learning algorithms for classification. Common features include statistical measures (e.g., mean, variance), spectral features (e.g., Fast Fourier Transform coefficients), and time-domain features (e.g., autocorrelation).

For example, Shapelet-based decision trees [34] combine shapelets (i.e., discriminative subsequences that capture distinctive patterns in time series data) within an ensemble architecture. This method extracts shapelets from the training data and constructs an ensemble of decision trees (e.g., random forest), where each estimator focuses on a different subset of shapelets, typically measured by Euclidean distance. While effective at capturing local patterns in multivariate time series data, this approach can be computationally expensive and may struggle to identify relevant shapelets in high-dimensional spaces that are both informative and broadly applicable. Additionally, shapelets extracted from one dataset might not generalize well to other datasets with different dimensionalities, characteristics, and patterns.

To mitigate these issues, the Generalized Random Shapelet Forest (gRSF) [19] improves upon the original shapelet-based method by measuring distances between randomly selected time series and others within a threshold distance of the representative shapelet. Similarly, the Time Series Forest (TSF) [9] incorporates subseries, but instead of relying on distance measurements from learned subsequences, it derives summarized statistical features (such as mean, standard deviation, and gradient) within randomly divided intervals of the univariate time series. This approach treats each time step as a separate component and constructs decision trees for each feature dimension to capture temporal relationships and reduce the high-dimensional feature space. However, this method may not fully capture the interrelationships between different components of the time series, leading to a potential loss of crucial inter-channel relationships and dependencies in multivariate data. The Canonical Interval Forest (CIF) [?] extends TSF by incorporating additional canonical characteristics of the time series and *catch22* [22] features extracted from each interval. This approach aims to capture both individual patterns within each time series component and relationships between different components. However, interpreting an ensemble of decision trees remains challenging, making it less intuitive to understand the combined effects of multiple trees on multivariate time series data compared to single decision trees.

Many of these methods focus solely on understanding how each feature behaves independently, without considering interactions between different features. A relationship within a single time series parameter might be significant for one specific feature but not necessarily for others. The connections between distinct features are often unknown upfront. Understanding feature dependencies in time series data is crucial for improving model interpretability and performance [29]. Techniques like Partial Dependence Analysis (PDA) are commonly used in quantifying these dependencies but can be challenging to explore and analyze in multivariate aspects. [2] proposes a conceptual framework that refines the computed partial dependences on combinatorial feature subspaces (i.e., all the possible combinations of features on all their domains) but still lacks the capability of capturing intercorrelations that differentiate between features.

In time series classification problems, selecting the most relevant time intervals is crucial when generating features that effectively distinguish our data, thereby ensuring a robust model. However, identifying these relevant intervals is difficult because they cannot be directly determined and typically require a computationally expensive search across the entire series. Extracting the underlying mutual information from these relevant intervals can enhance our understanding of the predictive process and accelerate the transition from research to practical applications in flare forecasting models. Our objective in this work is to establish a framework that can recognize these characteristics and offer deeper insights into the behavior of classifiers during prediction tasks.

3 Methodology

In this section, we will outline our approach, including the extraction of statistical features from time intervals, the introduction of the sliding-window time series forest, and the feature ranking technique we employ.

Our proposed sliding-window multivariate time series forest is an early fusion, interval-based ensemble classification method. Figure 1 illustrates our feature generation process using the sliding window operation. This method employs multiple short decision trees, similar to random forests, which utilize interval-based features extracted from univariate time series through multi-scale sliding windows. By combining features from univariate time series at an early stage, we aim to understand the relationships among these features, using an embedded feature ranking based on mutual information.

Interval Features. To extract well-structured and relevant intervals, we calculate statistical characteristics such as mean, standard deviation, and slope for each interval. Additionally, we derive transformed features like maximum, minimum, and mean through a localized pooling procedure applied to the individual interval features extracted from consecutive intervals after the sliding window operation. In this process, all potential interval sets originating from the same time series are collected, and pooling functions are applied for consolidation. Essentially, we consider the highest, lowest, and average values of statistical properties from each parameter of each subseries obtained through sliding window operations. Formal definitions and explanations for processing multivariate time series and extracting vectorized features and transformation are provided in our previous research [16].

Sliding Window Multivariate Time Series Forest. After extracting interval features from subsequences obtained through the sliding window operation and applying secondary transformations to these statistical attributes, we merge these two groups of derived features into an input vector. This vector serves as the foundation for creating a versatile time series classifier we refer to as *Slim-TSF*. Among the wide array of supervised learning models available for making predictions, we have chosen random forest classifiers for two reasons: (1) their effectiveness and resilience when dealing with noisy, high-dimensional data, and (2) their ability to select the most relevant features from a given dataset with respect to a target feature.

It is important to highlight that, depending on the chosen parameter settings, such as using smaller window and step sizes, the interval feature vectors' data space can expand considerably. Additionally, the process of vectorization based on the sliding window approach may generate data points that exhibit some degree of correlation and potential noise. Consequently, it is crucial to systematically identify and remove these features. This is achieved through the application of information-theoretic relevance metrics (e.g., Gini index or entropy).

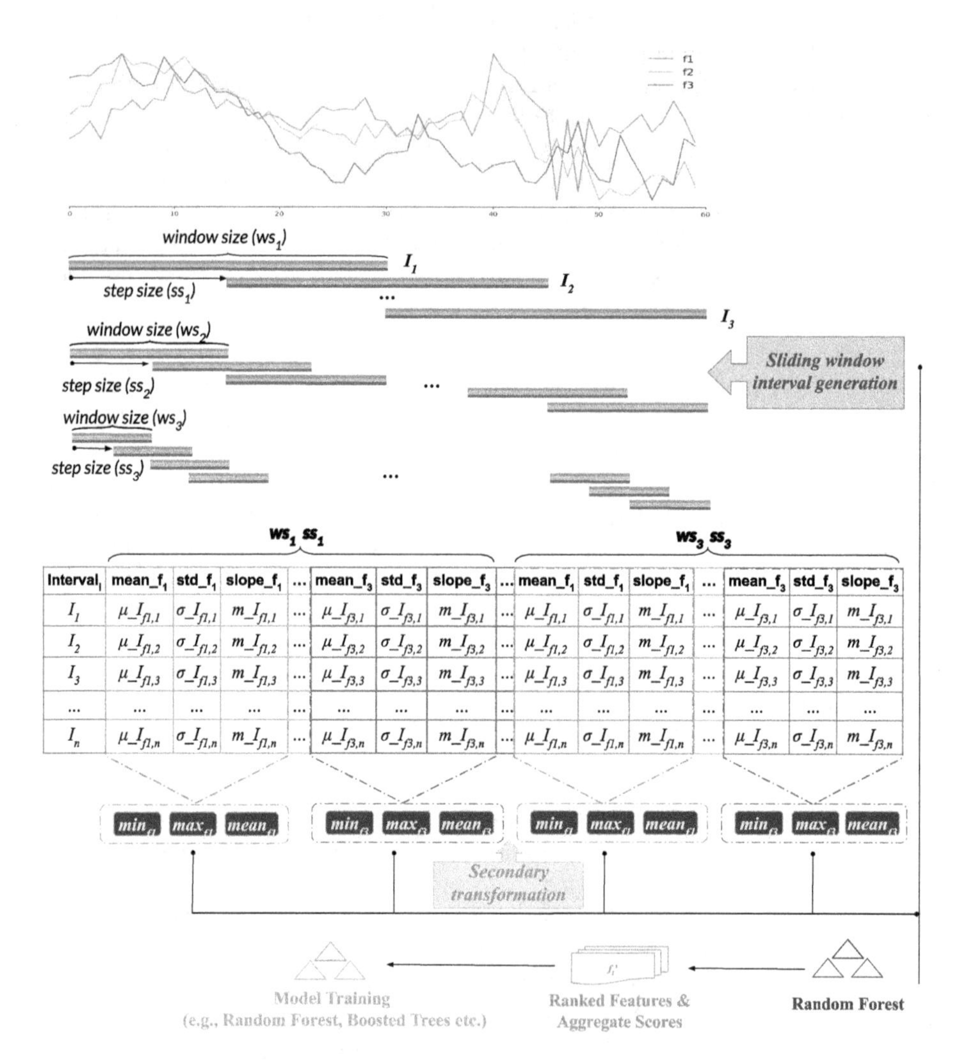

Fig. 1. An illustration of the sliding window-based statistical feature generation. We first generate subsequences (intervals) with a fixed-size sliding window and step size. Then, we create vectorized features from these intervals where these features can be used as input for the sliding window multivariate time series forest (a random forest built on multivariate time series features) and features are ranked with aggregated relevance scores.

This meticulous feature selection process ensures the efficacy of our approach by retaining only the most informative attributes while discarding redundant ones.

Feature Ranking. Our ranking methodology involves a systematic process conducted through multiple experiments, denoted by the total number N, each executed with distinct experimental configurations. This is analogous to a grid search process. The experiments with different configurations yield individual selected features, denoted as exp_j, where j signifies a specific experiment. The ranking denoted as r, is a mapping that assigns a rank i to each feature, reflecting its position within the ranking. In each experiment, the features are ranked to create a specific ordering, denoted as $r_{f,i}$, which designates that feature f has achieved the ith rank in that particular experiment. Subsequently, the top-k features selected for inclusion in the selected feature set, denoted as SFS_j, within each individual experiment j are determined from the ranking r (i.e., include features whose rank is less than or equal to k). This selected feature set is represented as $\{r_{f,1}, r_{f,2}, ..., r_{f,k}\}$. In the end, the selected feature sets across all experiments are aggregated by summing the sparse representation of top-k membership vectors $(\widehat{SFS_j})$ from each experiment (as in Eq. 1).

$$SFS = \sum_{j=1,N} \widehat{SFS_j} \tag{1}$$

This approach allows for a systematic and consistent method of selecting top features across multiple experiments, enhancing the robustness and reliability of the feature selection process. Furthermore, we create a counting vector per each interval of each parameter, denoted as ct_v to represent the value counts of individual intervals in the selected feature set SFS. This counting vector serves as a transformation function, indicating the frequency with which a given interval appears within the top-k selections of the feature set.

Hyperparameter Optimization. Hyperparameter optimization is the process of selecting the optimal hyperparameters to achieve the best performance for a classifier. This process is applied to determine the optimal hyperparameters of our slim-TSF classifier. Traditional grid search cross-validation (CV) is designed for tabulated data and assumes that instances are independent, meaning random assignment of instances to different training and testing folds does not risk overfitting or memorization for the trained models. This can lead to similar instances being included in both training and testing sets, resulting in models that tend to memorize rather than learn. While these models may initially show better results, this is due to sub-optimal sampling rather than stronger generalization capabilities [1].

In time series analysis, where instances are obtained with a sliding window, data partitions for training, testing, and validation need to be time-segmented. Traditional grid search CV cannot ensure that instances from consecutive overlapping segments are not placed in both training and testing sets, which undermines the reliability of time series classification performance evaluations. To address this issue, we implemented a customized CV schema that modifies the original grid search to split by the SWAN-SF partitions, maintaining continuous time segmentation instead of randomized sampling. Each time-segmented training partition dataset is assigned a partition label to ensure it is not included in both training and testing sets.

Additionally, we modified the scoring function for our CV, replacing classification accuracy with forecast skill scores, primarily the True Skill Statistic score and the Heidke Skill Score, which will be discussed further in the next section.

4 Experimental Evaluations

The experiments conducted in this study are designed with two primary objectives. Firstly, they aim to demonstrate the effectiveness of time series classifiers developed using distinct interval features and to perform a comprehensive performance comparison among them. Another key objective is to identify the intervals within the time series that hold the greatest relevance to the initial time series. This effort is primarily designed to offer interpretable insights into our model. It involves pinpointing the specific segments of the time series that exert significant influence on predictions and understanding the aggregation strategies that can lead to more accurate outcomes.

4.1 Data Collection

For predicting solar flares, we utilized the SWAN-SF dataset, an open-source multivariate time series dataset introduced in [3]. This dataset offers a comprehensive collection of space weather-related physical parameters derived from solar magnetograms, incorporating data from various solar active regions and flare observations. Our experiments, encompassing both classification and feature ranking, utilized all 24 active region parameters provided by the dataset. These parameters are widely recognized as highly representative of solar activity, with detailed descriptions available in [7] and [3].

The SWAN-SF dataset is organized into five distinct time-segmented partitions to ensure that data instances within each partition do not temporally overlap. Active regions within the dataset are segmented using a sliding observation window of 12-hour intervals across the multivariate time series. Each segment captures essential data, including an active region number (aligned with NOAA Active Region numbers and HMI Active Region Patches identifiers), a class label (indicating the maximum intensity flare expected from that region within the subsequent 24 h), and start and end timestamps for each segment.

Flare intensity is categorized by the logarithmic classification of peak X-ray flux into major flaring classes (X, M, C, B, or A). For our analysis, instances classified as M- and X-class flares are considered flaring (i.e., positive class), while those classified as C- and B-class flares, along with flare-quiet regions, are treated as non-flaring (i.e., negative class). This binary classification framework is applied to model the solar flare forecasting problem as a binary multivariate time series classification task.

4.2 Experimental Settings

To assess our model's performance, we employed a binary 2×2 contingency matrix, supplemented by other well-known evaluation metrics for evaluating forecasting accuracy. Within these metrics, the positive class corresponds to significant flare events ($\geq$M1.0 or M- and X-class flares), while the negative class includes smaller flares and regions with minimal flare activity (i.e., instances labeled below the M1.0 threshold). In this context, True Positives (TPs) and True Negatives (TNs) represent instances where the model accurately predicts flaring and non-flaring events, respectively. False Positives (FPs) are false alarms, where the model incorrectly predicts a flare and False Negatives (FNs) are misses, where the model fails to predict an actual flare event.

For rigorous evaluation, we utilize the True Skill Statistic (TSS) and a weighted version of TSS (ωTSS), detailed in Eq. 2 and Eq. 3 respectively. The TSS measures the difference between the Probability of Detection (recall for the positive class) and the Probability of False Detection (POFD), providing a robust indicator of model skill.

$$TSS = \frac{TP}{TP + FN} - \frac{FP}{FP + TN} \tag{2}$$

In essence, TSS can be reformulated as the sum of true positive rate (TPR) and true negative rate (TNR), offset by 1 ($TPR + TNR - 1$). The general purpose of TSS is a good all-around forecast evaluation method, especially for evaluating scores among datasets with different imbalance ratios. However, it focuses on a simpler, more understandable scoring schema where both TPR and TNR are treated equally. To change the importance given to each term in this equation we can use a regularization term $\alpha/2$, and create the following weighted TSS (ωTSS):

$$\omega TSS = \alpha TPR + (2 - \alpha) TNR - 1 \tag{3}$$

Here, $\alpha/2$ and $1-\alpha/2$ are regularization parameters that show how important each term is. For instance, if correctly predicting an M- or X-class flare is 3 times more important than correctly predicting a non-flaring class, then we can use $\alpha/2 = 0.75$.

The second skill score we employed is the Heidke Skill Score (HSS), which serves as a critical measure of the forecast's improvement over a climatology-aware random prediction. The HSS ranges from -1 to 1, where a score of 0 indicates that the forecast's accuracy is equivalent to that of a random binary forecast, based solely on the provided class distributions. The formula for calculating this metric is provided in Eq. 4. Here, P denotes the actual positives, which is the sum of true positives (TP) and false negatives (FN)), and N represents the actual negatives, the sum of false positives (FP) and true negatives (TN)).

$$HSS = \frac{2 \cdot ((TP \cdot TN) - (FN \cdot FP))}{P \cdot (FN + TN) + N \cdot (TP + FP)} \tag{4}$$

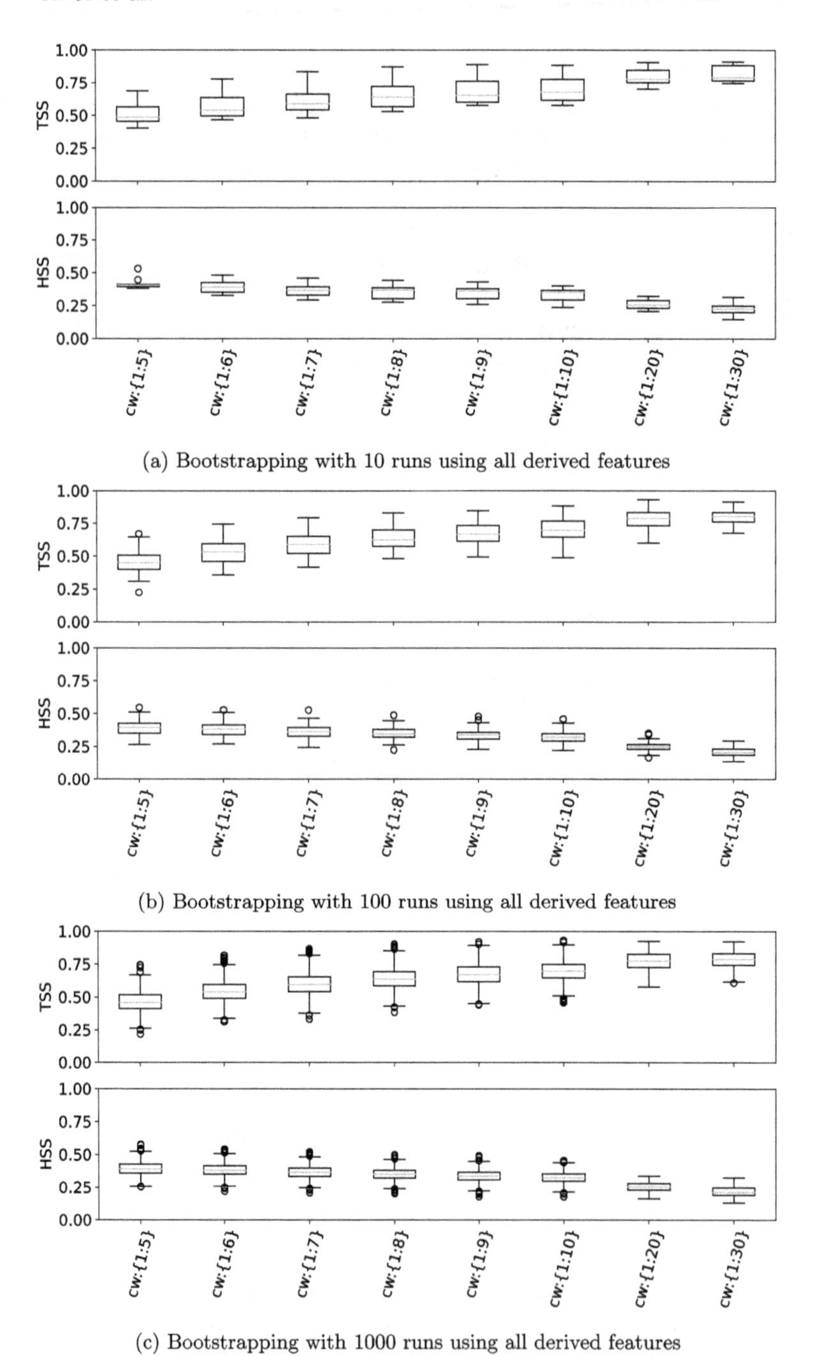

(a) Bootstrapping with 10 runs using all derived features

(b) Bootstrapping with 100 runs using all derived features

(c) Bootstrapping with 1000 runs using all derived features

Fig. 2. Error Bar representation of slim-TSF evaluation with ex-ante bootstrapping feature selection using different class weight (i.e., cw) ratio. The most relevant features are selected (per each model trained) across different class weights using the log-scale filter. The TSS and HSS scores are shown for each bootstrapping experiment.

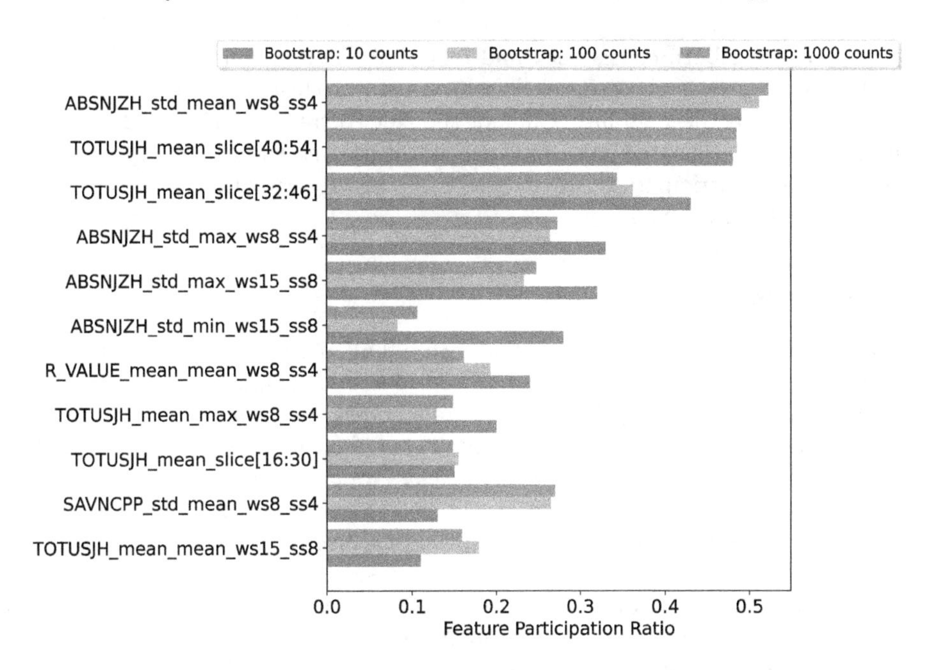

Fig. 3. A bar plot representation of feature participation ratio in three bootstrap evaluation counts. All features from sliding window intervals and transformed features are used.

4.3 Bootstrapping

In this study, we introduce a novel approach to feature selection that deviates from the methodology used in [16]. Instead of limiting our selection to only the top 5 highly ranked parameters from individual experiments, we base our feature selection on the cumulative results of the entire bootstrapping process. This involves compiling data from each iteration of the bootstrap subsampling and identifying the most relevant features based on their frequency of appearance throughout the random subsampling procedure. To further refine our feature selection and reduce the risks associated with an overly extensive feature set, we apply a filter $k = log2(N)$ to select the top k features, where N is the total number of features. The results of this refined process are illustrated in Fig. 2a, 2b, and 2c, demonstrating the use of selected important features from various window and step size configurations. This approach helps to mitigate the influence of outlier features, which might otherwise compromise the accuracy of our predictions.

This refined selection strategy enables us to achieve results comparable to those of our initially proposed Slim-TSF model, but with a significantly smaller set of parameters and features. Consequently, we can maintain average testing scores of approximately 60% in TSS and 35% in HSS while utilizing fewer inputs. Despite these reductions, the robustness of our feature selection process is confirmed through its repeated application with random subsamples of the original dataset, ensuring both consistency and reliability. Throughout our studies, certain features, such as those derived from the Total Unsigned Current Helicity (TOTUSJH) and the Absolute Value of the Net Current Helicity (ABSNJZH), are consistently selected across multiple iterations, as shown in Fig. 3. These features have a participation rate of over 40%, underscoring their critical role in predicting solar flare events.

4.4 Remarks

In the results, we demonstrate that the Slim-TSF models, using only the top cumulative important features selected from our bootstrapping iterations, perform comparably, ensuring efficiency and robustness. These models achieve similar outcomes using fewer but more significant features from the original 24 parameters. Notably, models with lower class weights show an average performance improvement of 5% over previous research [17]. This improvement occurs as we reduce redundancy by limiting the use of extensive derived features, thereby increasing feature relevancy. Consequently, the models can concentrate more effectively on key factors by minimizing the redundancy found in less informative features, ultimately enhancing performance significantly.

The outcomes of our study systematically evaluate the performance of our Slim-TSF models, incorporating an additional filter during feature selection. Specifically, our findings reveal that these models improve when utilizing only the top k (from a log-scale) most significant features. This highlights the principle that quality often outweighs quantity in feature selection, as these streamlined models achieve results comparable to their more complex counterparts that utilize all 24 features. Additionally, it is worth noting that adjusting the class weight hyperparameter significantly enhances model performance by reducing the imbalance ratio.

5 Conclusions

This study builds upon our previous work, which utilized interval-based features generated from sliding window operations in multivariate time series classifiers, also useful for ranking key features, intervals, and transformed pooling features. The primary goal of this work is to enhance the interpretability of high-dimensional multivariate time series classifiers. By employing a comprehensive and methodical approach to feature selection, our research not only improves the predictive accuracy and efficiency of the Slim-TSF model but also offers valuable insights into solar flare prediction, especially under the constraints of limited observational data. This advancement marks significant progress in the field of solar weather forecasting, underscoring the importance of adaptability and innovation in addressing the challenges of data scarcity.

Acknowledgment. This work is supported in part under two grants from NSF (Award #2104004) and NASA (SWR2O2R Grant #80NSSC22K0272).

References

1. Ahmadzadeh, A., et al.: Challenges with extreme class-imbalance and temporal coherence: a study on solar flare data. In: 2019 IEEE International Conference on Big Data (Big Data). IEEE (2019). https://doi.org/10.1109/bigdata47090.2019.9006505
2. Angelini, M., Blasilli, G., Lenti, S., Santucci, G.: A visual analytics conceptual framework for explorable and steerable partial dependence analysis. IEEE Trans. Vis. Comput. Graph. 1–16 (2024). https://doi.org/10.1109/tvcg.2023.3263739
3. Angryk, R.A., et al.: Multivariate time series dataset for space weather data analytics. Sci. Data **7**(1) (2020).https://doi.org/10.1038/s41597-020-0548-x
4. Bagnall, A., Lines, J., Bostrom, A., Large, J., Keogh, E.: The great time series classification bake off: a review and experimental evaluation of recent algorithmic advances. Data Min. Knowl. Disc. **31**(3), 606–660 (2016). https://doi.org/10.1007/s10618-016-0483-9
5. Baydogan, M.G., Runger, G., Tuv, E.: A bag-of-features framework to classify time series. IEEE Trans. Pattern Anal. Mach. Intell. **35**(11), 2796–2802 (2013). https://doi.org/10.1109/TPAMI.2013.72
6. Benz, A.O.: Flare observations. Living Rev. Solar Phys. **5** (2008). https://doi.org/10.12942/lrsp-2008-1
7. Bobra, M.G., Couvidat, S.: Solar flare prediction using SDO/HMI vector magnetic field data with a machine-learning algorithm. Astrophys. J. **798**(2), 135 (2015). https://doi.org/10.1088/0004-637x/798/2/135
8. Chen, Y., Ji, A., Babajiyavar, P.A., Ahmadzadeh, A., Angryk, R.A.: On the effectiveness of imaging of time series for flare forecasting problem. In: 2020 IEEE International Conference on Big Data (Big Data), pp. 4184–4191 (2020)
9. Deng, H., Runger, G., Tuv, E., Vladimir, M.: A time series forest for classification and feature extraction. Inf. Sci. **239**, 142–153 (2013). https://doi.org/10.1016/j.ins.2013.02.030
10. Georgoulis, M.K.: On our ability to predict major solar flares. In: Obridko, V., Georgieva, K., Nagovitsyn, Y. (eds.) The Sun: New Challenges. ASSSP, vol. 30, pp. 93–104. Springer, Heidelberg (2012). https://doi.org/10.1007/978-3-642-29417-4_9
11. Geurts, P.: Pattern extraction for time series classification. In: De Raedt, L., Siebes, A. (eds.) PKDD 2001. LNCS (LNAI), vol. 2168, pp. 115–127. Springer, Heidelberg (2001). https://doi.org/10.1007/3-540-44794-6_10
12. Homayouni, H., Ghosh, S., Ray, I., Gondalia, S., Duggan, J., Kahn, M.G.: An autocorrelation-based LSTM-autoencoder for anomaly detection on time-series data. In: 2020 IEEE International Conference on Big Data (Big Data). IEEE (2020). https://doi.org/10.1109/bigdata50022.2020.9378192
13. Hong, J., Ji, A., Pandey, C., Aydin, B.: Beyond traditional flare forecasting: a data-driven labeling approach for high-fidelity predictions. In: Wrembel, R., Gamper, J., Kotsis, G., Tjoa, A.M., Khalil, I. (eds.) DaWaK 2023. LNCS, vol. 14148, pp. 380–385. Springer, Cham (2023). https://doi.org/10.1007/978-3-031-39831-5_34
14. Ji, A., Arya, A., Kempton, D., Angryk, R., Georgoulis, M.K., Aydin, B.: A modular approach to building solar energetic particle event forecasting systems. In: 2021 IEEE Third International Conference on Cognitive Machine Intelligence (CogMI), pp. 106–115 (2021)
15. Ji, A., Aydin, B.: Active region-based flare forecasting with sliding window multivariate time series forest classifiers. In: 2023 IEEE 5th International Conference on Cognitive Machine Intelligence (CogMI), pp. 196–203 (2023). https://doi.org/10.1109/CogMI58952.2023.00036

16. Ji, A., Aydin, B.: Active region-based flare forecasting with sliding window multivariate time series forest classifiers. In: The Fourth IEEE International Conference on Cognitive Machine Intelligence. IEEE (2023)
17. Ji, A., Aydin, B.: Interpretable solar flare prediction with sliding window multivariate time series forests. In: 2023 IEEE International Conference on Big Data (BigData), pp. 1519–1524 (2023)
18. Ji, A., Aydin, B., Georgoulis, M.K., Angryk, R.: All-clear flare prediction using interval-based time series classifiers. In: 2020 IEEE International Conference on Big Data (Big Data), pp. 4218–4225 (2020)
19. Karlsson, I., Papapetrou, P., Boström, H.: Generalized random Shapelet forests. Data Min. Knowl. Disc. **30**(5), 1053–1085 (2016). https://doi.org/10.1007/s10618-016-0473-y
20. Kusano, K., Iju, T., Bamba, Y., Inoue, S.: A physics-based method that can predict imminent large solar flares. Science **369**(6503), 587–591 (2020)
21. Lines, J., Bagnall, A.: Time series classification with ensembles of elastic distance measures. Data Min. Knowl. Disc. **29**(3), 565–592 (2014). https://doi.org/10.1007/s10618-014-0361-2
22. Lubba, C.H., Sethi, S.S., Knaute, P., Schultz, S.R., Fulcher, B.D., Jones, N.S.: catch22: CAnonical time-series CHaracteristics. Data Min. Knowl. Disc. **33**(6), 1821–1852 (2019). https://doi.org/10.1007/s10618-019-00647-x
23. Nanopoulos, A., Alcock, R., Manolopoulos, Y.: Feature-Based Classification of Time-Series Data, pp. 49–61. Nova Science Publishers, Inc. (2001)
24. Pandey, C., Angryk, R.A., Aydin, B.: Solar flare forecasting with deep neural networks using compressed full-disk HMI magnetograms. In: 2021 IEEE International Conference on Big Data (Big Data), pp. 1725–1730 (2021)
25. Pandey, C., Angryk, R.A., Aydin, B.: Explaining full-disk deep learning model for solar flare prediction using attribution methods. In: De Francisci Morales, G., Perlich, C., Ruchansky, N., Kourtellis, N., Baralis, E., Bonchi, F. (eds.) ECML PKDD 2023. LNCS, vol. 14175, pp. 72–89. Springer, Cham (2023). https://doi.org/10.1007/978-3-031-43430-3_5
26. Pandey, C., Ji, A., Angryk, R.A., Aydin, B.: Towards interpretable solar flare prediction with attention-based deep neural networks (2023)
27. Pandey, C., Ji, A., Angryk, R.A., Georgoulis, M.K., Aydin, B.: Towards coupling full-disk and active region-based flare prediction for operational space weather forecasting. Front. Astronomy Space Sci. **9** (2022). https://doi.org/10.3389/fspas.2022.897301
28. Priest, E., Forbes, T.: The magnetic nature of solar flares. Astron. Astrophys. Rev. **10**(4), 313–377 (2002)
29. Saeed, W., Omlin, C.: Explainable AI (XAI): a systematic meta-survey of current challenges and future opportunities. Knowl.-Based Syst. **263**, 110273 (2023). https://doi.org/10.1016/j.knosys.2023.110273
30. Sakoe, H., Chiba, S.: Dynamic programming algorithm optimization for spoken word recognition. IEEE Trans. Acoust. Speech Signal Process. **26**(1), 43–49 (1978). https://doi.org/10.1109/TASSP.1978.1163055
31. Shibata, K., Magara, T.: Solar flares: magnetohydrodynamic processes. Living Rev. Sol. Phys. **8**(1), 6 (2011)
32. Silva, D.F., Giusti, R., Keogh, E., Batista, G.E.A.P.A.: Speeding up similarity search under dynamic time warping by pruning unpromising alignments. Data Min. Knowl. Discov. **32**(4), 988–1016 (2018). https://doi.org/10.1007/s10618-018-0557-y

33. Song, H., Tan, C., Jing, J., Wang, H., Yurchyshyn, V., Abramenko, V.: Statistical assessment of photospheric magnetic features in imminent solar flare predictions. Solar Phys. **254**(1), 101–125 (2008). https://doi.org/10.1007/s11207-008-9288-3
34. Ye, L., Keogh, E.: Time series shapelets: a novel technique that allows accurate, interpretable and fast classification. Data Min. Knowl. Disc. **22**(1–2), 149–182 (2010). https://doi.org/10.1007/s10618-010-0179-5

Evaluation of High Sparsity Strategies for Efficient Binary Classification

Laura Erhan[1(✉)], Lucia Cavallaro[2], Mattia Andrea Antinori[1], and Antonio Liotta[1]

[1] Faculty of Engineering, Free University of Bozen-Bolzano, Bolzano, Italy
laura.erhan@unibz.it

[2] Data Science Department, Institute for Computing and Information Sciences, Faculty of Science, Radboud University, Nijmegen, The Netherlands

Abstract. In the dynamic landscape of Artificial Intelligence (AI) advancements, particularly in the development of compact and highly efficient models for space-constrained environments, the strategic sparsification of neural networks takes center stage. In this work, we investigate creating, training, and evaluating Convolutional Neural Network (CNN), DenseNet, and ResNet models taking advantage of sparse neural networks with the help of the Sparse Evolutionary Training (SET) approach. This paper extends the existing framework, while also considering critical performance metrics, such as accuracy, precision, recall, inference time, execution time, memory usage, and energy consumption. The proposed script facilitates exploration across diverse model architectures, sparsity levels, quantization options, and the number of training epochs, alongside a recording of these performance metrics throughout both the training and inference phases. We analyse and evaluate the performance of the chosen models and parameters on a classic Malaria dataset. Results confirm that highly sparse ANNs (*e.g.*, 70%) can achieve results comparable to their fully-connected counterparts while allowing for smaller models and possibly more energy-efficient systems. Our contribution advances the ongoing discussion on optimizing embedded systems for efficient AI applications.

Keywords: Sparse Artificial Neural Networks · Sparse Evolutionary Training · Binary Classification

1 Introduction

Embedded Machine Learning (ML) connects embedded systems with ML models, facilitating the deployment of Artificial Neural Networks (ANNs) on microcontrollers and embedded subsystems, namely low-power devices such as mobile devices, wearables, and IoT devices. Consequently, ML models can be executed locally, minimizing the impact on device battery life and eliminating the need to transmit data to cloud servers for processing [1]. This paradigm shift promotes faster and more efficient operations at the edge. Addressing challenges

R. Wrembel et al. (Eds.): DaWaK 2024, LNCS 14912, pp. 106–111, 2024.
https://doi.org/10.1007/978-3-031-68323-7_8

arising from the escalating size of ANNs and the imperative for deployment on resource-constrained devices, literature advocates for the adoption of sparse ANNs [10]. Various techniques, such as magnitude pruning, variational dropout, l0 regularization, and random pruning baseline, have been proposed to induce sparsity in these models [7]. This sparsity allows for the identification and utilization of zero-valued weights, facilitating skipped computations and compact storage through sparse matrix formats. A notable contribution in this context is the SET method proposed by Mocanu *et al.* [11], addressing limitations in training efficiency associated with traditional sparsity induction methods. SET considers data distributions and introduces sparse bipartite layers to replace fully-connected bipartite layers in any type of ANN. This innovation results in reduced memory requirements and the potential for quadratically faster computational times in both the training and inference phases.

In the pursuit of advancing the domain of miniaturization for binary classifiers, this research extends the groundwork established in our previous study [3]. The authors enrich the comprehension of SET concerning three distinct binary classifier architectures: a conventional Convolutional Neural Network (CNN), a Densely Connected Neural Network (DenseNet), and a Residual Network (ResNet). The objective is to showcase the efficacy of SET in reducing the dimensions of these models while preserving their accuracy and enabling deployment on embedded devices. The principal aim of this study was to acquire a deeper understanding of the effectiveness of the SET approach in model miniaturization by considering performance metrics, such as accuracy, precision, recall, inference time, execution time in seconds, memory usage, and energy consumption. This research argues that the SET approach, when compared with the dense one, can achieve comparable performance in terms of accuracy while benefitting from a decrease in size and needed computational resources. This positions the SET approach as an optimal solution for the development of embedded ML models.

2 Related Work

Within the realm of deep artificial neural networks, the concept of sparsity entails the strategic adjustment of a subset of network parameters to zero, fostering a more compact and efficient model architecture [4]. This adjustment involves setting a percentage of weights in the artificial neural networks to zero, facilitating skipped multiplications, and enabling compressed storage through the utilization of sparse matrices. Sparsity induction in neural networks often involves pruning techniques, wherein parameters that do not significantly impact network accuracy are selectively removed either during or after the training process [8]. While effective in compressing Convolutional Neural Network (CNN) models, pruning presents challenges associated with hyperparameter tuning. To tackle the crucial challenge of sustaining accuracy in pruned models, the scientific community has delved into sparse training algorithms, strategically designed to alleviate the risk of encountering local minima. These algorithms, as outlined in [6], prioritize the removal of connections with the smallest magnitudes while activating connections with the highest gradients. Alternatively, Evolutionary Algorithms (EAs) have emerged as a promising solution, seamlessly integrating self-adaptive mechanisms with search-based approaches to pinpoint optimal network topologies [2]. Within this landscape of innovative solutions, we leveraged the SET method [11] to miniaturize binary classification models. This application signifies a pioneering use of sparse training algorithms, contributing to the ongoing discourse on efficient model deployment in resource-constrained environments.

3 Materials and Methods

Dataset. We conducted our analysis using the publicly available Malaria dataset [12], chosen for its curated data, balanced distribution, and representative collection of images suitable for addressing the complexity of our research problem. The dataset comprises 27,558 images, evenly divided into two classes representing infected and uninfected cells. The images were normalised to maintain values within the [0, 1] range. The dataset was partitioned into training, testing, and validation sets respecting the following percentages: 80% of the data for training and 20% for testing; the training data was further split into training and validation sets, maintaining the 80% and 20% split.

The SET Implementation. In our implementation, we utilized Keras and TensorFlow APIs to establish a training pipeline facilitating the generation of machine learning models at different sparsity levels. Network sparsification is achieved through the SET approach, inducing sparsity within the network during the model training process, rather than removing neurons and connections on the final trained model. This results in a reduction of computations and faster inference time, because weights and activation functions that are set to zero are skipped.

Chosen Models belong to the Convolutional Neural Network (CNN) architecture. CNNs employ a hierarchical architecture to systematically analyze visual data, an approach that enables CNNs to identify and prioritize relevant features, facilitating heightened precision and efficacy in tasks such as image recognition, classification, and segmentation [9]. The chosen models include a regular CNN, a DenseNet, and a Residual Neural Network (ResNet). For image resolution, we considered three levels: 8×8, 16×16, and 32×32.

Quantization. Miniaturization techniques play a vital role in adapting previously defined model architectures to fit computationally limited end devices by reducing their size and complexity. Dynamic range quantization is applied directly to the trained outputs. The initial Keras models are transformed into compact and portable TensorFlow Lite models, suitable for deployment across a range of compatible devices.

Experimental Setup. The user initiates the workflow by selecting a machine learning model (CNN, DenseNet, or ResNet), specifying the color mode (RGB or grayscale), the desired image resolution (*i.e.*, 8×8, 16×16, 32×32), and the number of epochs to be tried. Furthermore, the sparsity levels to be tested are chosen. If the sparsity level is set to zero, the model progresses without parameter reduction. In contrast, a non-zero sparsity level triggers the sparsity induction process.

The model enters a dynamic training and evaluation loop, iterating over the defined number of epochs. During each epoch, the model goes through all training samples, updates its parameters (weights and biases), and adjusts them to minimize the training loss. At each iteration, metrics such as execution times, and peak memory usage are recorded. Upon completing the training and evaluation loop, the model undergoes TensorFlow Lite (TFLite) conversion, facilitating deployment in resource-constrained environments.

Further, we offer a brief overview of the **evaluation metrics. Accuracy** represents the model's efficiency as the percentage of accurately classified instances out of the total inferences made. **Precision** quantifies the ratio of truly identified infected cells to the total number of cells predicted as infected. **Recall** shows the model's capability to accurately identify infected cells.

Inference Time delineates the temporal interval between tensor allocation and output formulation. A lower inference time aligns with our objective to craft sleek and efficient models. For **energy consumption**, we looked at the inference phase of a machine learning model and estimated it as $inference\_energy = inference\_time * power\_consumption$ measured in watt-hours. For inference tasks, the power consumption is generally lower than for tasks such as training. For mid-range to high-end CPUs like an Intel i7, it might be in the range of 15 to 50 W, depending on the workload and the efficiency of the hardware [5]. For our estimations, a value of 50 W was considered.

Memory usage provides insight into the memory utilised by the model and the resource requirements for optimizing deployment across diverse environments. It is monitored using the *tracemalloc* module.

4 Results and Discussion

A primary focus of our study implied investigating sparsity levels, ranging from the benchmark (dense network with 0% sparsity) up to a sparsity level of 95%.

The *epochs_to_try* parameter varied between 0 and 30, values chosen due to hardware constraints. A value of 0 means that the model will not undergo any training epochs; otherwise, the model is trained for the indicated number of epochs. The aim is to observe how the model's performance changes with varying amounts of training. To this end, in Fig. 1 we can visually examine and compare how the accuracy of the CNN and DenseNet machine learning models evolves with varying sparsity levels and number of epochs. It can be noticed that sparse ANNs have comparable accuracy to the fully-connected ANNs. This is also observed in the ResNet model. Quantized models consistently outperformed their non-quantized counterparts, a trend that persisted for higher image resolutions such as 16×16 and 32×32. However, the improvement curve for these resolutions was less pronounced, suggesting that 16×16 and 32×32 models exhibited

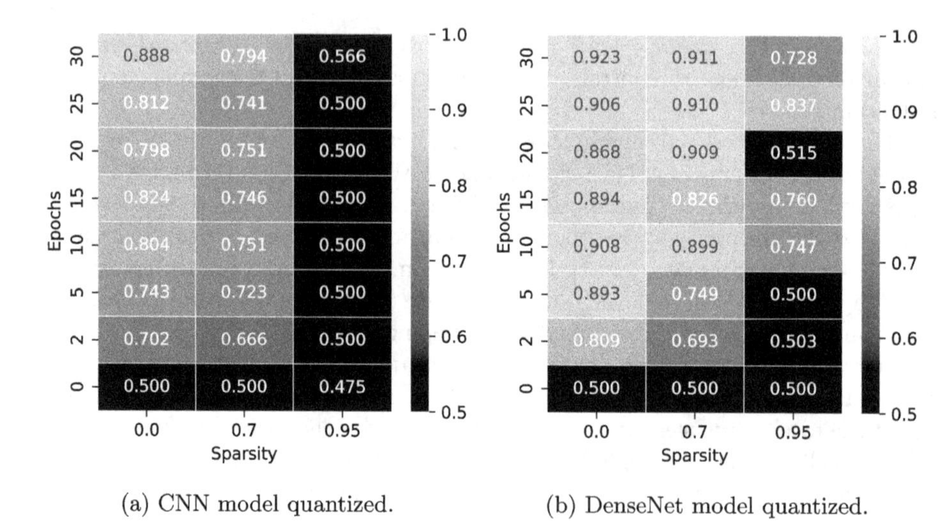

(a) CNN model quantized. (b) DenseNet model quantized.

Fig. 1. Accuracy heatmaps for the 2 models: CNN, DenseNet quantized versions.

stability and did not show significant improvement due to the increased clarity of the images. Beyond 70% sparsity, the accuracy undergoes a decline, noticeable for the 95% sparsity level. The use of a sparse ANN, more lightweight in comparison to its fully-connected counterpart, opens the avenue for possible deployment in edge IoT scenarios. In terms of inference time (μs), values of around 0.4 were noticed for the CNN model, 1.3 for the DenseNet model, and 0.75 for the ResNet model with corresponding estimated energy consumption levels (mW) of 6, 19, and 11 respectively. There were no significant variations noticed between the different scenarios, but an investigation into training time and the associated energy consumption would be of interest.

5 Conclusions and Future Work

As part of this work, an experimental matrix, wherein parameters were systematically varied to gauge their impact on outcomes, was created. Parameters of interest included architectural choices (*i.e.*, CNN, DenseNet, and ResNet), color preferences (*i.e.*, grayscale and in RGB color), image resolutions (*i.e.*, 8×8, 16×16, and 32×32), and sparsity levels. A script was developed to automate the execution of experiments, encompassing diverse parameter combinations, as well as collecting metrics of interest. Our initial findings suggest that sparse ANNs can have a performance similar to that of fully-connected ANNs while allowing for more lightweight models. An in-depth analysis is required to compare the required training times for the two and if significant energy saving can be achieved with the use of sparse ANNs.

As next steps, the experimental matrix is to be deployed on computationally powerful hardware (as opposed to a personal laptop *i.e.*, a Huawei Matebook 16 s,

12th Gen Intel(R) Core (TM) i7-12700H 2.30 GHz processor) to exhaustively test the various parameter combinations to validate the initial findings presented herein and to offer a better insight into the time, energy, and memory footprint. For future experiments, it is of interest to explore an early stopping function that allows the script to automatically determine the optimal number of epochs to be run. Furthermore, a promising investigation includes exploring more complex datasets and integrating other network architectures. Moreover, an analysis can be carried out to investigate the feasibility of employing such techniques on computationally limited devices, such as edge IoT devices.

Acknowledgments. This work was supported by the PRIN 2020 project COMMON-WEARS (grant number I53C21000210001) and by the STEADIER Project (grant number I55F21001900005).

References

1. Ajani, T.S., Imoize, A.L., Atayero, A.A.: An overview of machine learning within embedded and mobile devices-optimizations and applications. Sensors **21**(13), 4412 (2021)
2. Bartz-Beielstein, T., Branke, J., Mehnen, J., Mersmann, O.: Evolutionary algorithms. Wiley Interdisc. Rev. Data Min. Knowl. Discov. **4**(3), 178–195 (2014)
3. Cavallaro, L., Serafin, T., Liotta, A.: Miniaturisation of binary classifiers through sparse neural networks. Numer. Comput. Theory Algorithms NUMTA **2023**, 74 (2023)
4. Changpinyo, S., Sandler, M., Zhmoginov, A.: The power of sparsity in convolutional neural networks. arXiv preprint arXiv:1702.06257 (2017)
5. Daghero, F., Pagliari, D.J., Poncino, M.: Energy-efficient deep learning inference on edge devices. In: Advances in Computers, vol. 122, pp. 247–301. Elsevier (2021)
6. Evci, U., Gale, T., Menick, J., Castro, P.S., Elsen, E.: Rigging the lottery: making all tickets winners. In: International conference on machine learning, pp. 2943–2952. PMLR (2020)
7. Gale, T., Elsen, E., Hooker, S.: The state of sparsity in deep neural networks. arXiv preprint arXiv:1902.09574 (2019)
8. Hoefler, T., Alistarh, D., Ben-Nun, T., Dryden, N., Peste, A.: Sparsity in deep learning: pruning and growth for efficient inference and training in neural networks. J. Mach. Learn. Res. **22**(241), 1–124 (2021)
9. Li, Z., Liu, F., Yang, W., Peng, S., Zhou, J.: A survey of convolutional neural networks: analysis, applications, and prospects. IEEE Trans. Neural Netw. Learn. Syst. **33**(12), 6999–7019 (2021)
10. Mocanu, D.C., et al.: Sparse training theory for scalable and efficient agents. In: Proceedings of the 20th International Conference on Autonomous Agents and MultiAgent Systems, pp. 34–38. AAMAS 2021, International Foundation for Autonomous Agents and Multiagent Systems, Richland, SC (2021)
11. Mocanu, D.C., Mocanu, E., Stone, P., Nguyen, P.H., Gibescu, M., Liotta, A.: Scalable training of artificial neural networks with adaptive sparse connectivity inspired by network science. Nat. Commun. **9**(1), 2383 (2018)
12. Rajaraman, S., et al.: Pre-trained convolutional neural networks as feature extractors toward improved malaria parasite detection in thin blood smear images. PeerJ **6**, e4568 (2018)

Incremental SMOTE with Control Coefficient for Classifiers in Data Starved Medical Applications

Wan D. Bae[1], Shayma Alkobaisi[2(✉)], Siddheshwari Bankar[1], Sartaj Bhuvaji[1], Jay Singhvi[1], Madhuroopa Irukulla[1], and William McDonnell[1]

[1] Computer Science, Seattle University, Seattle, WA, USA
{baew,sbankar,sbhuvaji,jsinghvi,mirukulla,mcdonn11}@seattleu.edu
[2] College of Information Technology, United Arab Emirates University, Al Ain, UAE
shayma.alkobaisi@uaeu.ac.ae

Abstract. Prediction models for data-starved medical applications lag behind general machine learning solutions, despite their potential to improve early interventions. This is largely due to the assumption that optimization approaches are applied on a balanced distribution of events, yet medical data often has an imbalanced distribution within classes. The curse of dimensionality is further exacerbated by small samples and a high number of features in individual-based risk prediction models. In this paper, we propose a data augmentation system to gradually create synthetic minority samples with a control coefficient, which improves the quality of generated data over time and consequently boosts prediction model performance. This system incrementally adjusts to the data distribution, avoiding overfitting. We evaluate our approach using four synthetic oversampling techniques on real asthma patient data. Our results show that this system enhances classifiers' overall performance across all four techniques. Specifically, applying the incremental data augmentation approach to three oversampling methods led to an increase in sensitivity of 4.01% to 7.79% in deep transfer learning-based classifiers.

Keywords: class imbalance problem · synthetic minority oversampling technique · rare event prediction · data starved contexts · control coefficient

1 Introduction

Recent advances in machine learning play a pivotal role in clinical decision-making, enabling early disease detection and enhancing patient care across various medical conditions. For example, avoidable asthma exacerbations account for 63% of the total annual asthma costs. Preventive approaches to predict the likelihood of symptom changes and risk levels can improve healthcare quality and achieve significant economic savings [8].

R. Wrembel et al. (Eds.): DaWaK 2024, LNCS 14912, pp. 112–119, 2024.
https://doi.org/10.1007/978-3-031-68323-7_9

However, building robust predictive models in healthcare often faces the challenge of class imbalance within medical datasets. Class imbalance occurs when certain health conditions, demographics, or outcomes are rare or underrepresented, resulting in the minority class having significantly fewer samples than the majority class. This imbalance leads to biases machine learning algorithms and compromises performance, as accurately identifying minority class data is crucial for effective healthcare delivery.

Synthetic minority oversampling (SMOTE) techniques [1,4–6,11], are actively researched to improve prediction models. SMOTE variants and boosted SMOTE methods provide a wide range of solutions for addressing class imbalance. However, their success in data-starved contexts, common in medical applications with few daily observations per patient, is limited. In our individual asthma risk prediction modeling, the average dataset size is 168 and the imbalance ratio is 3.98, hence traditional SMOTE methods may not suffice.

Given the need to generate a substantial amount of synthetic data to rebalance classes in data deprived medical applications, we introduce a new meta-algorithm to improve the quality of synthetic data produced by SMOTE methods and develop a system that can be extended to other data generation techniques. Our research unfolds in several stages. First, we conduct data-level analysis to assess the quality of synthetic data generated by SMOTE variants within the incremental data generation system. Next, we assess the effectiveness of the proposed system on conventional classifiers. Finally, we explore deep transfer learning (TL) classifiers to enhance classifier performance.

2 Related Work

Synthetic Minority Oversampling Technique (SMOTE) is a widely used method for addressing class imbalance by generating synthetic samples similar to minority class data, thus achieving a more balanced class distribution. The first SMOTE method [1] generates synthetic samples using k-nearest neighbors and linear interpolation. This often results in data lying on the same line, motivating researchers to explore alternative data distributions.

Gaussian SMOTE (G-SMOTE) [6] generates synthetic samples similar to SMOTE [1] but incorporates a Gaussian distribution to produce more varied data, avoiding the duplication common with SMOTE. Gamma SMOTE (Gamma-SMOTE) [5] uses the Gamma distribution to create new samples in a non-linear manner. Due to the Gamma distribution's asymmetry, new minority samples are generated close to the existing minority data samples. Sample Density Distribution SMOTE (SDD-SMOTE) [11] considers the overall dataset distribution and local sample density to reduce fuzzy classification boundaries and control the randomness of the SMOTE method. It identifies the k-nearest neighbors of minority samples, measures their density, and generates synthetic samples with controlled coefficients to balance class distribution. Integrating AdaBoost with SMOTE, as in SMOTEBoost [2], is a natural way to enhance the SMOTE methods by combining oversampling with boosting algorithms. Another approach [9] adjusts the weights of synthetic data to mitigate data noise.

Generative Adversarial Networks (GANs) and autoencoders, typically used for synthetic image generation, have been applied to tabular data. Techniques like conditional tabular GAN [13] and the SMOTified GAN technique [10] show potential in augmenting tabular data and improving prediction models. Despite these advancements, they have not been successful with small training datasets.

3 Method

"Incremental boosting" typically refers to a technique where boosting algorithms are applied sequentially in multiple stages, each building on the results of the previous stage. The goal is to iteratively correct the mistakes made by previous learners and to improve the overall accuracy of the model. In our context, incremental boosting is applied in multiple iterations to enhance the performance of the SMOTE methods. After each iteration, new synthetic samples are generated using the augmented training data, and the boosting process is reapplied to further refine the model's ability to classify minority class instances.

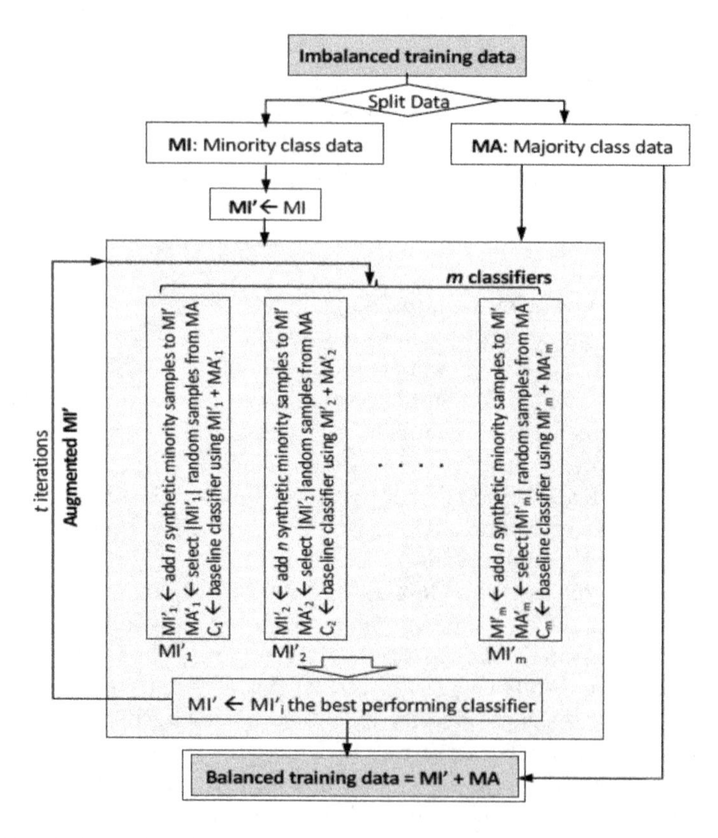

Fig. 1. Overview of an incremental synthetic data generation system

3.1 An Incremental Synthetic Data Generation System

The incremental synthetic data generation system divides the SMOTE data generation process into several iterations, as illustrated in Fig. 1. In each iteration, a SMOTE method generates n synthetic minority class samples and adds them to the current minority training dataset. The system creates m such minority training datasets. Then m classifiers are trained and validated using these datasets. The system selects the subset of synthetic data that produces the best-performing classifier and adds it into the current training data.

To select the best subset, we utilize a weighted evaluation function that incorporates performance metrics, focusing on correctly predicting minority class events. In subsequent iterations, the method generates minority class samples by using the augmented training data. By including synthetic data from previous iterations, the SMOTE method can generate more diverse data, thus reducing overfitting in classification models.

For further improvement, we adopt the Control Coefficient (CC) from the SDD-SMOTE algorithm [11] to consider the distribution of synthetic data points relative to the dense area of minority samples. The CC value is calculated during data preprocessing and applied when the SMOTE_{incrCC} method generates synthetic data points. A random probability distribution function is used for SDD and SDD-SMOTE methods, Gaussian function for G-SMOTE, and Gamma function for Gamma-SMOTE.

4 Experiments

4.1 Datasets and Experiments Setup

Our experiments used datasets from 20 non-smoking asthma patients, collected through a case study [12]. Each data sample consists of 27 variables including the patient' peak expiratory flow rate (PEFR) and indoor/outdoor environmental exposure data, and a binary label indicating risk (0 or 1). The variables and measurements are summarized in Table 1.

Patients' exposures to environmental variables were estimated using a 24-h time window for PEFR measurement. The high-risk zone is defined as a PEFR below a patient's critical cutoff (PEFR_C), set at the 20% quantile PEFR value of the patient's dataset, with samples below PEFR_C being minority samples. Patients' datasets contain between 88 and 210 samples (average 168), with minority class samples ranging from 16 to 38 (average 35) and majority class samples from 72 to 172 (average 132). The class imbalance ratio ranged from 2.32 to 5.52 (average 3.98), higher than the ratios in the SDD-SMOTE [11].

All our data generation methods were implemented in Python 3.8 using the Keras framework, with data analysis and model performance evaluation conducted using scikit-learn. Model hyperparameters were selected through extended training and validation with k-fold cross validation (CV) to avoid overfitting and enhance performance. For TVAE, we used the Adam optimizer (learning rate = 0.001), 200 epochs, 16 batch size, and a validation split rate of 0.25.

Table 1. Variables and measurements in 20 asthma patients' datasets

Data Category	Major variables	Measurement
Physiological data	yesterday's PEFRs	twice a day
Indoor air quality	$PM_{2.5}$, CO_2, temperature, humidity	every 60 s.
Outdoor air quality	SO_2, CO, O_3 , NO_2, PM_{10}, temperature, humidity	every 30 min.
Home location	home distance from major roads	level 1–level 5
Life style	income level	level 1–level 9
Cooking habit	frequency of frying	level 1–level 7

Table 2. Statistical summary on the generated synthetic data

Method	Mean diff.	STD diff.	KL	KDE area	Gretel score
SMOTE	**0.27%**	13.08%	0.008074	1.02	92.63
SMOTE$_{incrCC}$	0.33%	**12.31%**	**0.006867**	**0.85**	**92.95**
G-SMOTE	1.33%	14.17%	**0.012217**	**1.32**	**91.88**
G-SMOTE$_{incrCC}$	**0.96%**	**10.32%**	0.015719	1.44	91.58
Gamma-SMOTE	0.69%	14.73%	0.011598	1.17	90.25
Gamma-SMOTE$_{incrCC}$	**0.56%**	**12.03%**	**0.004292**	**0.69**	**92.84**
SDD-SMOTE	2.24%	12.95%	**0.005129**	**0.82**	**93.74**
SDD-SMOTE$_{incrCC}$	**1.84%**	**11.03%**	0.006220	0.85	91.52
TVAE synthesizer	3.84%	33.21%	0.070124	2.84	77.34

Conventional classifiers used 3-fold CV, while TL classifiers used 5-fold CV for the source model and 3-fold CV for the target model, with the Adam optimizer (learning rate = 0.001) and 100–1,000 epochs for both models. Each dataset was split into 80% training/validation and 20% testing. The training/validation data was augmented with synthetically generated data, while the testing data retained the original imbalance ratio.

4.2 Statistical Analysis

Synthetic data generation for rebalancing classes focuses on three factors: (1) maintaining similar probability density functions of variables within the augmented training dataset, (2) preserving class boundaries, and (3) increasing data diversity to reduce overfitting.

Factor (1) is measured by the difference between the means of the original and generated datasets, with a smaller difference indicating better maintenance of data distribution. Factor (2) is assessed by the difference in standard deviations (STD), where a large increase suggests boundary distortions. Factor (3) is measured by probability density functions, such as Kullback-Leibler (KL) divergence and Kernel density estimation (KDE). We also used an open platform called "Gretel" [3] that assesses the overall quality of synthetically generated

Table 3. Conventional classifiers with SDD-SMOTE and incremental SDD-SMOTE

Classifier	Method	Accuracy	Sensitivity	Specificity	Precision	F_1 score	AUC ROC
DT	SDD-SMOTE	0.5955	0.3979	**0.7931**	0.5849	0.5772	0.5908
	SDD-SMOTE$_{incrCC}$	**0.5987** (+0.54%)	**0.4194**(+5.40%)	0.7780 (−1.90%)	**0.5884** (+0.60%)	**0.5783** (+0.19%)	**0.5974** (+0.11%)
KNN	SDD-SMOTE	0.5974	0.5255	0.6722	0.5783	0.5615	0.5851
	SDD-SMOTE$_{incrCC}$	**0.6177** (+3.40%)	**0.5287** (+0.61%)	**0.7127** (+6.02%)	**0.5904** (+2.09%)	**0.5710** (+1.69%)	**0.5976** (+2.13%)
LR	SDD-SMOTE	0.6180	0.5290	0.7089	0.5894	0.5805	0.5997
	SDD-SMOTE$_{incrCC}$	**0.6453** (+4.42%)	**0.5447** (+2.97%)	**0.7459** (+5.22%)	**0.6097** (+3.44%)	**0.6043** (+4.10%)	**0.6178** (+3.02%)
NB	SDD-SMOTE	0.6011	0.3880	**0.8141**	0.5881	0.5876	0.5992
	SDD-SMOTE$_{incrCC}$	**0.6080** (+1.15%)	**0.4095** (+5.54%)	0.8065 (−0.93%)	**0.5951** (+1.19%)	**0.5890** (+0.24%)	**0.6070** (+1.30%)

data. Gretel measures data quality by comparing the distributional distance between the principal components in the original and synthetic data, with closer principal components indicating higher data quality.

Table 2 summarizes the statistics of data generated by the incremental data generation system compared to existing SMOTE methods and the TVAE synthesizer [7]. The mean differences between the generated and original data using the incremental generation system are relatively small, ranging from 0.33% to 1.84%. G-SMOTE, Gamma-SMOTE, and SDD-SMOTE reduced the mean values, while SMOTE increased it. All SMOTE methods reduced the STD values: 5.9% in SMOTE, 27.2% in G-SMOTE, 18.3% in Gamma-SMOTE, and 14.8% in SDD-SMOTE. The proposed system decreased KL divergence and KDE area in SMOTE and Gamma-SMOTE but increased them in G-SMOTE and SDD-SMOTE, a trend also seen in Gretel scores. Further analysis is needed to relate these metrics to actual performance in risk prediction. In contrast, data generated by TVAE showed the highest mean and STD differences from the original data, 3.84% and 33.21%, respectively. This resulted in the highest KL and KDE values and the lowest Gretel score.

4.3 Performance Evaluation on Classifiers

We evaluated classifiers using common binary classification metrics: weighted accuracy, F_1-score average, and Receiver Operating Characteristic Area Under the Curve (ROCAUC). While all these metrics are important, we focused on improving sensitivity, representing the model's ability to predict health risks correctly. The effectiveness of the incremental data generation system on each SMOTE method was tested using conventional classifiers and TL classifiers. The conventional classifiers include: (1) Decision Tree (DT), (2) K-Nearest Neighbors (KNN), (3) Logistic Regression (LR), and (4) Naive Bayes (NB). TL classifiers train the source model with population data from 19 asthma patients' datasets and retrain the target model with a target patient's dataset, and then the target model predicts the patient's health risk.

Table 3 shows that the proposed system with SDD-SMOTE$_{incrCC}$ improved classifier performance on augmented data in all metrics except specificity, with improvements of 0.54%-4.42% in accuracy, 0.61%-5.54% in sensitivity, 0.60%-3.44% in precision, 0.19%-4.10% in F1 score, and 0.11%-3.02% in ROCAUC.

Table 4. TL classifiers with SMOTEs, incremental SMOTEs and TVAE

Method	Accuracy	Sensitivity	Specificity	Precision	F_1 score	AUC ROC
SMOTE	0.6697	**0.5449**	0.7846	0.6503	0.6489	0.6647
SMOTE$_{incrCC}$	**0.6762**(+0.97%)	0.5371 (−3.21%)	**0.8153** (+3.91%)	**0.6599** (+1.48%)	**0.6572** (+1.26%)	**0.6762** (+1.72%)
G-SMOTE	0.6592	0.5461	0.7723	0.6308	0.6282	0.6592
G-SMOTE$_{incrCC}$	**0.6973** (+5.79%)	**0.5886** (+7.79%)	**0.8060** (+4.37%)	**0.6549** (+3.81%)	**0.6597** (+5.01%)	**0.6973** (+5.79%)
Gamma-SMOTE	0.6737	0.5505	0.7969	0.6508	0.6501	0.6770
Gamma-SMOTE$_{incrCC}$	**0.6975** (+3.54%)	**0.5726** (+4.01%)	**0.8225**(+3.21%)	**0.6708** (+3.08%)	**0.6704** (+3.12%)	**0.6975** (+3.04%)
SDD-SMOTE	0.6853	0.5720	0.7986	0.6556	0.6543	0.6853
SDD-SMOTE$_{incrCC}$	**0.6982** (+1.88%)	**0.5953** (+4.07%)	**0.7998** (+0.15%)	**0.6591** (+0.54%)	**0.6598** (+0.84%)	**0.6975** (+1.78%)
TVAE Synthesizer	0.6653	0.5062	0.8233	0.6474	0.6461	0.6650

Due to space limitations, we present results using SDD-SMOTE but note similar performance was achieved with other SMOTE methods.

Table 4 presents the performance improvement in TL models trained using augmented datasets generated by the existing SMOTE methods, the incremental SMOTE and TVAE. The improvement in sensitivity for TL models, +4.01% with SDD-SMOTE$_{incrCC}$, +4.07% with Gamma-SMOTE$_{incrCC}$, and +7.79% with G-SMOTE$_{incrCC}$ while it decreased by 3.21% with SMOTE$_{incrCC}$. G-SMOTE$_{incrCC}$ showed the highest improvement across all metrics while sensitivity decreased by 3.21% with SMOTE$_{incrCC}$. Classifiers trained on data generated by SMOTE variants and incremental SMOTE methods outperformed those trained on data by TVAE, with sensitivity improvements ranging from 6.08% to 17.5%.

5 Conclusions

In medical applications where predicting rare disease events or exacerbations is the main concern, class imbalance can affect the accuracy of prediction models. This study systematically evaluated various SMOTE variants and proposed an incremental data generation system to enhance these variants, which generate better-quality training data. We compared our incremental SMOTE methods to the original SMOTE variants using real asthma patients' datasets with four conventional and TL-based classifiers. The findings demonstrate that the incremental SMOTE methods improved prediction accuracy by 4.01% to 7.79% in the sensitivity of TL models using three SMOTE variants. However, with very few minority samples, it may not generate enough non-duplicate data points to balance the dataset. Open challenges include developing flexible, scalable SMOTE variants robust to different imbalanced ratios and data sizes, and TL architectures with other classifiers.

Acknowledgement. This study received support from United Arab Emirates University under UAEU NFRP grant (Grant No. G00004281) and Seattle University.

References

1. Chawla, N.V., Bowyer, K.W., Hall, L.O., Kegelmeyer, W.P.: Smote: synthetic minority over-sampling technique. J. Artif. Intell. Res. **16**, 321–357 (2002)
2. Chawla, N.V., Lazarevic, A., Hall, L.O., Bowyer, K.W.: SMOTEBoost: improving prediction of the minority class in boosting. In: Lavrač, N., Gamberger, D., Todorovski, L., Blockeel, H. (eds.) PKDD 2003. LNCS (LNAI), vol. 2838, pp. 107–119. Springer, Heidelberg (2003). https://doi.org/10.1007/978-3-540-39804-2_12
3. Gretel: Gretel. https://gretel.ai/. Accessed 4 May 2024
4. Hoens, T.R., Chawla, N.V.: Imbalanced datasets: from sampling to classifiers. Foundations, Algorithms, and Applications. Wiley, Imbalanced Learning (2013)
5. Kamalov, F., Denisov, D.: Gamma distribution-based sampling for imbalanced data. Knowl.-Based Syst. **207**, 106368 (2020)
6. Lee, H., Kim, J., Kim, S.: Gaussian-based smote algorithm for solving skewed class distributions. Int. J. Fuzzy Logic Intell. Syst. **17**(4), 229–234 (2017)
7. MIT: The synthetic data vault. https://sdv.dev. Accessed 4 May 2024
8. Raghupathi, W., Raghupathi, V.: Big data analytics in healthcare: promise and potential. Health Inf. Sci. Syst. **2**(1), 1–10 (2014)
9. Sağlam, F., Cengiz, M.A.: A novel smote-based resampling technique trough noise detection and the boosting procedure. Expert Syst. Appl. **200**, 117023 (2022)
10. Sharma, A., Singh, P.K., Chandra, R.: SMOTified-GAN for class imbalanced pattern classification problems. Ieee Access **10**, 30655–30665 (2022)
11. Wan, Q., Deng, X., Li, M., Yang, H.: Sddsmote: synthetic minority oversampling technique based on sample density distribution for enhanced classification on imbalanced microarray data. In: The 6th International Conference on Compute and Data Analysis, pp. 35–42 (2022)
12. Woo, J., Rudasingwa, G., Kim, S.: Assessment of daily personal pm2. 5 exposure level according to four major activities among children. Appl. Sci. **10**(1), 159 (2020)
13. Xu, L., Skoularidou, M., Cuesta-Infante, A., Veeramachaneni, K.: Modeling tabular data using conditional GAN. In: Advances in Neural Information Processing Systems, vol. 32 (2019)

Exploring Evaluation Metrics for Binary Classification in Data Analysis: the Worthiness Benchmark Concept

Mohammad Shirdel[1(✉)], Mario Di Mauro[2], and Antonio Liotta[1]

[1] Free University of Bozen-Bolzano, Piazzetta dell'Universitá, 1, Bolzano, Italy
{mshirdel,antonio.liotta}@unibz.it
[2] University of Salerno, Via Giovanni Paolo II, 132, Fisciano, Italy
mdimauro@unisa.it

Abstract. In binary data classification, the main goal is to determine if elements belong to one of two classes. Various metrics assess the efficacy of classification models, making it essential to analyze and compare these metrics to select the most appropriate one. Despite significant research, a comprehensive comparison of these metrics has not been adequately addressed. The effectiveness of classification models is typically represented by a confusion matrix, detailing the count of correct and incorrect predictions for each class. Evaluating changes in the confusion matrix is crucial to discern model superiority, but different metrics may yield varying interpretations of the same matrices. We propose the *Worthiness Benchmark* (γ), a novel concept characterizing the classification metrics' principles for ranking classifiers and is useful to select the best metric for a given problem.

Keywords: Classification · Evaluation metrics · Worthiness Benchmark

1 Introduction and Related Research

In binary data classification problems, new metrics must be evaluated against established benchmarks [5,12]. One main purpose of a metric is to rank models accurately, but differences in these rankings often lead to debates on the most effective metric for determining model superiority. The confusion matrix plays a crucial role in defining most evaluation metrics. It illustrates a classifier's performance as follows: $\begin{pmatrix} TP\ FN \\ FP\ TN \end{pmatrix}$ where True Positives (TP) and True Negatives (TN) indicate correct predictions, and False Negatives (FN) and False Positives (FP) mark prediction errors.

To illustrate the challenge of identifying the optimal metric for ranking classifiers, consider the *Profit Analysis Example*. Suppose we evaluate two classifiers, A and B, within the context of binary classification involving 100 instances.

R. Wrembel et al. (Eds.): DaWaK 2024, LNCS 14912, pp. 120–125, 2024.
https://doi.org/10.1007/978-3-031-68323-7_10

Classifier A has a confusion matrix of $\begin{pmatrix} 40 & 20 \\ 15 & 35 \end{pmatrix}$ and Classifier B has a confusion matrix of $\begin{pmatrix} 30 & 30 \\ 10 & 30 \end{pmatrix}$. We evaluate the classifiers by creating two betting scenarios focused on soccer game results to examine profits from positive forecasts.

Scenario 1: A correct prediction earns \$16 (a \$6 gain plus the stake), while an incorrect one loses \$10. Profits are $6 \times TP - 10 \times FP$, resulting in \$90 for Classifier A and \$80 for Classifier B.

Scenario 2: A correct prediction earns \$14 (a \$4 gain plus the stake), while an incorrect one loses \$10. Profits are $4 \times TP - 10 \times FP$, leading to \$10 for Classifier A and \$20 for Classifier B.

This analysis reveals the limitations of relying solely on confusion matrix-devised metrics. Since these metrics produce a single ranking from different matrices, the best metric choice depends on the specific scenario. This insight guides us towards developing a new approach for metric analysis, focusing on their applicability across various contexts. Understanding how metrics rank classifiers is critical as it directly impacts model selection.

With this motivation, our paper addresses the research question: *How do the metrics rank the classifiers?* Despite numerous studies on classification metrics [1,5], this problem remains underexplored. This study proposes a methodology for evaluating binary classification metrics, introducing the *Worthiness Benchmark* (γ) that represents the principle of a metric to rank two classifiers.

There have been attempts in the technical literature in metrics analysis, as outlined below. Authors in [5] focus on the Area Under Curve (AUC) metric, scrutinizing its application of a misclassification cost function and its alignment with real-world costs. This debate, including critiques by [8] and responses by [6], underscores the contentious nature of achieving reliable evaluation metrics. New frameworks like *BenchMetrics* [1] and *PToPI* [2] suggest fresh ways to test the robustness and utility of metrics across different situations and classification problems. Additionally, [3] analyze metrics by assessing the correlation in classifier rankings across different metrics to group similar metrics into clusters, providing insights into the consistency of evaluation metrics. Further analyses, such as those by [9] and [4], highlight the potentially misleading nature of metrics like *Accuracy* in imbalanced class distributions. Studies like [7] provide insights into metric behavior in response to changes in the confusion matrix. Despite extensive exploration, there is a gap in understanding how metrics rank classifiers. Our contribution aims to address this gap, highlighting principles for selecting evaluation metrics to ensure a more accurate and meaningful assessment of classifier performance. Readers interested in examining some mathematical details in depth (here omitted for space constraints) are invited to refer to [11].

2 Methodology

Before illustrating the main methodology, it is useful to introduce some basic definitions. We recall that to evaluate classification models, the confusion matrix is crucial. For simplicity, we use the terms confusion matrix and classifier interchangeably (e.g., classifier A yields confusion matrix A). In a given dataset, the number of Actual Positive (AP) and Actual Negative (AN) samples are constant, therefore, we can rewrite the confusion matrix as $\begin{pmatrix} TP & AP - TP \\ FP & AN - FP \end{pmatrix}$.

A key component of our analysis is the *Reward Matrix*, which quantifies classification outcomes through rewards and costs for TP and FP. It is defined as $\begin{pmatrix} reward_P(TP) & cost_P(TP) \\ cost_N(FP) & reward_N(FP) \end{pmatrix}$, where $reward_P(TP)$ and $reward_N(FP)$ are benefits for correct predictions, and $cost_P(TP)$ and $cost_N(FP)$ are penalties for misclassifications. The goal is to maximize reward and minimize cost.

Another crucial quantity is the Profit function, $P(TP, FP)$, which defines the net gain of a classifier, considering the dynamics between accurate and inaccurate predictions. For example in the exemplary Scenario 1 presented in Sect. 1, $Reward\, Matrix = \begin{pmatrix} 6\,TP & 0 \\ -10\,FP & 0 \end{pmatrix}$ and the Profit function is:

$$P(TP, FP) = \sum Elements\, of\, Reward\, Matrix = 6\,TP - 10\,FP. \quad (1)$$

We also introduce the *Isoprofit Line*, a curve whose points have the same Profit value, given by TP and FP from confusion matrices. The function *isoprofit(A)* represents the profit value of confusion matrix A and its isoprofit line. For example, given matrix $A = \begin{pmatrix} 40 & 20 \\ 15 & 35 \end{pmatrix}$ in Scenario 1, we have *isoprofit(A) = 90.*

Key assumptions for our analysis of classification metrics are: (1) $P(TP, FP)$ is defined on a single confusion matrix, and *Accuracy* and $F1$-*score* can be expressed in terms of $P(TP, FP)$. (2) $P(TP, FP)$ is assumed to be continuous, treating TP and FP as continuous despite their discrete nature. (3) $P(TP, FP)$ increases with TP and decreases with FP.

In this study, we introduce the *Worthiness Benchmark* (γ) which is our main contribution. It evaluates if changes in a confusion matrix's elements by a new classifier are beneficial from the metric's viewpoint. We study how changes in TP and FP influence the Profit function, showing the directions in FP-TP space that enhance the Profit. If TP and FP are increasing, then:

$$\nabla P(TP, FP) > 0 \Leftrightarrow \frac{d(TP)}{d(FP)} > -\frac{\frac{d(P(TP,FP))}{d(FP)}}{\frac{d(P(TP,FP))}{d(TP)}}. \quad (2)$$

To prove, we define D as the unit vector (u, v) with u corresponding to changes in TP and v to changes in FP, and adhering to $u^2 + v^2 = 1$ such that

$u = \frac{d(TP)}{\sqrt{d(TP)^2+d(FP)^2}}$ and $v = \frac{d(FP)}{\sqrt{d(TP)^2+d(FP)^2}}$. we set:

$$\nabla P(TP, FP) \cdot D > 0 \Leftrightarrow \frac{d(P(TP, FP))}{d(TP)} u + \frac{d(P(TP, FP))}{d(FP)} v > 0 \tag{3}$$

$$\Leftrightarrow \frac{u}{v} > -\frac{\frac{d(P(TP,FP))}{d(FP)}}{\frac{d(P(TP,FP))}{d(TP)}} \Leftrightarrow \frac{d(TP)}{d(FP)} > -\frac{\frac{d(P(TP,FP))}{d(FP)}}{\frac{d(P(TP,FP))}{d(TP)}}. \tag{4}$$

This inequality introduces the *Worthiness Benchmark*, γ, assessing when confusion matrix changes enhance performance, setting a benchmark for improvement. The *Worthiness Benchmark*, γ, is defined as:

$$\gamma(TP, FP) = -\frac{\frac{d(P(TP,FP))}{d(FP)}}{\frac{d(P(TP,FP))}{d(TP)}}. \tag{5}$$

Matching γ signifies equivalent performance, exceeding it indicates improvement, and falling short suggests a decline. γ is invariably positive.

The differential equation below describes the *isoprofit lines*, illustrating the ratio between changes in TP and FP that maintains constant profit:

$$d(TP) = \gamma \cdot d(FP), \tag{6}$$

$$d(TPR) = \frac{\gamma}{\alpha} \cdot d(FPR). \tag{7}$$

where, TPR and FPR are True Positive Rate ($\frac{TP}{AP}$) and False Positive Rate ($\frac{FP}{AN}$) respectively and $\alpha = \frac{AP}{AN}$. Therefore, if TP and FP in *isoprofit lines* have linear relationship (e.g. when γ is constant), we can use $\frac{\Delta TP}{\Delta FP} > \gamma$ instead of Eq. 2. To illustrate the *Worthiness Benchmark*, let's revisit the *'Profit Analysis Example'*. Classifier A achieves 40 TP and 15 FP, while Classifier B gets 30 TP and 10 FP. A's advantage stems from an additional 10 TP and 5 FP over B. The comparison between A and B boils down to whether 10 extra TP compensates for 5 more FP. Applying the *Worthiness Benchmark*, we find $\gamma = \frac{10}{6}$ for Scenario 1, signifying a classifier must gain $\frac{10}{6}$ more TP than each FP to be superior. Hence, A surpasses B as $\frac{TP_A - TP_B}{FP_A - FP_B} = 2 > \frac{10}{6}$. In Scenario 2, B is favored because $\frac{\Delta TP}{\Delta FP} = 2 < \gamma = \frac{10}{4}$, illustrating the benchmark's application in evaluating classifier performance.

The *Worthiness Benchmark* (γ) is useful for examining how binary classification metrics rank classifiers against a baseline. By conceptualizing metrics (M) as Profit functions, we aim to identify the γ value each metric conforms to: $\gamma_M = -\frac{\frac{d(M)}{d(FP)}}{\frac{d(M)}{d(TP)}}$. Originally, γ was defined based on the Profit function, reflecting specific problem attributes. Nonetheless, γ_M is inferred with M acting as the Profit function, albeit without full consideration of problem specifics in metric definition and choice. γ_M illustrates the characteristics of the problems where M is suitable for ranking classifiers. To separate the γ associated with a metric and the γ determined by the problem's characteristics, we use a subscript.

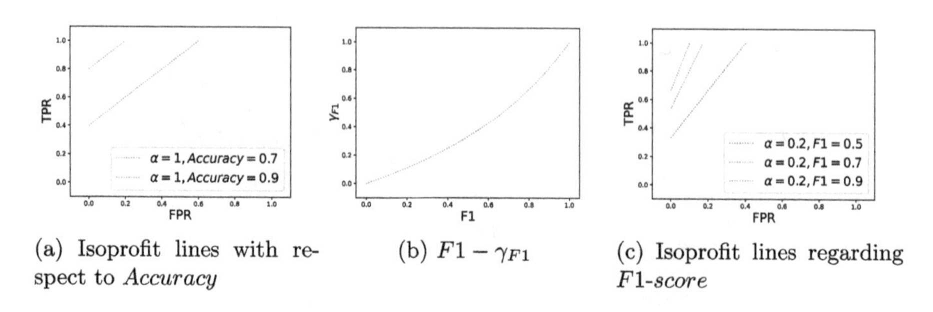

(a) Isoprofit lines with respect to *Accuracy*

(b) $F1 - \gamma_{F1}$

(c) Isoprofit lines regarding $F1$-*score*

Fig. 1. γ-Analysis of *Accuracy* and $F1$-*score*

3 Discussion and Conclusion

The primary contribution of our research was to introduce a new mathematical method known as γ-analysis to evaluate metrics used in binary classification. At the heart of γ-analysis lies the newly proposed concept of the *Worthiness Benchmark*. This concept is crucial for assessing the trade-off between changes in TP and FP from the perspective of the metrics. In this section, we γ-analyze the *Accuracy* and $F1$-*score* metrics.

Accuracy is a common metric for binary classification defined as: $Accuracy = \frac{TP+TN}{N} = \frac{TP+AN-FP}{N}$. While often used for balanced datasets, it can be applied to imbalanced datasets considering class gain-loss relationships. The *Worthiness Benchmark* helps determine when *Accuracy* reliably ranks models. $\gamma_{Acc} = 1$, indicating a constant value regardless of *Accuracy*. This shows that increasing TP relative to FP enhances *Accuracy*. Isoprofit lines for *Accuracy* are given by $TP = FP + c$ (by solving Eq. 6), showing a linear TP and FP relationship as depicted in Fig. 1a. Parallel lines suggest that γ remains constant. Higher lines denote better performance, confirming our assumptions when FP is constant and TP increases (assumption 3).

The $F1$-*score* is the harmonic mean of *Precision* and *Recall*, designed to maximize positive case identification while maintaining high *Precision*: $F1 = \frac{2 \cdot Precision \cdot Recall}{Precision + Recall} = \frac{2TP}{AP+TP+FP}$. Grasping how the $F1$-*score* ranks classifiers is crucial, particularly regarding the TP and FP trade-offs it involves. The *Worthiness Benchmark* for $F1$-*score*, γ_{F1}, is: $\gamma_{F1} = \frac{TP}{AP+FP}$, with $\gamma_{F1} < 1$ signifies that an addition of one TP and one FP boosts the $F1$-*score*. Unlike the constant γ_{Acc}, γ_{F1} varies, defined by: $\gamma_{F1} = \frac{F1}{2-F1}$. This dependence on $F1$, shown in Fig. 1b, means improving $F1$-*score* becomes harder at higher values. Isoprofit lines for $F1$-*score* are given by: $TP = c \cdot FP + d$, where c and d are constants. This linear relationship, shown in Fig. 1c, enables substituting $\frac{\Delta TP}{\Delta FP}$ for $\frac{d(TP)}{d(FP)}$ in Eq. 2. Unlike *Accuracy*, these lines are not parallel, and higher slopes at higher $F1$-*scores* indicate increasing difficulty in improving $F1$.

Metrics are reliable for ranking only when their *Worthiness Benchmarks* match the problem's benchmark. This benchmark is an intrinsic property of the problem itself. Thus, no metric is universally superior without considering the problem domain. γ-analysis and the *Worthiness Benchmark* help identify the most suitable metric, shifting metric evaluation from intuition to methodology. This approach also aids in developing new metrics and tuning parameters, such as β in F_β-*score* [10].

Our study's limitations highlight areas for future exploration. The *Worthiness Benchmark* is based on specific assumptions that may not apply universally. Future research could broaden its applicability or adapt it to different conditions.

Disclosure of Interests. This work was supported by the PRIN 2020 project COMMON-WEARS (grant number I53C21000210001) and by the STEADIER Project (grant number I55F21001900005).

References

1. Canbek, G., Taskaya Temizel, T., Sagiroglu, S.: Benchmetrics: a systematic benchmarking method for binary classification performance metrics. Neural Comput. Appl. **33**(21), 14623–14650 (2021)
2. Canbek, G., Taskaya Temizel, T., Sagiroglu, S.: Ptopi: a comprehensive review, analysis, and knowledge representation of binary classification performance measures/metrics. SN Comput. Sci. **4**(1), 13 (2022)
3. Ferri, C., Hernández-Orallo, J., Modroiu, R.: An experimental comparison of performance measures for classification. Pattern Recogn. Lett. **30**(1), 27–38 (2009)
4. Gösgens, M., Zhiyanov, A., Tikhonov, A., Prokhorenkova, L.: Good classification measures and how to find them. Adv. Neural. Inf. Process. Syst. **34**, 17136–17147 (2021)
5. Hand, D.J.: Measuring classifier performance: a coherent alternative to the area under the roc curve. Mach. Learn. **77**(1), 103–123 (2009)
6. Hernández-Orallo, J., Flach, P., Ferri Ramírez, C.: A unified view of performance metrics: translating threshold choice into expected classification loss. J. Mach. Learn. Res. **13**, 2813–2869 (2012)
7. Luque, A., Carrasco, A., Martín, A., Lama, J.R.: Exploring symmetry of binary classification performance metrics. Symmetry **11**(1), 47 (2019)
8. Parker, C.: An analysis of performance measures for binary classifiers. In: 2011 IEEE 11th International Conference on Data Mining, pp. 517–526. IEEE (2011)
9. Powers, D.: Evaluation: From precision, recall and f-measure to ROC, informedness, markedness and correlation. J. Mach. Learn. Technol. **2**(1), 37–63 (2011)
10. van Rijsbergen, C.J.: Information Retrieval. Butterworths, 2 edn. (1980)
11. Shirdel, M., Di Mauro, M., Liotta, A.: Worthiness benchmark: a novel concept for analyzing binary classification evaluation metrics. Inf. Sci. (article in press)
12. Shirdel, M., Di Mauro, M., Liotta, A.: Relative Information Superiority (RIS): a novel evaluation measure for binary rule-based classification models. In: Procweedings of International Conference on Embedded Wireless Systems and Networks (EWSN). ACM (2023)

Machine Learning Methods and Applications

Exploring Causal Chain Identification: Comprehensive Insights from Text and Knowledge Graphs

Ziwei Xu[1(✉)] and Ryutaro Ichise[1,2]

[1] National Institute of Advanced Industrial Science and Technology, Tokyo, Japan
xu.ziwei@aist.go.jp

[2] Tokyo Institute of Technology, Tokyo, Japan
ichise@iee.e.titech.ac.jp

Abstract. During real-world reasoning, the logic path is generally not explicitly articulated. An appropriate causal chain can offer abundant informative details to depict a logical pathway, which is also beneficial in preventing ambiguity problems during text generation. However, most causal chains tend to lose their causal meaning after multiple hops, also this phenomenon occurs in other chains of relations. To discriminate the broken linkage in chain detection task, we introduce the **CK-CEVAE** model, **C**hained domain **K**nowledge in **C**ause **E**ffect **V**ariational **A**uto**E**ncoder, which integrates knowledge into the representation of causal assumptions within chains, employing sequential probabilistic distributions for cause-effect estimation. Our model demonstrates an improvement of around 4% in F1-score over LLM-based and neural-based models in identifying causal chains originating from text. Furthermore, to investigate the semantic continuity of chains within established knowledge graphs, we curate a chain-structured dataset, highlighting both causal relations and multiple non-causal relations, i.e. *used for*, *synonym* and *similar to*, termed **ConceptNet-CC dataset**. We noticed that the longer the chains, the fewer instances of existence. However, contrary to our intuitions, models perform better at identifying longer chains than shorter ones in uni-directional relations like *causes* and *used for*.

Keywords: Causal Chain · VAE · Knowledge Graph

1 Introduction

In real-world reasoning, clear expression of the logic path becomes indispensable, especially when engaging in complex reasoning with multiple factors. Causality forms the cornerstone of reasoning, with widespread research focused on textual causal pairs. However, relying on a single causal pair is generally inadequate for capturing the full scope of logical reasoning. There is a growing focus on constructing causal chains because of their potential to provide detailed insights into logical progressions. Unfortunately, in most cases, causal chains fail to maintain meaningful connections from beginning to end. For example, while causal

R. Wrembel et al. (Eds.): DaWaK 2024, LNCS 14912, pp. 129–146, 2024.
https://doi.org/10.1007/978-3-031-68323-7_11

pairs <yoga, flexibility> and <flexibility, problem> make sense independently, their causal chain <yoga, flexibility, problem> loses its causal semantics. This phenomenon is not limited to causality but extends to non-causal relationships as well.

This study centers on identifying effective causal chains and investigating semantic continuity in non-causal sequences. Introducing the **CK-CEVAE** model—**C**hained **K**nowledge in **C**ause **E**ffect **V**ariational **A**uto**E**ncoder, we address disruptions caused by cross-domain concepts by employing chained domain knowledge to detect and address broken linkages. Our model highlights the potential to enhance Cause Effect VAE models (CEVAE) [10] for sequential cause-effect estimations in textual data, distinct from CEVAE's conventional emphasis on single-variable estimation in structured datasets. In causal chains originating from text, CK-CEVAE outperforms other LLM-based models and neural-based models such as GPT-4 [12], BiLSTM-CRF [5], and CVAE-SRNN [17], achieving significantly higher F1 and accuracy scores. Despite these successes, the model's recall performance on structural resources remains a challenge.

Additionally, for exploring non-causal chains, we created a structured dataset named **ConceptNet-CC** from ConceptNet [14]. We analyzed information strength differences between uni-directional and bi-directional relational chains. In our investigation, uni-directional relations such as *causes* and *used for* demonstrated more robust semantic continuity in message passing than bi-directional relations such as *synonym* and *similar to*. Moreover, we noted a low conversion rate from pairs to chains, ranging from 0.53% to 2.16% across various relationships. Longer chains tend to have fewer instances. Contrary to our expectations, models demonstrate superior performance in identifying longer chains over shorter ones in uni-directional relations like *causes* and *used for*.

This paper initiates by discussing relevant literature in Sect. 2. The methodology outlined in Sect. 3 comprises several pivotal elements, including the transformation of in-chain domain knowledge, the structure of sequential models, and the algorithms for chained prediction units. Following this, Sect. 4 introduces the utilized datasets in chains, with which we conduct an experimental analysis. Moreover, Sect. 5 presents a case study that investigates the semantic continuity for different relational chains. To conclude, the paper wraps up with an exhaustive discussion.

2 Related Work

Causal inference from text involves two main methodologies: measuring the causality of causal pairs by LLM-based methods and estimating cause effects of interest given a certain assumption among variables by neural-based methods.

Causal Inference by LLM-Based Methods. The earliest approaches relied on finding the explicit causality in sentences. The explicit clues are a collection of causative verbs, causation adv./adj., causal links, and resultative constructions [1]. Getting rid of the crafty procedures at feature engineering, many language models like BERT [3] were examined and explored to detect causality directly from the labeled sentences or word pairs. To enhance causality detection from text, DTGNN [15] incorporated dependency relation features from a sentence to concatenate with the BERT encoder. Instead, ilab-FinCau [18] injected the causal knowledge injections into the encoders to significantly prevent the wrong identification of cause and effect parts. Despite advances in single causality detection, the ability to infer causal relationships when events are linked in a chain (like A causing B, and B causing C) is not well-developed, neither from text analysis nor from knowledge graph research. There is a study aiming to connect all causal tuples following the acquisition of causal pairs, leading to the creation of the causality knowledge graph FinCaKG [19]. However, validating their causal chains remains notably challenging.

In fact, given the widespread adoption of LLM-based models across diverse applications, framing causal inference as a classification problem offers a pragmatic approach: 1) training a sentence classifier using BERT by transforming concatenated causal chains into individual sentences; 2) embedding the linguistic information to DNN [8] for label prediction task on chains; 3) implementing the sequence labeling models like BiLSTM-CRF [5] for chained connection detection; 4) querying generative models like GPT-4 [12] for question and answer tasks. However, to the best of our knowledge, the application of these prevailing methods for causal chain detection is limited, and there continues to be a deficiency in comparing those implementations within this field.

Causal Inference by Neural-Based Methods. Treatment effect estimation refers to a set of neural-based methods used to assess whether a treatment has a significant outcome based on the covariates of the observation data and to quantify the size of that effect [2]. A practical example, evaluating the treatment effects of two different medications for one disease, is the typical application of binary treatments. This conventional task has been extensively studied in observational studies. For example, TARNet [6] addresses the challenge of estimating treatment effects from observational data where treatments are not randomly assigned. Meanwhile, conditional VAE [13] presents a flexible framework that can be applied to diverse tasks needing to condition on additional information. From the observation data, covariates could be collected in different types, e.g. observed confounders and hidden confounders. The observed confounders are directly induced from the observation data, however, identifying all of the confounders is impossible in practice. We need further assumptions (hidden confounders) to estimate the treatment effects on the outcomes. CEVAE [10] was proposed to simultaneously estimate the unknown hidden space summarizing the hidden confounders and the cause effect. It is a probabilistic model that jointly modeling treatment, potential outcomes, and observed outcomes, and aims to provide more accurate and robust estimates of causal effects compared to traditional methods. However, this model is primarily designed for binary outcomes and may not perform chained predictions. Comparatively, CVAE-SRNN [17] represents hidden confounders directly from causal pairs with conditional VAE and utilizes the structural neural networks for chained reasoning in causal chains. However, its precision suffers due to the strong assumption of treating all contextual information as pivot exogenous variables.

3 Methodology

For a causal chain of N components, e.g. A-B-C-D-E, we retrieve the embeddings of each component individually and combine the embeddings from the same causal pair by concatenating them in the final dimension, forming input variables X_n. With the presence of a causal chain instance, we depict this process in the left of Fig. 1. Subsequently, our proposed CK-CEVAE model is introduced in the middle of the figure, followed by a demonstration of how the outcomes of chained CEVAEs are employed in the chain detection task.

3.1 In-Chain Domain Knowledge

We posit that the comparison in domain knowledge between each component reflects the semantic alterations through chains. We would like to investigate the consistency of domain distribution inside each chain, denoting the components in chains by their sequence, e.g. head, 2nd, 3rd, etc. Due to the established semantic certainty of adjacent causal pairs in chains, we skip the first pair but start to compare the subsequent components, e.g. 3rd, 4th ..., to the head term one by one, termed as 'indirect causal pairs'. Analyzing these indirect causal pairs highlights the coherence of domain distribution within chains, namely **in-Chain domain Knowledge (CK)**. In this context, our investigation is restricted to 12 domains categorized according to Wikipedia classification[1].

For each indirect causal pair within a chain, we identify the top-k domains most closely associated with each component, with k set to 2 in our case. If both components of a pair share at least one common domain, we label this indirect causal pair as demonstrating positive domain consistency, unless negative. Thus, we pre-calculate the representation of in-chain domain knowledge for each input chain. The specifications for the domain detection model are discussed in Sect. 4.2. In the subsequent module, we will integrate the insights gained from analyzing domain distribution within chains into the proposed model.

[1] https://en.wikipedia.org/wiki/Wikipedia:Contents/Categories, access date: 2024/03/26.

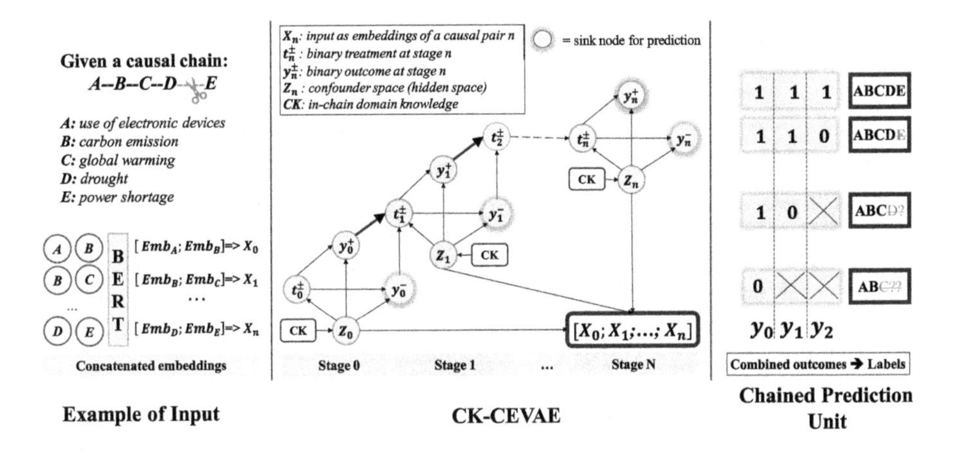

Fig. 1. The framework of CK-CEVAE for the chain detection task.

3.2 CK-CEVAE

The vanilla Cause Effect VAE (CEVAE) [10] architecture is designed for structural data, aiming to simultaneously capture hidden confounders while estimating a solitary effect. To extend its applicability to text data, the preprocessing operation is essential to convert textual information into numerical representations. This transformation relies on the assumption that each causal pair in a causal chain signifies independent causality, facilitating the use of independent numerical representations for these pairs. As depicted in the left component of Fig. 1, our approach proposes to pre-process a causal chain into distinct causal pairs, representing them as concatenated embeddings X_n, which serve as variables supporting the causality of each pair. The aggregation of these causal pair embeddings constructs the representation of a causal chain, illustrated as $[X_0; X_1; ...; X_n]$, which serves as the input data for CK-CEVAE model, positioned at the center of Fig. 1.

From the representation of causal chains, utilizing vanilla CEVAE enables the extraction of independent hidden confounders Z_n, independent treatments $t_n^{\pm}$, and independent outcomes $y_n^{\pm}$. However, this method only yields binary outcomes for each causal chain, which is unsuitable for our purposes of identifying a cohesive causal sequence within a lengthy chain. Moreover, it neglects the transitivity aspect in a chain, wherein the causal linkage from preceding pairs (AB) to subsequent pairs (BC) remains unaddressed. To incorporate this kind of chaining information effectively, we aim to establish connections between treatments and outcomes across causal pairs. Given that, within the same model, treatments should be independent of their potential outcomes ($(y^+, y^-) \perp t|X$), we adopt independent models for sequential estimation tasks while still preserving chained connections between these models. As depicted at the lower center of Fig. 1, we name each estimation task of a causal pair as a *Stage* along the entire sequential tasks.

In *Stage* 0, the causal graph equipped with hidden confounder Z_0 and its treatment t_0 results in the binary outcomes y_0^+ and y_0^-. In the next stage, these outcomes y_0^+ and y_0^- are assigned as the observed treatments t_1 for each instance. Based on the same input embedding, the estimation procedure in *Stage* 1 will generate the dedicated hidden confounder Z_1 and yield the outputs y_1^+ and y_1^-. In this manner, we connect n estimation tasks from *Stage* 0 to *Stage* n together, by passing through the precedent outcomes to the following treatments, while keeping the hidden confounders unique and dedicated for local embeddings.

At startup, we initialize the treatment t_0 with all 1 by assuming the precedent outcomes are all positive. In the sequential stages, we bold the paths with positive outcomes or treatments to highlight the forwarding positive influences. For the termination of each path, we signify the sink of positive influences with the shaded nodes, i.e. $y_0^-, y_1^-, y_n^-, y_n^+$.

Generative Network. CEVAE assumes the observations factorize conditioned on the latent variables in the generative process, and uses an inference network that follows a factorization of the true posterior. In the generative procedure, we adhere to the core concept of the original CEVAE, but we enhance it by incorporating the generation of extra knowledge $\mathbf{g_i}$ conditioned on the hidden space. To simplify comparison, we use **bold** to highlight differences from the original formula.

Here the notations of stage number s_n are omitted, i.e. $z_i^{(s_n)}$ is simplified as z_i. For each stage, the model first generates the probability of hidden confounder z_i following normal distribution as $p(z_i)$. Conditioning on the generated confounder z_i, on the one hand, we calculate the distribution of extra knowledge $\mathbf{p(g_i|z_i)}$ and the distribution of input embeddings $p(x_i|z_i)$, on the other hand, we have the probability of binary treatment $p(t_i|z_i)$. Conditioning on the treatments and hidden confounder, $p(y_i|t_i, z_i)$ express the probabilities of the outcome y_i, where they both follow the Bernoulli distribution:

$$p(z_i) \sim \prod_{j=1}^{D_z} \mathcal{N}(z_{ij} \mid 0, 1); \quad \mathbf{p(g_i \mid z_i) \sim \prod_{j=1}^{D_g} p(g_{ij} \mid z_i)} \tag{1}$$

$$p(x_i \mid z_i) \sim \prod_{j=1}^{D_x} p(x_{ij} \mid z_i); \quad p(t_i \mid z_i) \sim Bern(\sigma(f_1(z_i))) \tag{2}$$

$$p(y_i \mid t_i, z_i) \sim Bern(\pi = \hat{\pi}_i), \quad \hat{\pi}_i = \sigma(t_i f_2(z_i) + (1 - t_i) f_3(z_i)) \tag{3}$$

with $\mathbf{p(g_i|z_i)}$ and $p(x_i|z_i)$ being the appropriate probability distributions. Notice that i represents the number of instances, j is the covariate for different probabilities, i.e. in the case of $p(z_i)$, j ranges to D_z the dimension of hidden space z; for $\mathbf{p(g_i|z_i)}$, j ranges to $\mathbf{D_g}$ the number of features in extra knowledge $\mathbf{g}$; for $p(x_i|z_i)$, j ranges to D_x the dimension of x. For functions, $\sigma(\cdot)$ refers to the logistic function and $f_n(\cdot)$ for neural network parametrized by its parameters.

Inference Network. The hidden space z_i performs as the true posterior depends on x_i, t_i and y_i, the posterior approximation $q(z_i|x_i, t_i, y_i)$ calculated by x_i and y_i and conditioned on treatment t. This approximation architecture is similar to the classical inference network of CEVAE, here we omit the details.

Instead, to preserve the chaining effect of sequential estimation, we focus on embedding the precedent information to the latter one through loss function formation, enhancing both training and inference processes with a focus on information transfer.

In each stage, we could form a single objective for both generative and inference networks, the variational lower bound is given by:

$$\mathcal{L}^{(s_n)} = \sum_{i=1}^{C} \mathbb{E}_{q(z_i{}^{(s_n)}|x_i, \boldsymbol{y_i}{}^{(s_n-1)}, y_i{}^{(s_n)})}[\boldsymbol{\log p(g_i \mid z_i{}^{(s_n)})} + \log p(x_i, \boldsymbol{y_i}{}^{(s_n-1)} \mid z_i{}^{(s_n)}) \\ + \log p(y_i \mid \boldsymbol{y_i}{}^{(s_n-1)}, z_i{}^{(s_n)}) + \log p(z_i{}^{(s_n)}) - \log q(z_i{}^{(s_n)} \mid x_i, \boldsymbol{y_i}{}^{(s_n-1)}, y_i{}^{(s_n)})] \tag{4}$$

with C representing the instance count, (s_n) denoting objects served for computation in stage n. We highlight $\boldsymbol{y_i}^{(s_n-1)}$, the output from stage $n-1$, which replaces $t_i^{s_n}$ in the original loss function to address the absence of model chaining information across stages.

Concerning out-of-sample predictions (unseen instances), theoretically, access to the treatment assignment t_i and its corresponding outcome y_i is essential for each stage before inferring the distribution over z_i. However, these values are unobservable except for the observed input x_i^*. Therefore, two auxiliary distributions are introduced to predict t_i' and y_i' for the masked instances during training:

$$q(t_i' \mid x_i^*) \sim Bern(\pi = \sigma(f_4(x_i^*))); \quad q(y_i' \mid x_i^*, t_i') \sim Bern(\pi = \overline{\pi_i}) \tag{5}$$

In our notation, variables to predict are marked with $'$, while observed variables in the current stage are denoted with $*$. Due to the model's chaining effects, treatments can always obtained from the outcomes of the precedent stage. Therefore, for stage n, we adjust the variational lower bound to estimate the parameters of these two auxiliary distributions as:

$$\begin{aligned} \mathcal{L}_{predict}^{(s_n)} &= \mathcal{L}^{(s_n)} + \sum_{i=1}^{C} [\log q(t_i'^{(s_n)} \mid x_i^*) + \log q(y_i'^{(s_n)} \mid x_i^*, t_i'^{(s_n)})] \\ &= \mathcal{L}^{(s_n)} + \sum_{i=1}^{C} [\log q(\boldsymbol{y_i}^{*(s_n-1)} \mid x_i^*) + \log q(y_i'^{(s_n)} \mid x_i^*, \boldsymbol{y_i}^{*(s_n-1)}] \end{aligned} \tag{6}$$

with $y_i^{*(s_{n-1})}$ being observed outcomes from stage $n-1$ and $y_i'^{(s_n)}$ being outcomes to predict in stage n only.

3.3 Chained Prediction Unit

Variational Auto-Encoders (VAEs) [7], as a generative model, leverages probabilistic distributions, requiring numerous samples for a comprehensive assessment of its outcomes. In our case, the outcomes produced by CK-CEVAE are more complex than those of a single model: we employ sequential models across different stages (i.e. *N stages*) with each model executed in distinct random

Function *Sequential-Prediction* ($y^{sn}_{i,m,r}$, θ, N, R, M)

Input : $y^{sn}_{i,m,r}$ - An individual value of instance i in stage n as m-th sample at r-th randoms;
θ - A threshold of variance tolerance among all sampling outcomes;
N, R, M - The number of models' stages, random states, and sampling results;

Output: y_i - An overall outcome of instance i;

for *n in N stages* **do**
  for *r in R random states* **do**
    for *m in M samples* **do**
      if $y^{sn}_{i,m,r} > 0.5$ **then** $y^{sn}_{i,m,r} = 1$ **else** $y^{sn}_{i,m,r} = 0$; // Standardize initial values
    end
    if $\sum_{m=0}^{M} y^{sn}_{i,m,r}$ ¿ $M * \theta$ **then** $y^{sn}_{i,r} = 1$ **else** $y^{sn}_{i,r} = 0$; // Aggregate sampling results
  end
  if $\sum_{r=0}^{R} y^{sn}_{i,r}$ ¿ $R * 0.5$ **then** $y^{sn}_{i} = 1$ **else** $y^{sn}_{i} = 0$; // Obtain agreements among randoms
end
return $y_i = \left[y^{s0}_i, y^{s1}_i ... y^{sn}_i\right]$

end

Algorithm 1: The algorithm for deriving consolidated predictions from diverse sampling outcomes.

states (i.e. *R states*) to eliminate stochastic influences. Due to the models' probabilistic nature, the results for each sampling operation (i.e. *M samples*) may exhibit considerable diversity even for the same input. To tackle this complex challenge, we introduce the sequential prediction algorithm to obtain a consolidated predictive output for each input. As outlined in Algorithm 1, except for the given outcomes $y^{sn}_{i,m,r} \in \mathbb{R}^{N*R*M}$, our algorithm also takes into account the threshold θ of variance tolerance during sampling as well as the predetermined parameters N, R, M. These inputs lead to the prediction unit producing the overall outcomes y_i, presented as a sequence of binary labels for each instance i, which can serve as categorical labels for evaluation procedures.

To facilitate understanding, we furnish an example showcasing the mapping from sequential binary to categorical labels, situated on the right side of Fig. 1. Each colored box represents the binary prediction sequence for its respective stage. For instance, the orange box corresponds to the negative sink node y_0^- terminated on stage 0 without progressing further, resulting in an outcome at y_0 only, while y_1 and y_2 remain blank. This yields the combined outcome of '0 null null'. Adjacent to the orange box, these outcomes are interpreted into predicted labels 'AB', indicating a causality break from component B to C, with this discontinuity extending to subsequent unknown components marked by '?'. Conversely, assuming positive outcomes persist until the last stage N, the lower green box corresponds to the sink node y_n^-, signifying a break in the causal chain at the final component. The upper green box corresponds to the sink node y_n^+, highlighting that the 5-length chain has successfully maintained causality.

4 Experiments

4.1 Chains Acquisition

Relational chains, like causal chains, consist of sequentially connected word pairs with the same relation. Causal chains are a subset of them, connected by the rela-

Table 1. The instances of dataset **CCR-EN** and **ConceptNet-CC**.

Relations	Datasets	Examples of Relational Chains	Labels
causes	CCR-EN	[yoga, flexibility, problems, stress, marital problems]	Len2
		[excess estrogen, hormonal imbalance, acne, bacteria, gingivitis]	Len5
	ConceptNet-CC	[maintaining muscle strength, maintaining good health, living for long time, art objects, non believer]	Len3
		[taking final exams, passing class, graduating, getting job, needing to pay taxes]	Len4
used for		[stick eraser, pencil, erasing, self conscious, polynymous]	Len3
synonym		[hold, have, accept, bear, disinter]	Len4
similar to		[sentimental, tender, young, little, emotional]	Len5

tion 'causality'. Actually, it is not easy to acquire relational chains, because current knowledge graphs are designed to include and present all types of relations together. In this section, we talk about what kinds of relations are interesting and how to acquire the required relational chains from different resources.

Chains from Text. The chain-structured dataset **CCR-EN**[2], proposed by Xiong et al. [17], exclusively concentrates on causality from text in chains. Their causal chains, spanning 5 terms, draw from the corresponding causal pairs of referenced sentences, showcasing natural mentions reflective of real-world scenarios. Yet, there are no explicit tags for these terms to be associated with any Knowledge Graphs (KGs). The causality depicted within such causal chains is regarded as implicit knowledge lacking structured representation, posing a significant challenge for existing models.

For instance, Table 1 lists the relations and examples alongside their ground truth labels. Each chain comprises 5 terms in sequence, and the label reflects the extent to which the relation is successfully conveyed from the head term. 'Len2' denotes the relation extends to the 2nd term, e.g. 'yoga' causes 'flexibility' but does not continue to result in 'problems'. While 'Len5' signifies the relation successfully extends to the 5th term, e.g. 'excess estrogen' causes 'hormonal imbalance', leading to 'acne', then 'bacteria', and finally resulting in 'gingivitis'. Moreover, the dataset contains the original sentences where the causal pairs originated, serving as the textual context for those terms. It is important to note that only this dataset has context for causal chains, not the subsequent dataset.

[2] https://github.com/waste-wood/reco, access date: 2024/03/26.

Table 2. The statistics of different relations in dataset **CCR-EN** and **ConceptNet-CC**. *Notes: 'uni' for 'uni-directional' category and 'bi' for 'bi-directional' category.*

Resources	Relations		Pairs	Len3	Len4	Len5	Chain Ratio (Len3/Pairs)
CCR-EN	uni	causes	4,640	305	254	543	**6.57%**
ConceptNet-CC	uni	causes	33,481	682	384	264	2.04%
		used for	79,574	1,715	2,172	3,267	2.16%
	bi	synonym	400,609	2,117	1,433	1,443	0.53%
		similar to	60,560	551	468	350	0.91%

Chains from Knowledge Graphs. Obtaining relational chains is not straightforward, as current knowledge graphs are crafted to encompass a multitude of relations in unison. In this section, we explore the types of relations and the method to extract the relational chains from existing knowledge bases.

We seek to understand how semantic information is conveyed through different kinds of relational chains. Our exploration starts by two relation categories: **bi-directional**, exemplified by *synonym, similar to*, and **uni-directional**, represented by *used for, causes.* We choose these relations because they connect terms with similar characteristics to those connected by 'causes' relations like the CCR-EN dataset, specifically focusing on **concepts**. Put simply, we avoid relations linking **name entity** terms. Consider, for example, "Barack Obama born in Honolulu", where both subject and object terms are recognized as named entities. However, the relationship 'born in' is notably less compelling when investigating its chaining effects. Besides, the selected relations are typical examples of 'uni-directional' or 'bi-directional' relations in commonsense knowledge graphs. To locate examples of the aforementioned relational chains, please refer to Table 1.

Though relational chains exist in knowledge graphs, most chains encounter difficulties in passing continuous information from head to tail. The challenge lies in selecting chains that reliably convey information. In the extraction of positive relational chains, we adhere to this rule: a relational chain is deemed positive if its head and tail are directly connected by the same relation. ConceptNet [14] is a well-established knowledge base that incorporates crowd-sourced knowledge from sources like DBPedia [9], WordNet [11], Wikipedia[3], and more. Utilizing ConceptNet-5.7 as the main resource, we form the relation chains from connected pairs and generate the dataset **ConceptNet-CC**. As shown in Table 2, we illustrate the quantity of its related pairs in the **Pairs** column. Additionally, we present an analysis of the statistics for positive relational chains identified at varying lengths-3, 4, and 5-alongside the **Chain Ratio** column, which indicates the likelihood of different relations forming a chain.

Furthermore, we provide the statistics of the CCR-EN dataset in the same table. We noted that bi-directional pairs are generally plentiful, yet their chain

[3] https://www.wikipedia.org/, access date: 2024/03/26.

ratios are consistently low. Regarding uni-directional relations, *used for* and *causes* exhibit relatively high chain ratios. Comparatively, uni-directional relations demonstrate enhanced chaining capabilities. Particularly for *causes* in CCR-EN, the chain ratio is approximately three times higher compared to the uni-directional relations from ConceptNet-CC. This is mainly because the CCR-EN [17] dataset was specifically created for discovering causality from text. Moreover, we observe a general decrease in most relations from short to long chains, except for *used for* and *synonym*. This deviation can be attributed to their long chains sharing a common short chain as a foundation in most cases, with *used for* being a notable example. Relevant information of ConceptNet-CC is available on the web site[4].

4.2 Domain Detection Model

DeBERTa [4] stands out in capturing intricate semantic associations, making it particularly suitable for domain classification. It achieves an accuracy rate of around 90.6% on the MNLI-m dataset [16] and holds a prominent position on the leaderboard[5]. We employ this model[6] for zero-shot classification on 12 Wikipedia categories. After iterating through all indirect causal pairs within a sequence, we convert the aforementioned binary container into decimal digits, representing the domain distribution information for an entire chain.

4.3 Models Configurations

During training, in the presetting of sequential models structure of CK-CEVAE, we set the number of stages N to 3 considering the length of chains being 5, the number of sampling M for each operation is 100 and each model runs with 10 different random states R. For the technical parameters of each model, we list the value of each element: epoch to 500, latent dimension for confounder space to 50, hidden dimension to 200, number of neural layers to 2, learning rate to 5e-4, and weight decay to 1e-4. As for the comparison models, using some prevailing LLM-based and neural-based models straight out of the box isn't ideal for causal chain detection. Here, we outline the distinct preprocessing steps for each model to demonstrate how they can be effectively employed with the relational chains:

GPT-3.5/4 [12]: A LLM-based model using provided prompts to generate responses tailored to the given context or question. Employing few-shot prompting on both GPT-3.5-turbo and GPT-4, for each prompt, we randomly sample three chains (and their context if have) accompanied with gold labels from the training set of the same relation and ask for the label given a relational chain from the corresponding test set.

---

[4] https://www.ai.iee.e.titech.ac.jp/CKceVAE/.
[5] https://gluebenchmark.com/leaderboard/, access date: 2024/03/26.
[6] https://huggingface.co/MoritzLaurer/DeBERTa-v3-base-mnli-fever-anli, access date: 2024/03/26.

Table 3. The overall results in the CCR-EN dataset and ConceptNet-CC dataset on *causes* relation.

	Models	CCR-EN				ConceptNet-CC			
		P	R	F1	Acc	P	R	F1	Acc
Prompting	GPT-3.5-turbo	19.64	26.59	21.78	25.71	30.27	31.25	26.59	31.25
	GPT-4	29.46	29.37	24.12	25.71	44.63	43.75	39.53	43.75
Fine-Tuning	BERT	71.66	81.17	76.12	66.53	90.38	**97.92**	**94.00**	**91.67**
	Sequential DNN	**76.25**	75.31	75.78	68.36	73.58	81.25	77.23	68.06
	BiLSTM-CRF	74.42	79.01	76.65	68.36	95.56	89.58	92.47	90.28
	CVAE-SRNN	66.3	**94.75**	78.01	64.90	70.17	83.33	76.19	65.28
	CK-CEVAE (ours)	74.16	88.58	**80.73**	**72.21**	**96.67**	80.56	87.88	85.19

Table 4. The selected examples showcase varying predictions between CK-CEVAE and comparative models.

Sources	Selected examples	Ground Truth	Prediction	
			CK-CEVAE	Other models
CCR-EN	[precursor tropical wave, thunderstorms, severe weather, flooding, extensive damage]	**Len3**	**Len3**	Len5
ConceptNet-CC	[visiting museum, seeing artifacts, learning about past, undefeatable, polygenesis]	**Len2**	Len3	**Len2**

BERT [3]: A feedforward neural network on the pre-trained language models for the sequence classification task. We use it to categorize a sequence of five terms of a chain into a label denoting the different successful lengths of a chain. The training parameters like epochs are the same as CK-CEVAE but with an early stop.

Sequential DNN [8]: Following the sequential models structure of CK-CEVAE, instead, in each stage, we train an independent Deep Neural Network (i.e. two-layer neural network) to predict the binary outcomes. Without the requirements of a sequential prediction unit, we simply combine the binary predictions in each stage for categorical evaluation. For a better comparison, we apply with the same parameters as CK-CEVAE, e.g. epochs, learning rate, and hidden dimensions. The lack of causality linkage between different stages is the main difference to CK-CEVAE.

BiLSTM-CRF [5]: Bidirectional LSTM with a Condition Random Field layer is practically sufficient for sequence tagging tasks, it not only efficiently uses both past and future features of adjacent terms but also considers the tag information in a whole sequence. We transform our chain identification tasks into sequence tagging tasks by tagging terms with the BIO scheme. For instance, for each chain with length 5, we assign "BIOOO" for each term for a positive instance at Len2, and assign "BIIII" for a positive instance at Len5. In this way, each chain is tagged and each term in a chain will be predicted with a tag.

CVAE-SRNN [17]: This model uses conditional VAE to generate hidden space to represent the causality of each causal pair and propose the structural causal RNN to predict the causality of a subsequent term in a chain. Comparatively, CVAE-SRNN applied VAE independently for each causal pair but embedded the chaining connections in the stage of the proposed structural RNN.

4.4 Overall Analysis

We summarized the overall results of all comparative models in the CCR-EN dataset and ConceptNet-CC dataset on *causes* relations in Table 3. We observed that prompting models significantly lag behind fine-tuning models in both datasets, underscoring the limited capability of prevailing models like the GPT family in diagnosing causal chains. Particularly, they perform worse on text-originated causal chains (CCR-EN) than knowledge graph-originated ones (ConceptNet-CC).

For the CCR-EN dataset, our model can reach the highest F1 score and accuracy with a 2–4% margin. Sequential DNN shows good precision but suffers a lot in recall, in contrast, CVAE-SRNN achieves the highest recall but the precision is the lowest. For a much more detailed analysis, in the comparison of sequential DNN and CK-CEVAE, the inclusion of chaining connections between sequential models increases a lot, showing the benefits of our coupling technique. As for BERT, it has been proven by its best chaining performance but CK-CEVAE generalizes a rather better-hidden space representation, resulting in the higher recall when the precision is close. Additionally, in the ConceptNet-CC dataset, our model reaches the highest precision among others but suffers from low recall as well. The effectiveness of sequence tagging models, such as BiLSTM-CRF and BERT, is evident in this dataset. It seems that we showed a good way to generalize this chaining identification task into a sequence tag recognition task.

Regarding the different performances of the two datasets, we believe that the CCR-EN dataset performs like open-world puzzles and the ConceptNet-CC dataset like close-world puzzles. Please check the different predictions of selected examples in Table 4. The open-world instances (CCR-EN) are shown with the causality between concepts from natural phenomena to weather changes, while closed-world instances (ConceptNet-CC) are comprised of the multiple actions predefined in the knowledge graphs. In these instances, our model correctly identifies the causal link from 'precursor tropical wave' to 'severe weather' rather than disasters, but incorrectly generalizes the action "visiting museum" causing "learning about the past". For the structured causality in the ConceptNet-CC

Table 5. The ablation study on the CCR-EN dataset ("w/o DK" notes for "without domain knowledge", "Emb. CTX" for term embedding with context information, and "Emb. SENT" for term embeddings replaced by its sentence embedding that a target term occurs).

	P	R	F1	Acc
CK-CEVAE	**74.16**	**88.58**	**80.73**	**72.21**
- w/o DK	73.75	86.73	79.72	70.99
Embedding Replacement				
* Emb. CTX	73.63	87.04	79.77	70.99
* Emb. SENT	73.83	87.96	80.28	71.60

dataset, their knowledge is well captured from the large language model embeddings and the typical approaches are capable enough to identify them. But for the CCR-EN dataset, it turns out to be challenging. We believe the probabilistic models could represent well the causality assumptions from real-world text and use it for chaining identification.

4.5 Ablation Study

In order to show the effectiveness of knowledge injection in the proposed model, we employ the ablation study on different injection manners, including domain knowledge injection in hidden space, context knowledge fusion in embeddings, and sentence knowledge fusion in embeddings in Table 5.

We first remove the domain knowledge operation in the model, so as to delete the influence of domain knowledge in the hidden space distribution. The results of row *w/o DK* indicate a noticeable decline in all metrics. Keeping the absence of domain knowledge, we replace the original input embeddings with context embeddings and sentence embeddings. The former technique is to extract the term embedding from the mentions in its context, e.g. a causal sentence, the latter is to extract this sentence embedding directly. We observe a decline in performance for both experiments. This leads us to the conclusion that these embeddings introduce more uncertainty or chaos to our model in representing their causality assumptions. Furthermore, when we attempt to incorporate both domain knowledge injection and either sentence or context embedding fusion, we observe an even greater decline in performance. We infer that domain knowledge possesses abundant and unambiguous information for identifying chain effects, demonstrating the effectiveness of our approach in domain knowledge injection.

Table 6. The F1-score of relational chains detection in ConceptNet-CC dataset.

Relations	Size		Intruder Discrimination			Length Bias			
	Train	Test	SameRel	DiffRel	NoRel	Len2	Len3	Len4	Len5
causes	816	32	0.92	0.85	0.91	0.55	0.73	**0.93**	**1.00**
used for	5,396	636	**0.95**	**0.96**	**0.93**	**0.86**	**0.86**	0.91	0.95
synonym	3,216	32	0.61	0.69	0.65	0.59	0.41	0.22	0.22
similar to	952	80	0.62	0.65	0.69	0.20	0.32	0.18	0.26

5 Case Study: Understanding Semantic Continuity in Knowledge Graphs

Theoretically, there is broad consensus that, for one specific relation, the impact of semantic continuity during messaging passing decreases linearly with an increase in the number of hops. In reality, this consensus is not applicable since the information strength of different relations varies significantly. In this case study, besides causality, we investigate all other relations to understand the semantic continuity within current knowledge graphs, employing ConceptNet-CC as the underlying dataset.

Once we have positive chains, we propose to apply three different techniques to introduce the intruder components for those relational chains. Suppose we have a 3-length positive chain: A-B-C, we would like to replace the final word (i.e. C) with other terms (i.e. N) to have a negative chain A-B-N. The selection of replaced terms dominates whether this example is an easy negative or a difficult negative. We define three different techniques to replace the C term by considering its relation to the precedent term B. It is presented as the *intruder discrimination* column in Table 6. For example, if B and N hold the same relation as in the target chain, it is a difficult negative, tagged as **SameRel**. If B and N hold a relation but not the same as the target chain, tagged as **DiffRel**. If N has no relation to B but just is randomly selected, we consider it an easy negative and tag it as **NoRel**. In this way, we add up the intruder terms to the tail of positive examples until all positive chains reach length 5.

We analyze the semantic continuity of different relations by examining the challenges associated with their chain detection performance, utilizing the most typical model, BERT. Additionally, we deployed CK-CEVAE and regarded its results as additional information to affirm the overall performance of various relations. As discussed earlier, its performance, not outlined here, places it in a subordinate position compared to BERT concerning relations originating from the knowledge graph.

Generally, we summarized the performance of relational chain detection in Table 6. In the columns of *Intruder Discrimination* of Table 6, we observe that the uni-directional relations, e.g. *causes* and *used for* are less sensitive to three different intruders and turns to be easy to be detected. While bi-directional relations vary in a low F-1 score, conforming to our intuitions, the detection of chains is strongly influenced by the intruders with the same relations, but not others. As for intruders with different relations and no relations, their performance varies depending on different kinds of relations. Generally, we can conclude that uni-directional relations have higher information strength than bi-directional relations, owing to their steady and proficient performance, demonstrating reduced vulnerability to intruders. As for CK-CEVAE, we confirm this conclusion but with lower performance.

Except for information strength, we also examine the challenges associated with detecting chains of varying lengths. In the columns of *Length Bias* of Table 6, as the length grows, surprisingly the detection of uni-directional relation chains becomes smoother, contrasting with the increased difficulty observed in bi-directional relation chains. We posit that the information strength of bi-directional chains diminishes significantly after a few hops, whereas for uni-directional chains, the information becomes more resilient and concrete. Therefore, we deduce that, within the context of the knowledge graph, uni-directional chains demonstrate stronger semantic continuity.

6 Discussion

To address the current limitations of ambiguity in text generation, exemplified by ChatGPT, the inclusion of causal or relational chains could be a vital recommendation before generation. If one seeks to request detailed statements or evidence of a causal pair, ChatGPT is often inclined to produce ambiguously all-purpose remarks. However, if the sequential keywords for the corresponding causal chain are provided before, the generated content becomes spot-on and situation-specific. We insist that extracting concrete relational or logical chains from the text, the primary manifestation of knowledge, reveals deeper insights than the constrained resources available in knowledge graphs. The introduction of a new algorithm is deemed necessary to effectively capture hidden causal chains in text, enabling their application to chain identification issues.

The strength of our model lies in its ability to identify the causal chains sourced from text, yet it falls behind the typical sequence tagging models when the causal chains are derived from knowledge graphs. In addition, the primary emphasis of this model lies in representing the hidden causal connections between variables and outcomes, but its versatility is constrained for other relations.

Nevertheless, it's essential to highlight that in this study, our experiments practically reform diverse input formats to examine various comparative models on the specific chain detection task and report their performance across different datasets and relationships. We anticipate that this discussion will inspire other studies in chain detection, either originating from text or knowledge graphs.

7 Conclusion

We suggest that domain knowledge could be effective characteristics for distinguishing the continuity of chained relations. We propose CK-CEVAE to inject the in-chain domain knowledge for a better representation of the hidden causality in each stage and the effective sequential connections between different stages in chains. The results show that our model proves to be proficient in identifying causality originating from text but its effectiveness diminishes when faced with causality sourced from a knowledge graph. Furthermore, we offer the chain-structured ConceptNet-CC, designed for the relational chains detection task incorporating different levels of difficulty in handling intruder challenges. We conduct a thorough examination of the differences in semantic continuity for various relational chains of knowledge graphs. This investigation points to the conclusion that, in message passing, uni-directional relations have greater strength than their bi-directional counterparts. In the future, we discern that the intriguing aspect centers on the selection or ranking of the most suitable chains from different candidates derived from the same relational pairs. The chains deemed most fitting are those appropriate for specific events or scenarios.

Acknowledgment. This study is partially based on the results obtained from a project JPNP20006, commissioned by the New Energy and Industrial Technology Development Organization (NEDO).

References

1. Ali, W., Zuo, W., Ali, R., Zuo, X., Rahman, G.: Causality mining in natural languages using machine and deep learning techniques: a survey. Appl. Sci. **11**, 10064 (2021)
2. Chu, Z., Huang, J., Li, R., Chu, W., Li, S.: Causal effect estimation: recent advances, challenges, and opportunities (2023)
3. Devlin, J., Chang, M.W., Lee, K., Toutanova, K.: BERT: pre-training of deep bidirectional transformers for language understanding. In: Proceedings of the 2019 Conference of the North American Chapter of the Association for Computational Linguistics, pp. 4171–4186 (2019)
4. He, P., Liu, X., Gao, J., Chen, W.: DeBERTa: decoding-enhanced BERT with disentangled attention. In: International Conference on Learning Representations (2021)
5. Huang, Z., Xu, W., Yu, K.: Bidirectional LSTM-CRF models for sequence tagging. arXiv (2015)
6. Johansson, F.D., Shalit, U., Sontag, D.: Learning representations for counterfactual inference. In: Proceedings of the 33rd International Conference on International Conference on Machine Learning, vol. 48, pp. 3020–3029 (2016)
7. Kingma, D.P., Welling, M.: Auto-encoding variational bayes. In: Proceeding of the 2nd International Conference on Learning Representations (2014)
8. LeCun, Y., Bengio, Y., Hinton, G.E.: Deep learning. Nature **521**(7553), 436–444 (2015)
9. Lehmann, J., et al.: DBpedia - a large-scale, multilingual knowledge base extracted from Wikipedia. Semant. Web **6**(2), 167–195 (2015)
10. Louizos, C., Shalit, U., Mooij, J., Sontag, D., Zemel, R., Welling, M.: Causal effect inference with deep latent-variable models. In: Proceedings of the 31st International Conference on Neural Information Processing Systems, pp. 6449–6459 (2017)

11. Miller, G.A.: WordNet: a lexical database for English. In: Human Language Technology: Proceedings of a Workshop held at Plainsboro, New Jersey (1994)
12. OpenAI, et al., J.A.: GPT-4 technical report (2024)
13. Sohn, K., Yan, X., Lee, H.: Learning structured output representation using deep conditional generative models. In: Proceedings of the 28th International Conference on Neural Information Processing Systems, pp. 3483–3491 (2015)
14. Speer, R., Chin, J., Havasi, C.: Conceptnet 5.5: an open multilingual graph of general knowledge. In: Proceedings of the 31st Conference on Artificial Intelligence, pp. 4444–4451 (2017)
15. Tan, F.A., Ng, S.K.: NUS-IDS at FinCausal 2021: dependency tree in graph neural network for better cause-effect span detection. In: Proceedings of the 3rd Financial Narrative Processing Workshop, pp. 37–43. Association for Computational Linguistics (2021)
16. Williams, A., Nangia, N., Bowman, S.: A broad-coverage challenge corpus for sentence understanding through inference. In: Proceedings of the 2018 Conference of the North American Chapter of the Association for Computational Linguistics: Human Language Technologies, pp. 1112–1122 (2018)
17. Xiong, K., et al.: ReCo: reliable causal chain reasoning via structural causal recurrent neural networks. In: Proceedings of the 2022 Conference on Empirical Methods in Natural Language Processing, pp. 6426–6438 (2022)
18. Xu, Z., Nararatwong, R., Kertkeidkachorn, N., Ichise, R.: iLab at fincausal 2022: enhancing causality detection with an external cause-effect knowledge graph. In: Proceedings of the 4th Financial Narrative Processing Workshop, pp. 124–127 (2022)
19. Xu, Z., Takamura, H., Ichise, R.: A framework to construct financial causality knowledge graph from text. In: 2024 IEEE 18th International Conference on Semantic Computing, pp. 57–64 (2024)

Towards Regional Explanations with Validity Domains for Local Explanations

Robin Cugny[2,4](✉), Julien Aligon[1], Max Chevalier[3], Geoffrey Roman Jimenez[4], and Olivier Teste[2]

[1] Université Toulouse 1, IRIT, Toulouse, France
julien.aligon@irit.fr
[2] Université Toulouse 2, IRIT, Toulouse, France
{robin.cugny,olivier.teste}@irit.fr
[3] Université Toulouse 3, IRIT, Toulouse, France
max.chevalier@irit.fr
[4] SolutionData Group, Toulouse, France
groman-jimenez@solutiondatagroup.fr

Abstract. The field of explainability in machine learning has become very prolific and numerous explanation methods have emerged during the last decade. Local explanations are of major interest because they are intelligible and claim to be locally faithful to the machine learning model. However, there is no guarantee that these explanations can apply to new similar observations. This raises questions about the usefulness of acquired knowledge through such explanations. In this paper, we propose a new concept, which we call *the validity domain of explanations*. This consists in searching for a region around a given local explanation allowing users to determine the extent to which the explanation generalizes. We experimentally show that our approach is more precise regarding the machine learning model and has a better coverage of the dataset than Anchors, a known competitor that paved the way for this issue.

Keywords: Explainable machine learning · Evaluation of explainability · Quality of explanations · Evaluation metrics · Information system

1 Introduction

Our ability to model complex phenomena has improved with the advent of Artificial Intelligence and especially with Machine Learning (ML). While the models produced this way are more accurate, we are not able to understand their reasoning. EXplainable Artificial Intelligence (XAI) tries to tackle the lack of understanding of ML models' inner working or their results. XAI methods are accepted as effective tools to explain ML-based decision-making systems in many application fields (medical, judicial, digital security, or autonomous driving) [4]. Explanations can either be global and explain the whole model or local and

R. Wrembel et al. (Eds.): DaWaK 2024, LNCS 14912, pp. 147–161, 2024.
https://doi.org/10.1007/978-3-031-68323-7_12

explain one or several predictions [9]. In this paper, we focus on a popular family of local XAI methods, named *additive feature attribution methods* [8]. These latter approximate the original model and produce simpler linear explanation models that can obtain the same prediction by summing the feature attributions.

Usually, users look for explanations to improve their understanding of someone or something so that they can derive a stable model that can be used for prediction and control [7]. However, as pointed out by [15], since these linear explanations are in some way local, it is not clear whether they apply to an unseen instance, i.e. new similar observations that the user discovers and has to process could have a different explanation. Indeed, as shown in Fig. 1, similar observations can have dissimilar explanations and conversely. Evaluation measures for explanations provide a score for continuity [10], but not for the area in which the explanation applies. To fill this gap, it would therefore be interesting to define the location of observations with similar explanations. However, these areas can be difficult to model, because of their arbitrary or discontinuous shape. [15] propose an algorithm to compute boxes that give an area where an explanation applies, these explanations, called anchors, are decision rules. However, rules cannot properly model non-linear or discontinuous shapes, as can be seen in Fig. 1d.

In this paper, we introduce a new concept we call the *validity domain of explanations*. It indicates to the user how generalizable the explanation is, so it is still locally faithful while extending its usage to regional explanations for potential new observations. We provide an algorithm to compute validity domains jointly with existing explanations. It is agnostic to the explanation method as long as the explanation is part of additive feature attribution methods. We show that it beats Anchors algorithm [15] in precision and coverage while still being interpretable. We show that it can handle discontinuities in the data space if there are separate areas with similar explanations while Anchors algorithm can only produce one continuous explanation area at a time. Also, we experimentally show that our proposal can cover more observations without false positives. Eventually, we conduct a sensitivity analysis to show the influence of our hyperparameters on validity domain properties.

The rest of this paper is organized as follows. Related work is introduced in Sect. 2. A toy example is shown in Sect. 3. Our proposal is formally described in Sect. 4. Results from experiments are detailed in Sect. 5. The limits of this work are identified in Sect. 6. Lastly, we conclude and discuss future work in Sect. 7.

2 Related Work

In this paper, we study explanation methods, especially local feature summary explanations from additive feature attribution methods. Explanations can be evaluated with metrics that assess if a quality property is met. Eventually, local explanations can be treated as tiny surrogate models that mimic the behavior of the black box model locally. Thus, we take inspiration from the literature on validity domains of models to propose validity domains of explanations.

2.1 Explanation Methods

Explanation methods produce one or several explanations for an existing model or learn an interpretable model. As for the form of the explanation, authors of [2] list several types of explanation: model internals, data point, surrogate intrinsically interpretable model, rule sets, explanations in natural language, question-answering, and feature summary, those studied in this paper. As mentioned in Sect. 1, explanations can either be global or local. Local explanations are a popular kind of explanation as they avoid the accuracy/interpretability trade-off by describing the black box model with locally faithful explanations. Among these methods, LIME [14] is an explanation method that trains interpretable models to mimic the black box model on perturbations around a given observation. In its implementation, the weights of a linear model capture the importance of features. The explanation consists of these weights. Meanwhile, authors of [8] propose to unify several of these explanation methods as *Additive Feature Attribution Methods*. This family of explanation methods gathers the *explanation models* that they define as any interpretable approximation of the original model. LIME belongs to this family of methods for instance. The authors also propose SHAP, an explanation method of this family that captures feature interaction using Shapley values from game theory. In this family, authors of [12] propose MAPLE, a predictive model that provides faithful self-explanations based on local linear modeling techniques and a dual interpretation of random forests.

Taking a different approach, authors of [15] propose a rule-based explanation method that can explain one prediction and other observations that have similar explanations regarding a subset of predicates. Authors call these explanations anchors and propose an algorithm to obtain anchors with a minimal coverage and the best precision, which are evaluation metrics for their explanations.

It is necessary to assess the quality of explanations to compare them and improve them. Hence, while authors of [15] propose metrics, many other metrics exist and they can evaluate other quality criteria.

2.2 Explanation Evaluation Metrics

Explanations can be evaluated subjectively by humans or objectively by metrics. For the latter, as there is no universal definition of what a good explanation is, we refer to a set of quality criteria called explanation properties [10]. For each property, there is a set of corresponding evaluation metrics and each of them is compatible with a set of explanation methods. We focus on the **Continuity**, **Correctness**, and **Completeness** properties as they apply to our context.

Continuity property describes how continuous and generalizable the explanation function is [10]. To do so, authors of [1] propose to assess the continuity of explanation with a neighborhood-based local Lipschitz continuity metric. It captures changes in the explanation w.r.t. small changes in the input so that small changes in the input should not lead to important changes in the output.

Correctness describes how faithful the explanation is, w.r.t. the black box [10]. Hence, authors of [18] propose to assess both continuity and correctness respectively with sensitivity; a metric that works similarly to [1] and infidelity, a metric that observes if the explanation can capture the prediction changes when the input changes. To evaluate correctness, authors of [15] propose precision, which is the percentage of observation in the anchor that has the same prediction as the reference observation.

Completeness describes how much of the black box behavior is described in the explanation [10]. In [15], authors propose to compute coverage, which is a completeness evaluation. The coverage is the percentage of observations belonging to the anchor in a given data distribution.

Evaluation measures produce a score for a property that often summarizes errors, thus losing information. Indeed, a continuity score does not allow the user to know in which direction or at which distance the explanation holds. To know where the explanation holds, rather than looking for a score, we are interested in non-aggregated information. To do so, we turn to the work on validity domain of models because the proposed methods could be adapted to the explanations.

2.3 Validity Domain of Models

The value of an approximated model depends on the consciousness of its descriptive limits and on the precise estimation of its parameters [13]. We believe that the same applies to explanations. Authors of [3] have first proposed to build a convex hull of the training data points to describe the validity domain. Although this method is efficient, it fails to capture non-convex domains of validity. This is why authors of [13] propose to compute non-linear classifiers for complex validity domains. They propose using a one-class Support Vector Machine (SVM) to learn the data distribution, which can also capture discontinuous validity domains. Eventually, authors of [17] propose to study the data space to select the most suitable algorithm to determine the validity domain.

To summarize, local explanations with feature importance are popular but to guarantee specific quality criteria, evaluation measures have been proposed. Our goal is to know to which extent an explanation applies to other observations. However, the evaluation measures give a score, but the score does not indicate the areas where the explanation applies. Therefore, we turn to the studies on validity domains of models that could help us to meet our goal if adapted to work on local explanations. That's the objective of this paper.

3 Toy Example

We illustrate the need for validity domains of explanations with the following toy example. We first train a Multilayer Perceptron (MLP) on Flame [6], a synthetic 2D dataset, and then we explain the predictions with SHAP [8]. Figure 1a displays the decision boundaries of the MLP model.

Figure 1b shows an example of a validity domain for one explanation. In this example, we define two observations as similar using a radius of 0.1 as in [1], and explanations are considered similar for a threshold of 0.2 w.r.t. Euclidean distance between explanations (the darkest blue in the color scale here). The

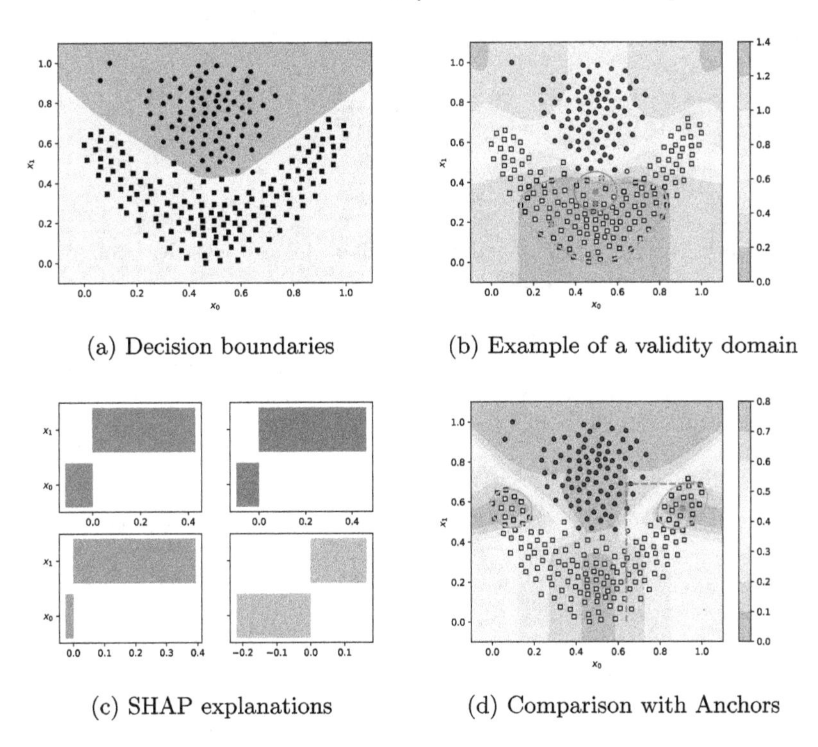

(a) Decision boundaries

(b) Example of a validity domain

(c) SHAP explanations

(d) Comparison with Anchors

Fig. 1. Multilayer Perceptron (MLP) trained on Flame dataset, explained with SHAP, with examples of validity domain and rules from Anchors. Figure a shows the decision boundaries of the MLP model. The colored areas correspond to the classes predicted by the model at the given coordinates, while shapes for the points represent their true label. Figure b shows an example of a validity domain for an explanation. The color of the data points refers to the predicted class. The red cross is the data point that is used to produce the *reference explanation* in red in Fig. c. The background color represents the Euclidean distance between the *reference explanation* and the explanation produced for a data point at given coordinates. The green contour corresponds to the validity domain of the explanation for the red cross observation. The red circle around it represents its neighborhood in data space within a radius of 0.1 like in [1], these observations are considered similar. Figure 1c represents the local explanations of the four data points computed by SHAP. The values are influences of each feature to the prediction. In Fig. 1d the green contours are the validity domains of the explanation for the red cross observation. The red dashed lines are rules from the anchor: ['x1 $\leq$ 0.69', 'x0 $>$ 0.64']. (Color figure online)

other colored crosses are observations of interest that will be described in the following paragraph.

In Fig. 1c, we show the local explanations produced by SHAP. They explain the predictions of an MLP for the data points whose colors match the crosses in Fig. 1b.

- The red one is the explanation whose area of validity is observed.

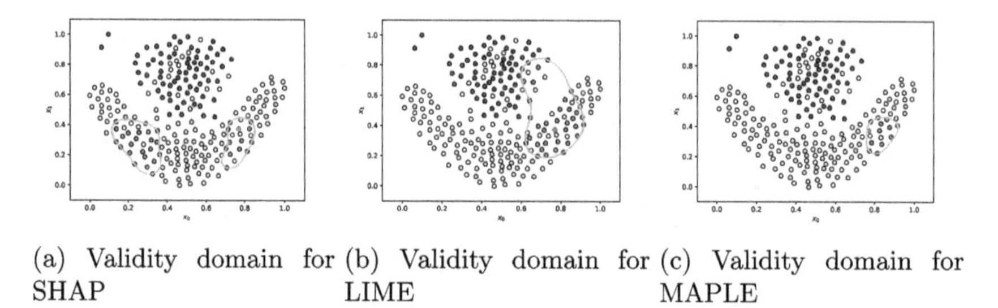

(a) Validity domain for SHAP (b) Validity domain for LIME (c) Validity domain for MAPLE

Fig. 2. Validity domains of a given explanation for different explanation methods. Yellow and purple points are predicted class, in addition, a contour is learned from the blue points (train set) and its generalization to new data points is observed with the white points (test set). (Color figure online)

- The blue one is a similar data point whose explanation belongs to the validity area.
- The green one is a similar data point while being outside the validity area, its explanation is considered dissimilar.
- The magenta data point has a similar explanation while being outside the neighborhood of the red data point.

The validity domain of an explanation is a region in the data domain where observations have an explanation similar to the reference one. We can see that observations can be similar and have dissimilar explanations, as well as the opposite situation: dissimilar observations with similar explanations. Hence, the validity domain of an explanation does not have a regular shape, and ignoring it can lead to approximations and errors. This motivates our approach, it is important to learn the contours of the validity domain of explanations with a non-linear model to reduce errors.

In Fig. 1d, it can be seen that Anchors includes observations that are predicted to be in the other class and a misclassified data point. In this example, similar SHAP explanations exist elsewhere in the dataset, and we can see that the validity domain manages to capture them. This means that one explanation is sufficient, whereas several explanations with Anchors would have been necessary to explain this set of observations. Figure 2 shows the observations inside the validity domain for different explanation methods as well as their learned contours. However, each of the explanation methods claims to reproduce the model's behavior, these figures show that for different definitions of an explanation, we obtain very different areas of validity. It is therefore important not to ignore the significance of the values returned by these methods, both to read the explanation and to judge its generalizability. In this paper we focus on SHAP: the explanations it returns integrate the relationships between attributes and the sum of SHAP values gives the prediction without requiring the data value. We will now detail our proposal more formally.

4 Our Proposal

4.1 Validity Domain

Validity areas should be accurate and guarantee (with a predefined margin) that the user does not need to recalculate other explanations within this area. Finally, these validity areas must be able to identify observations with the same explanations, even if they are far apart or there are several areas instead of one.

Here, we consider that a user has access to new observations that correspond to the same kind of data at the origin of a known dataset. The latter is defined as follows:

Definition 1 (Dataset). *X is a dataset so that $X = \{x_i\}_{i=1}^{n} | x_i \in \mathbb{R}^d$ are the observations. We also note $Y = \{y_i\}_{i=1}^{n} | y_i \in \mathbb{R}$ the labels of the datasets. n is the number of observations and d is the number of dimensions of the dataset.*

We consider an ML model, trained on the dataset X, addressing a given task to assist the user. It acts as a decision-support tool.

Definition 2 (ML model). *An ML model $\hat{f} : X \rightarrow \hat{Y}$ is an estimator whose parameters are learned from statistical relationships between X and Y. $\hat{Y} = \{\hat{y}_i\}_{i=1}^{n} | \hat{y}_i \in \mathbb{R}$ are the produced predictions.*

Including explanations completes the information input for the decision-support tool. This allows users to check whether they agree with the ML model's opinion.

Definition 3 (Explanations). *An explanation is a set of information about the ML model or its predictions. Here, we focus on local feature importance explanation $e_i \in \mathbb{R}^d | f_e(x_i) = e_i$, with f_e an explanation method.*

Validity domains of explanations are complementary information about the generalizability of these explanations. They make it possible to associate new observations with already-established explanations.

Definition 4 (Validity domain of a local explanation). *The validity domain of a given local explanation e_i is the region, in data space, in which the observations provide similar explanations. These explanations are similar to e_i w.r.t. a distance metric d and a tolerance threshold τ. $VD(e_i) \subseteq \mathbb{R}^d | \forall x_j \in VD(e_i), d(e_i, e_j) \leq \tau$, with $f_e(x_j) = e_j$.*

In this paper, we propose an algorithm to approximate the validity domain of local explanations. We proceed into three steps: **compute explanations**, **find observations with similar explanations**, and **compute the contour** of these observations. We compute explanations using any additive feature attribution method over the ML model predictions for each already known observation, i.e. data from the training set. One way to find similar explanations is by using a distance metric over explanations and a threshold. Intuitively, the choice for the distance metric depends mainly on the dimensionality of the dataset while the threshold is mainly the user's demand. We use a OneClassSVM [16] on the previously found observations. Based on this work, we propose Algorithm 1 to learn the contour of the validity domain for one explanation.

Algorithm 1. Learn Validity Domain

```
function LVD(x, E_train, Γ, τ)
    e ← f_e(x)
    ED ← {d(e, e_j) | e_j ∈ E_train}
    X_VD ← X_train[where(ED ≤ τ)]
    m_VD ← OneClassSVM.train(X_VD, Γ)
    return m_VD
```

Algorithm 1 shows how validity domains are learned for any observation x. It computes e the explanation for x using the explanation method f_e. Then it computes the distance between e and the explanations for the whole train dataset E_{train}. It selects the observations from the training dataset X_{train} for which the explanations are at a distance of less than the threshold distance τ. Hence, it obtains the data subset belonging to the validity domain X_{VD}. It can now learn the contour of X_{VD} using a OneClassSVM with an RBF kernel that relies on a Γ parameter. m_{VD}, the learned model, is a Boolean function that determines if a new unseen observation is inside the validity domain of e.

We propose an additional layer on the proposed algorithm. Subsets of E_{train} are created based on the predicted classes for each corresponding observation. It calculates the distances between explanations of each predicted class separately instead of the entire E_{train}. It then rescales distances them between 0 and 1. As the distances between explanations may be different for each class, it is possible to use a single intuitive threshold for all classes without introducing bias for any class. This choice means that observations that are not predicted by the same class are not used for contour learning.

Once several validity domains are learned from existing data, the user can use them to find one that matches a new observation.

Algorithm 2 tests each validity domain model from the set of learned models $\{m_{VD}\}$ until it finds one that predicts x to be inside the validity domain. It returns $\{m_{VD}\}[j]$: the corresponding validity domain model, $X[j]$: its original observation, and $E[j]$: its corresponding explanation. It doesn't need the explanation of the current observation, making it simple and fast.

Algorithm 2. Find corresponding Validity Domain

function FVD($x, X, E, \{m_{VD}\}$)
  $j \leftarrow 0$
  while $\{m_{VD}\}[j](x)$ is *False* and $j < \text{len}(\{m_{VD}\})$ **do**
    $j \leftarrow j + 1$
  return $X[j], E[j], \{m_{VD}\}[j]$

4.2 Model Summary

When several validity domains are learned, it is also possible to select the set of those that cover the greatest number of observations in the dataset. Hence, it is possible to approximate the model's behavior with a subset of local explanations. Such a summary eliminates the need to calculate new explanations as the dataset is covered. Moreover, the black box model is now easy to understand as it is summarized with several transparent models. When learning the contours, these models should have a minimal precision, otherwise it is easy to cover the whole dataset with bad explanations. Here is the algorithm we propose:

Algorithm 3. Model Summary

function MS($\{X_{VD}\}, K, precs, min_{prec}$)
  $idx \leftarrow \text{Filter}(precs, min_{precs})$
  $C \leftarrow \text{Combinations}(idx, K)$
  return $\text{argmax}_{c \in C} \text{Size}(\bigcup_{j \in K} \{X_{VD}\}[c_j])$

Algorithm 3 finds the combination of validity domains that covers the largest number of observations from the training dataset X_{train}. It uses $\{X_{VD}\}$, the set of observations subsets inside existing validity domains, and *precs*, the precisions of their corresponding OneClassSVM models $\{m_{VD}\}$. It keeps only the validity domains so that their models have a precision higher than min_{prec} to ensure reliable results. It computes every K-combinations of the remaining data subset from $\{X_{VD}\}$ (the K-length tuples from idx, no repeated elements). Eventually, it returns the indices of the data subsets whose union covers the largest number of observations.

4.3 Evaluation Metrics

To evaluate the quality of these regions, we observe correctness and continuity properties with precision and coverage respectively. We base our work on [15] metrics. We leave precision unchanged but improve coverage and complement with a new metric Maximum Distance between Observations (MDO).

Precision is computed as $\frac{tp}{tp+fp}$ where tp is the number of observations in the class of the given point and fp is the number of observations belonging to another class. This precision metric considers false positives as observations in the region while belonging to another class according to the model. Note that False negatives cannot be defined using the inverse reasoning here, otherwise every observation of the correct class should be in the region. This is why there is no recall metric.

The coverage that we propose corresponds to the percentage of observations belonging to the region out of all the dataset. It is based on existing data and not generated perturbations like the original metric. We use the whole dataset including the test set so that coverage reflects the region's capacity to integrate new observations. The scores tend to be smaller than the original coverage because the latter was calculated relatively to a smaller domain around the studied observation.

$$cov(R) = \frac{\text{Card}(X_R)}{\text{Card}(X)} \tag{1}$$

With X the whole dataset and X_R the observations inside the region R. Regions can be wide and discontinuous; non-intuitively, very dissimilar observations can have similar explanations depending on the ML model and the explanation method. Therefore, we also propose Maximum Distance between Observations to evaluate how dissimilar observations can be inside a region. It consists in computing the distance between every observation inside a region and keeping the maximum.

$$MDO(R) = \max d(X_R) \tag{2}$$

5 Experiments

The **first experiment** is to compare our proposal and Anchor [15] in terms of precision, coverage, and MDO, defined in Sect. 4.3. The results are shown in Table 2. The aim of the **second experiment** is to compare the total coverage of our proposal with Anchors for the model summary task. Eventually, we study the influence of hyperparameters of our proposal with a sensitivity analysis.

5.1 Protocol

We want to study whether the validity domains with non-linear contours lead to a gain in precision and coverage. Thus, we use these two measures to compare our proposal to Anchors on various datasets. We also measure the maximum distance between observations (MDO) in the validity domain and the zone defined by the anchors to evaluate their ability to fetch further observations. Although better scores are obtained for a higher distance threshold, we wish to have more similar explanations in the validity domains. As the distance between explanations is rescaled between 0 and 1, in Table 2 we only show here results for a distance threshold less or equal to the mid-range value: 0.5. Note that we also assess the sensitivity of the distance threshold in Sect. 5.4.

The datasets used are classification datasets detailed in Table 1. The ML model used for every dataset was a Multilayer Perceptron from scikit-learn [11]. To train it, we split the dataset into a train set (80%) and a test set (20%). This

Table 1. Characteristics of used datasets

Dataset	# Samples	# Features	# Classes
Flame [6]	240	2	2
Iris [5]	150	4	3
Wine [5]	178	13	3
Adult [5]	32561	14	2
Breast cancer [5]	116	9	2

Table 2. Average scores regarding Validity Domains and Anchors for different datasets.

	Precision		Coverage		MDO	
	VD	Anchors	VD	Anchors	VD	Anchors
flame	**1.00 ± 0.00**	0.98 ± 0.04	**0.19 ± 0.06**	0.18 ± 0.08	**0.79 ± 0.16**	0.52 ± 0.12
iris	**1.00 ± 0.00**	0.98 ± 0.05	**0.26 ± 0.03**	0.1± 0.08	**0.43 ± 0.05**	0.25 ± 0.18
wine	**1.00 ± 0.00**	**1.00 ± 0.00**	**0.12 ± 0.07**	0.12 ± 0.06	0.44 ± 0.21	**0.54 ± 0.12**
adult	**1.00 ± 0.00**	0.97 ± 0.08	**0.22 ± 0.08**	0.08 ± 0.10	**0.90 ± 0.04**	0.67 ± 0.33
breast cancer	**1.00 ± 0.00**	**1.00 ± 0.00**	**0.09 ± 0.05**	0.04 ± 0.03	**0.46 ± 0.10**	0.33 ± 0.22

allows us to see if the model handles new observations, and also to see if the explanation regions found can generalize to new points from the same distribution. The algorithms we evaluate for these experiments are Anchors [15] and our proposal. Hyperparameters are determined using Grid Search. For Anchors, it is done on δ, ϵ, and B, and for our validity domains, it is done on the distance threshold τ and Γ. For Anchors, the threshold for the desired confidence is the minimal precision required. Therefore, we set it using the average precision for the Validity Domains. It should result in more precise anchors. The explanation algorithm we use for our proposal is SHAP [8] for these experiments, but the method is compatible with other explanation algorithms as it can be seen in Fig. 2. To encourage reproducibility, the code, the hyperparameters and results of these experiments are available at https://github.com/RobinCugny/Validity-domains-of-explanations.

5.2 Evaluation of Methods

The results in Table 2 show the average scores and standard deviations for Precision, Coverage, and MDO of all regions (VD and Anchors) found in each dataset. Validity domains have a better average precision while having a better coverage most of the time. Indeed, our approach suffers less from the trade-off between precision and coverage thanks to the non-linear model. It fits more closely to the observations avoiding near observations belonging to other classes. Moreover, the average MDO is most of the time higher for validity domains. OneclassSVM models manage to fetch further observations with similar explanations without losing precision. Decision rules capturing regions with observations so far apart could lead to the capture of observations belonging to the wrong class.

5.3 Model Summary

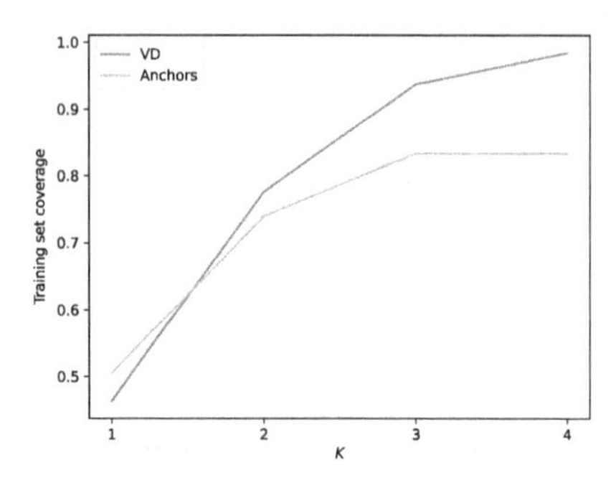

Fig. 3. Coverage of the training set for Flame dataset with k explanations regions. It shows the percentage of observations covered in the training set for k regions with both methods.

Figure 3 shows the percentage of observations covered in the training set for k regions with both methods. We impose the constraint that the regions used must have a precision of 1. The hyperparameters are $\tau = 0.2$ and $\Gamma = 18$. Our proposal achieves a coverage of 98.4% of observations covered with $k = 4$ regions while Anchors covers 83.3% of the training set. This means that the user needs more explanations with Anchors to explain the model which is less convenient. Otherwise, the user has to accept that a bigger fraction of the model cannot be explained which can mislead the user and have serious consequences in high-stakes decision domains.

Figure 4 shows how it is possible to reduce a model into a finite set of explanations. Although some observations are set aside, the 3 surrogate models proposed by SHAP allow a good understanding of the model as a whole. Note that this approach allows the approximation of any statistical model into explanation models such that a maximum number of observations are explained with guarantees on the approximation performed. These guarantees are the maximal margin between explanations and the identification of uncovered points.

5.4 Sensitivity Analysis

We illustrate the impact of the distance threshold τ and Γ hyperparameter on Iris and Flame dataset, with the MLP model, and SHAP. The tendency is similar for every tested dataset, see **link removed for anonymity** for more figures. In Fig. 5a and 5b, when τ increases, the precision decreases as the validity domain becomes more permissive. Regarding coverage and MDO, they increase as the validity domain gets bigger. We note that it is desirable to take a sufficiently large distance threshold so that there can be enough explanations in the validity domain, but that accuracy may be reduced if it is too large. For Γ, in Fig. 5c

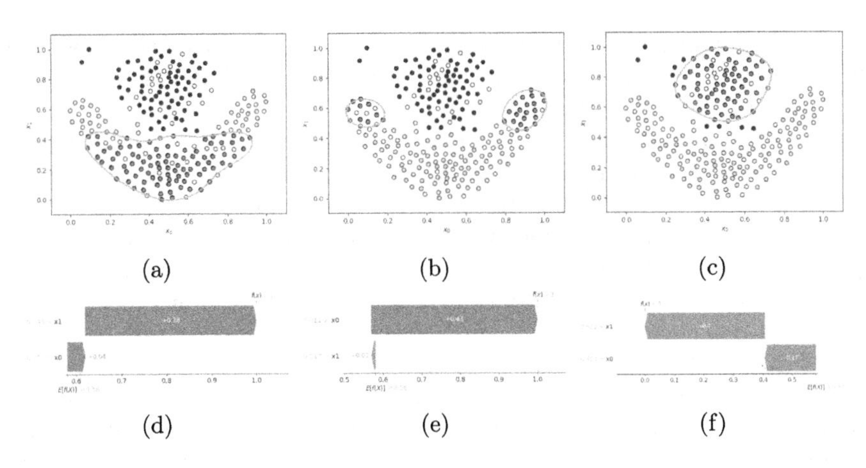

Fig. 4. Model summary applied on Flame dataset with validity domains of SHAP explanations. With $k = 3$, the algorithm finds (d), (e), and (f), 3 explanations that cover 93.8% of the training set. (a), (b) and (c) show the corresponding validity domains for the selected explanations.

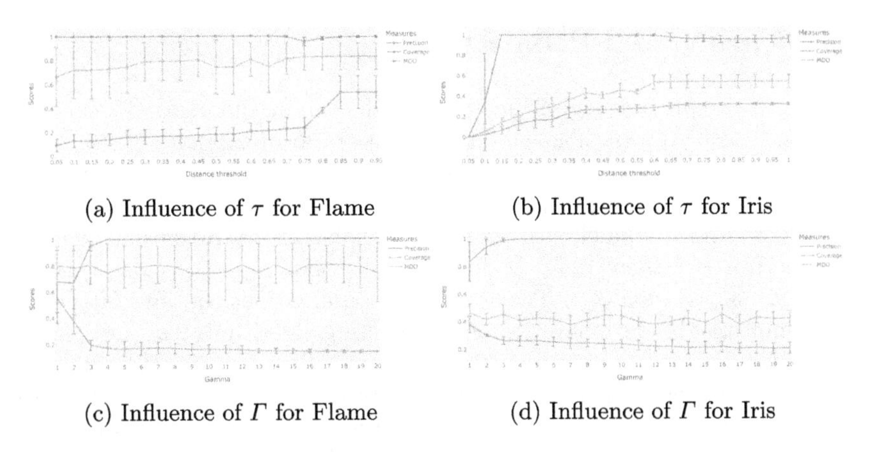

Fig. 5. Sensitivity analysis on Flame and Iris dataset regarding τ and Γ. The curves show the score for evaluation metrics and the corresponding standard deviation for each value of the studied hyperparameter.

and 5d, we observe that the higher the Γ, the more precise the validity domain is, but a plateau is quickly reached. Note that coverage decreases when Γ is too high as the OneClassSVM overfits data, hence the distance between observations decreases a little as boundaries are more precise.

6 Discussion and Limits

In this paper, we assume that the ML model will not have a huge discontinuity for a new instance inside the validity domain. Indeed, ML models become increasingly complex (non-linear) as the tasks are harder to solve, which could result in a local behavior different from the neighborhood and therefore a different explanation (depending on the method used to compute it). As it can be seen in Fig. 1, the area in which explanations are similar ($\tau \leq 2$) in dark blue goes beyond data distribution. This is because the ML model can give a prediction for any instance and the explanation method does the same. However, we choose to learn the validity domain only on existing observations. This approximates the data space, but we do it for two reasons. It would require synthesizing unrealistic artificial instances and mapping the data space and the explanation space requires producing as many explanations as there are coordinates in the data space. The complexity of such computation is exponential w.r.t. the number of dimensions which would make the method impractical. Also, the OneClassSVM model may occasionally fail to learn the domain of validity. This is due to the number of observations being insufficient for training. Regarding high-dimensional data, the choice of the distance metric remains to be studied. Moreover, high-dimensional data results in complex models that are more difficult to summarize with such explanations.

7 Conclusion and Perspectives

In this paper, we propose algorithms to learn the validity domains of local explanations and exploit them. The validity domains of local explanations show how generalizable these explanations are. They enable the knowledge gained from previous explanations to be applied to new observations. Finally, they can be used to summarize a model with a subset of explanations. Regarding the existing competitor, we show that validity domains offer a better coverage of the dataset with a precision of 1. They can reach observations that are further and even get multiple discontinuous regions. Clustering algorithms with medoids can group similar explanations with a representative. These approaches could give interesting ideas and their results could be alternatives to our model summary. A short-term work could be to adapt the method to another type of data such as data images and their corresponding explanations which raises the question of distances between such explanation types. Further work could be to conduct a user study and assess which hyperparameters or explanation methods are preferred regarding the context.

References

1. Alvarez-Melis, D., Jaakkola, T.S.: On the robustness of interpretability methods. arXiv preprint arXiv:1806.08049 (2018)
2. Carvalho, D.V., Pereira, E.M., Cardoso, J.S.: Machine learning interpretability: a survey on methods and metrics. Electronics **8**(8) (2019)
3. Courrieu, P.: Three algorithms for estimating the domain of validity of feedforward neural networks. Neural Netw. **7**(1), 169–174 (1994)

4. Cugny, R., Aligon, J., Chevalier, M., Roman Jimenez, G., Teste, O.: AutoxAI: a framework to automatically select the most adapted XAI solution. In: Proceedings of the 31st ACM International CIKM, New York, NY, USA, pp. 315–324 (2022)
5. Dua, D., Graff, C.: UCI machine learning repository (2017)
6. Fränti, P., Sieranoja, S.: K-means properties on six clustering benchmark datasets (2018)
7. Heider, F.: The Psychology of Interpersonal Relations. Wiley, Hoboken (1958)
8. Lundberg, S.M., Lee, S.I.: A unified approach to interpreting model predictions. In: Proceedings of the 31st International Conference on Neural Information Processing Systems, NIPS 2017. Curran Associates Inc., Red Hook (2017)
9. Molnar, C.: Interpretable Machine Learning. Lulu.com (2020)
10. Nauta, M., et al.: From anecdotal evidence to quantitative evaluation methods: a systematic review on evaluating explainable AI. arXiv preprint arXiv:2201.08164 (2022)
11. Pedregosa, F., et al.: Scikit-learn: machine learning in Python. J. Mach. Learn. Res. **12**, 2825–2830 (2011)
12. Plumb, G., Molitor, D., Talwalkar, A.S.: Model agnostic supervised local explanations. In: Advances in Neural Information Processing Systems, vol. 31 (2018)
13. Quaglio, M., Fraga, E.S., Cao, E., Gavriilidis, A., Galvanin, F.: A model-based data mining approach for determining the domain of validity of approximated models. Chemometr. Intell. Lab. Syst. **172**, 58–67 (2018)
14. Ribeiro, M.T., Singh, S., Guestrin, C.: "Why should I trust you?": explaining the predictions of any classifier. In: Proceedings of the 22nd ACM SIGKDD International Conference on Knowledge Discovery and Data Mining, KDD 2016, pp. 1135–1144. Association for Computing Machinery, New York (2016)
15. Ribeiro, M.T., Singh, S., Guestrin, C.: Anchors: high-precision model-agnostic explanations. In: Proceedings of the AAAI Conference on Artificial Intelligence, vol. 32, no. 1 (2018)
16. Schölkopf, B., Platt, J.C., Shawe-Taylor, J., Smola, A.J., Williamson, R.C.: Estimating the support of a high-dimensional distribution. Neural Comput. **13**(7), 1443–1471 (2001)
17. Schweidtmann, A.M., Weber, J.M., Wende, C., Netze, L., Mitsos, A.: Obey validity limits of data-driven models through topological data analysis and one-class classification. Optim. Eng. **23**(2), 855–876 (2022)
18. Yeh, C.K., Hsieh, C.Y., Suggala, A., Inouye, D.I., Ravikumar, P.K.: On the (in)fidelity and sensitivity of explanations. In: Advances in Neural Information Processing Systems, vol. 32. Curran Associates, Inc. (2019)

Analyzing a Decade of Evolution: Trends in Natural Language Processing

Richard A. A. Jonker(✉), Tiago Almeida, and Sérgio Matos

IEETA/DETI, LASI, University of Aveiro, Aveiro, Portugal
richard.jonker@ua.pt

Abstract. Natural Language Processing (NLP) stands at the forefront of the rapidly evolving landscape of Machine Learning, witnessing the emergence and evolution of diverse methodologies over the past decade. This study delves into the dynamic trends within the NLP domain, specifically spanning the years 2010 to 2022, through an empirical analysis of papers presented at conferences hosted by the Association for Computational Linguistics (ACL). We utilize ChatGPT in order to extract meaningful information from the data before performing an in depth analysis. Our investigation encompasses an exploration of several key aspects, namely **computational trends**, **research trends** and **geographic trends**. We further investigate the entry cost into NLP, the longevity of hardware and the environmental impact of NLP. The code to run our system is publicly available at https://github.com/ieeta-pt/nlp-trends.

Keywords: Data Analytics · Natural Language Processing · Data Extraction · NLP Trends

1 Introduction

The field of artificial intelligence (AI) has made remarkable strides in recent years, achieving significant milestones across various domains. These advancements range from breakthroughs in computer vision, enabling machines to interpret visual information, to innovations in drug discovery, where AI-driven systems are demonstrating their ability to predict molecular interactions and potential therapeutic effects.

One of the most notable areas of progress is in Natural Language Processing (NLP), where machines have evolved from basic rule-based and statistical systems to sophisticated models capable of understanding, interpreting, and interacting with human language. This evolution, driven by deep learning, seen in Fig. 1, has revolutionized NLP's capabilities over the past decade.

With the recent introduction of ChatGPT and other open-source Large Language Models (LLMs), the public has begun to recognize the transformative power of these technologies, which is rapidly reshaping how people worldwide work [4]. Although the effects of LLMs have only recently become apparent to

R. Wrembel et al. (Eds.): DaWaK 2024, LNCS 14912, pp. 162–176, 2024.
https://doi.org/10.1007/978-3-031-68323-7_13

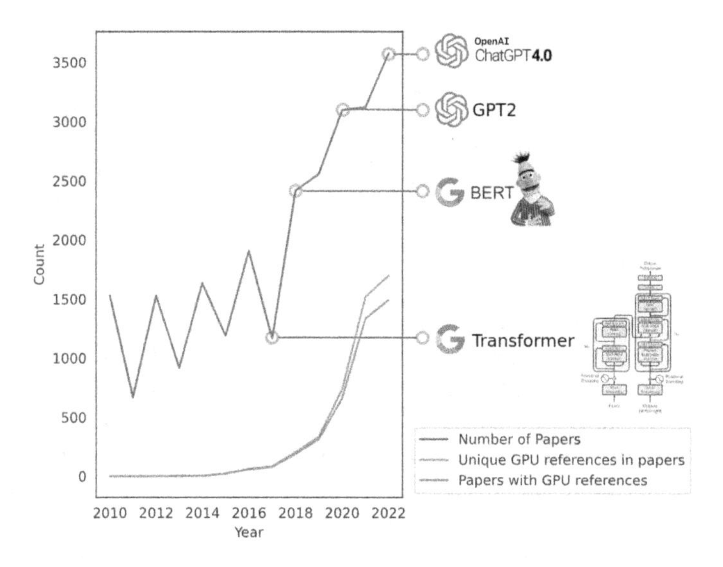

Fig. 1. Number of papers published per year in the various ACL conferences, as well as the number of papers with GPU references

the public, many have already experienced their impact. For instance, Google has employed BERT in its search engine since 2019 [10], and also uses NLP techniques to enhance search result comprehension with the use of passages in 2020 [13].

Although the impact of LLMs is mainly considered positive, there are negative aspects, namely the environmental impact to train these models. In 2019, it was estimated that up to 284,019 kg of CO_2e (carbon dioxide equivalents) was used to train a transformer using NAS [16]. This was later revealed to be overestimated by up to ×88 [11]. In 2019, it was estimated by both Nvidia [8] and Amazon [1] that up to 90% of the workload on machine learning is from inference alone. This iterates that the cost of training a model is, relatively speaking, not that harmful. However, in more recent years, it is becoming increasingly costly to train models. In order to train the Llama 2 model [18], it was estimated that 539,000 kg of CO_2e were used in the training of the model. It was estimated that the 70B model was trained for 1,720,320 GPU hours, using A100 80 GB GPUs, equating to 688,128 kWh in power consumption assuming a TDP of 400 W.

While many papers discuss trends of NLP, they mainly discuss the state of the art [7,20], tracking how methods and results evolved over time. Additionally, there are various blogs related to upcoming and trending techniques within the field [6,19]. To the best of our knowledge, there is no existing work that attempts to study the various trends using empirical methods. This study aims to address this gap, by examining the key developments in NLP over the last decade. Our analysis is guided by three research questions that aim to uncover the evolving trends in NLP. These questions are:

1. **Computational trends:** What are the dominant computational resources (hardware) in NLP research?
2. **Research trends:** How have NLP tasks, software frameworks and models evolved over the past decade?
3. **Geographic Trends:** How diverse are the publications in the field of NLP?

2 Methodology

To effectively study trends over the past few years, we analyzed papers from top conferences related to Natural Language Processing (NLP). We analyzed conference papers because conferences generally have a faster review process and overall turnover time for papers, indicating the latest work in the field. This is particularly important for our time-sensitive data analysis, where the most current publications is crucial. The Association for Computational Linguistics (ACL) stands out as one of the premier conferences for NLP-related research, achieving the highest h5-index in computational linguistics[1], with many of its other events ranking within the top 10 highest h5-indices for computational linguistics. We use the papers from these conferences spanning between the years 2010 and 2022 to perform analysis. The remaining methods are briefly summarized in Fig. 2.

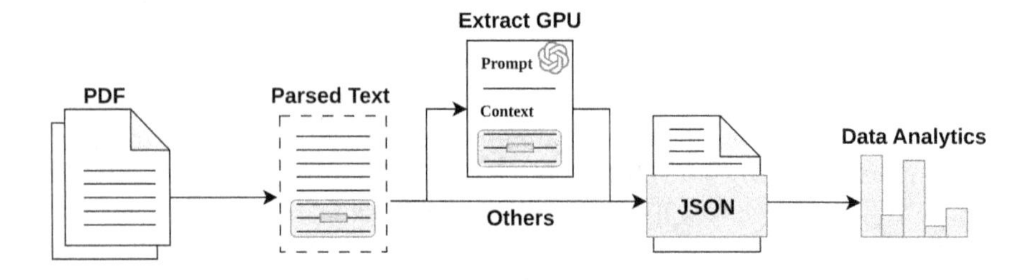

Fig. 2. Overview of the methodology used to extract information

From each conference, we exclusively download the main proceedings, excluding workshops, demonstrations, tutorials, student sessions, industry events, etc. This results in a dataset comprising the main proceedings of the conferences, encompassing both long and short papers. The list of conferences used in this study are as follows:

1. Annual Meeting of the Association for Computational Linguistics (ACL)
2. Conference on Empirical Methods in Natural Language Processing (EMNLP)
3. North American Chapter of the Association for Computational Linguistics (NAACL)

[1] https://scholar.google.com/citations?view_op=top_venues&hl=en&vq=eng_computationallinguistics.

4. International Conference on Computational Linguistics (COLING)
5. International Conference on Language Resources and Evaluation (LREC)
6. Conference on Computational Natural Language Learning (CoNLL)
7. European Chapter of the Association for Computational Linguistics (EACL)
8. International Joint Conference on Natural Language Processing (IJCNLP)

Further information regarding when the various conferences were held can be seen in Table 1.

Table 1. Conference years for ACL-related events.

Conference	2010	2011	2012	2013	2014	2015	2016	2017	2018	2019	2020	2021	2022
ACL	x	x	x	x	x	x	x	x	x	x	x	x	x
COLING	x		x		x		x		x		x		x
CoNLL	x	x	x	x	x	x	x	x	x	x	x	x	x
EACL			x		x			x				x	x
EMNLP	x	x	x	x	x	x	x	x	x	x	x	x	x
IJCNLP		x		x		x				x		x	x
LREC	x		x		x		x		x		x		x
NAACL	x		x	x		x	x		x	x		x	x

2.1 PDF Parsing

We download the papers in a PDF format, which is considered an unstructured format. When extracting text from a PDF, this can lead to the inclusion of unwanted data. The simpler approach involves extracting all text from a PDF in an unstructured format, using PyPDF2[2]. Alternatively, a more structured approach attempts to semantically parse the PDF data into sections visible to a viewer, using the SciPDF Parser[3]. However, the structured approach comes with the disadvantage of potential data loss, whereas the unstructured approach sacrifices structure and may include data that can interfere with subsequent data analysis. In this work, we employ both approaches.

We further collect citation information of all papers, with the use of Semantic Scholar API[4] (this was performed on July 24, 2022). When performing the match for citation results, we use Jaro similarity and set a threshold of 0.75 to account for inconsistencies of paper names. We were unable to find citation information for 20 of the papers.

The next step involved searching for specific keywords in the text that we aimed to extract. To successfully mine information about GPUs, we employed a two-stage pipeline. The initial step focused on identifying general GPU architectures in the text using keywords such as ‘rtx’, ‘gpu’, ‘nvidia’, ‘tesla’, ‘quadro’,

[2] https://pypi.org/project/PyPDF2/.
[3] https://github.com/titipata/scipdf_parser.
[4] https://www.semanticscholar.org/.

'geforce' and 'gtx'. Utilizing these keywords, we extracted a context of up to 500 characters surrounding the supporting word. This was achieved by chunking data into sentences and recursively adding these sentences to the context until the character limit was reached. We then aimed at extracting the exact GPU used, using exact dictionary matching.

Subsequently, we extracted exact GPU information from the context using ChatGPT. Pre-processing was performed on the GPU names, removing all keywords mentioned earlier. With regards to our prompt, our initial idea was to extract various different sources of information into a single JSON field. In order to do this we used the following prompt:

```
"You are a machine learning expert. Your
   goal is to extract correct information
   from a given CONTEXT and answer the
   QUESTION correctly. When in doubt, use
   the value -1.
CONTEXT: CONTEXT
QUESTION: What is the total training time of
    the models, explaining reasoning,
   return only a JSON: {total_time: NUMBER,
    unit: MINUTE/HOURS/DAYS, gpus:[{gpu:
   GPU_NAME, number_of_gpus: NUMBER},
   ...]}"]
```

Prompt 1.1. Prompt used to extract GPU information.

We attempted to extract information regarding the training time of the models, however this information was only correct in some cases, which is why we did not further investigate this. Similarly, we attempted to extract the number of GPUs used for training, which was correctly estimated when the number of GPUs is explicitly mentioned. However, we struggled to keep this information when merging data from the same source, and this could lead to incorrectly estimating the number of GPUs used.

We then compared the data extracted from ChatGPT with the data we manually extracted from the dictionary extraction methods. We conducted a validation of our information extraction methodology through data annotation, and the results are as follows: 100 samples were annotated by two annotators (authors of the paper). The agreement between the reviewers is 96%. Of these 96 samples, 86% (83) of the samples favoured ChatGPT annotations, 5% (5) of the samples favoured the dictionary based approach and 8% (8) were incorrectly labelled by both models. In most cases the dictionary extraction methods failed due to hyperparemeters being mistaken for GPU names such as '2000' and '680M'. This is why we used the data extracted from the ChatGPT analysis in the paper, as opposed to dictionary approaches.

We also collected some statistics regarding frameworks used, models used and general tasks authors aimed at solving. These were done on a simple dictionary matching scheme, whereby in cases where it makes sense, spaces were added/removed to maximize correct matches. We also limited the sections to analyze in the cases of architectures and NLP tasks, whereas for the frameworks, the entire text is searched.

3 Results

A summary of some general statistics are seen in Fig. 1, where the overall number of papers collected is presented, along with the number of papers with GPU references and the number of unique GPUs referenced in each paper. Analyzing first the number of papers, we notice some oscillation in the beginning, related to non-annual conferences, followed by an explosion in NLP papers from around 2018. In 2022 almost half the papers have mentioned specific GPUs used, which indicates that access to GPUs may be a limiting factor for publication in this field. We further note that only in more recent years do we see papers using multiple GPU architectures. In total 25,591 papers were collected of which 5,961 contain GPUs.

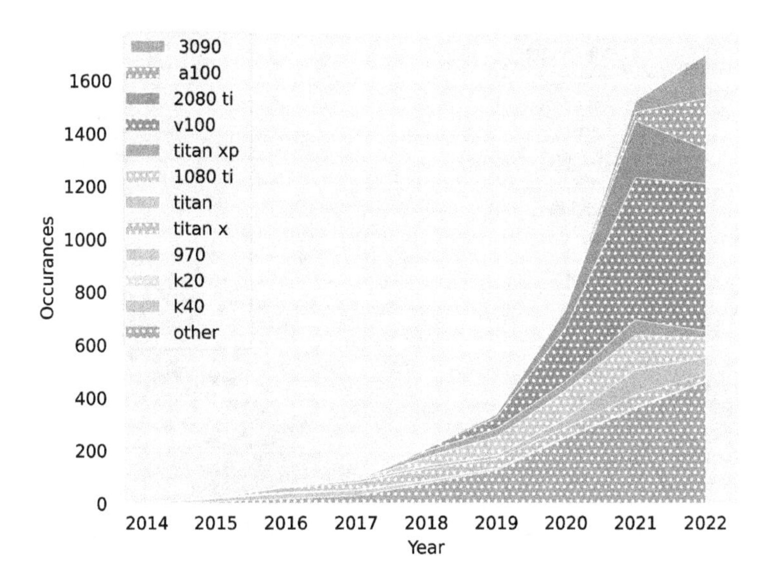

Fig. 3. Count of the top 3 GPUs from each year

In more recent years, the NVIDIA V100 emerged as the most popular GPU (Fig. 3), with newer GPUs such as the A100 and 3090 appearing to be on the rise, with a clear exponential growth in the number of GPUs. While older GPUs have a diminishing presence in later years, the counts of newer GPUs surge, indicating the transition to updated hardware. This shift is likely attributed to the escalating memory requirements essential for performing NLP. This further raises questions about the longevity of hardware, with subsequent models being released with increasing requirements. Although it appears that GPUs are required in order to produce meaningful work, a possible hypothesis is that there exist some interesting papers produced without a GPU. However, they are required to have comparisons to other state-of-the-art models, which necessitate GPUs to run. Although there is an apparent increase in entry cost to the field, there is still scope to conduct research without dedicated hardware. We further forecast that the use of openly available APIs, such as those presented by OpenAI, will benefit the research community, with many researchers opting to use these instead.

Analyzing the most popular frameworks used (Fig. 4), it is unsurprising that PyTorch and Hugging Face lead the field, primarily owing to their accessible APIs, facilitating rapid development. In close pursuit is TensorFlow, which has gradually lost popularity with NLP researchers, in more recent years. Overall, the framework counts are lower than the GPU frequencies, indicating that the frameworks used are not highly discussed. The substantial surge in Hugging Face's prevalence may also be attributed to the utilization of footnotes, as numerous authors reference models from the Hugging Face Hub in their text.

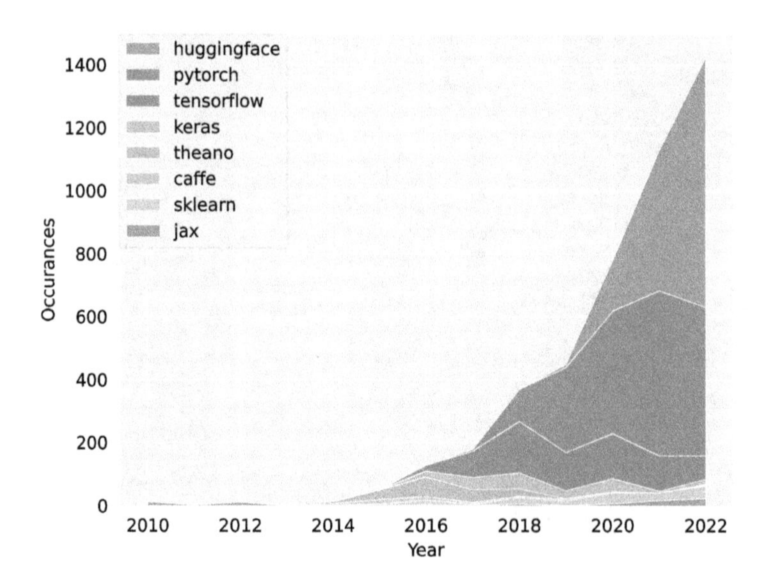

Fig. 4. Most popular NLP frameworks

We further analyzed the most popular algorithms and techniques presented in the conferences (Fig. 5). In the earlier years, there was a significant focus on Bayesian-based techniques with Support Vector Machines (SVMs). However, around 2013, the field shifted towards deep learning, and the use of deep neural networks and LSTMs emerged as dominant algorithms for solving tasks. We further date this to the use of embeddings such as GloVe [12] and Word2Vec [9].

Subsequently, there was an explosion in 2018 with the introduction of the transformer architecture, particularly BERT [3] and GPT [2,14,15], which we expect to further grow within upcoming years.

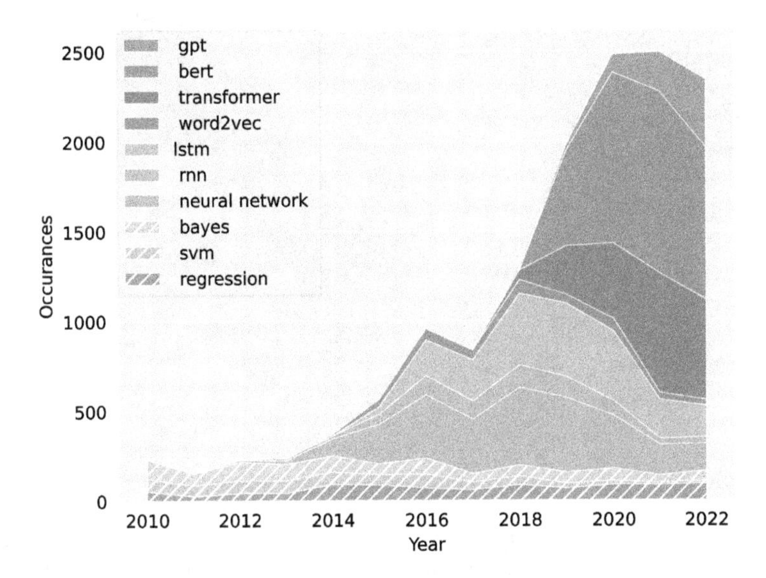

Fig. 5. Most popular NLP models. Pre-Deep learning machine learning approaches are marked with lines, deep learning approaches are unmarked.

Examining the most popular tasks (Fig. 6), it is evident that translation is the predominant task under investigation. Up until 2018, most trends remained consistent. However, with the introduction of transformer-based architectures, there is a substantial surge in the exploration of question answering tasks as well as text generation tasks, such as summarization, translation, paraphrasing and question answering, leading to significant progress in these tasks.

Examining the table containing the Top 5 most cited papers (Table 2), we observe that 4 out of 5 of the top papers include contributions from EMNLP. Notably, the third and fourth most popular papers are from 2014 as well. Comparing these values with Fig. 8, the results remain consistent, with ACL having a higher average number of citations in most cases (Fig. 9), corresponding to the various metrics used to rank these conferences.

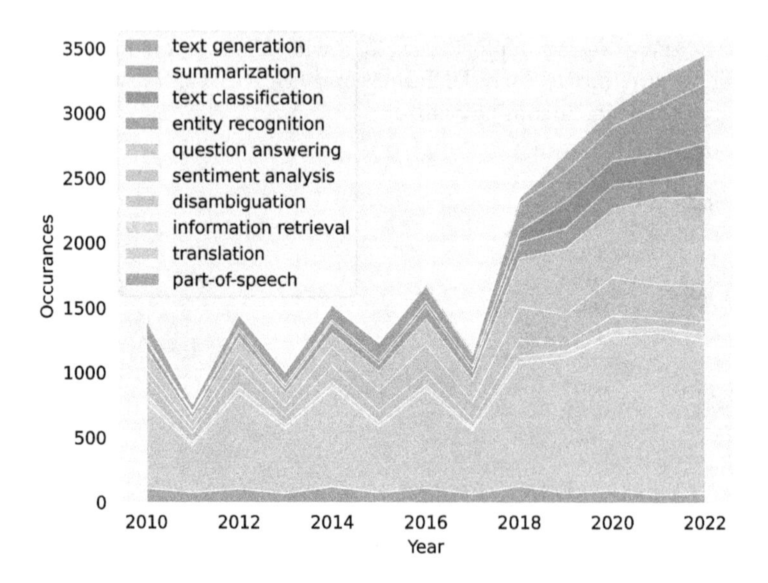

Fig. 6. Most popular NLP tasks

In Figs. 7 and 8, we observe the number of papers and citations from each conference per year. Beginning with the number of papers produced at each conference, there is a noticeable increase in the quantity of papers published in most conferences, while LREC remains relatively constant. The high value for NAACL in 2019, is due to the publication of the paper: BERT: Pre-training of Deep Bidirectional Transformers for Language Understanding [3]. Subsequently, there is a decline in the number of citations after 2019, which can be attributed to newer works having fewer citations due to having less 'visible' time.

Table 2. Top 5 cited papers from ACL conferences.

Paper Title	Year	Conference	Citations
BERT: Pre-training of Deep Bidirectional Transformers for Language Understanding	2019	NAACL	56,293
GloVe: Global Vectors for Word Representation	2014	EMNLP	27,236
Learning Phrase Representations using RNN Encoder-Decoder for Statistical Machine Translation	2014	EMNLP	18,698
Convolutional Neural Networks for Sentence Classification	2014	EMNLP	11,880
Deep Contextualized Word Representations	2018	NAACL	9,800

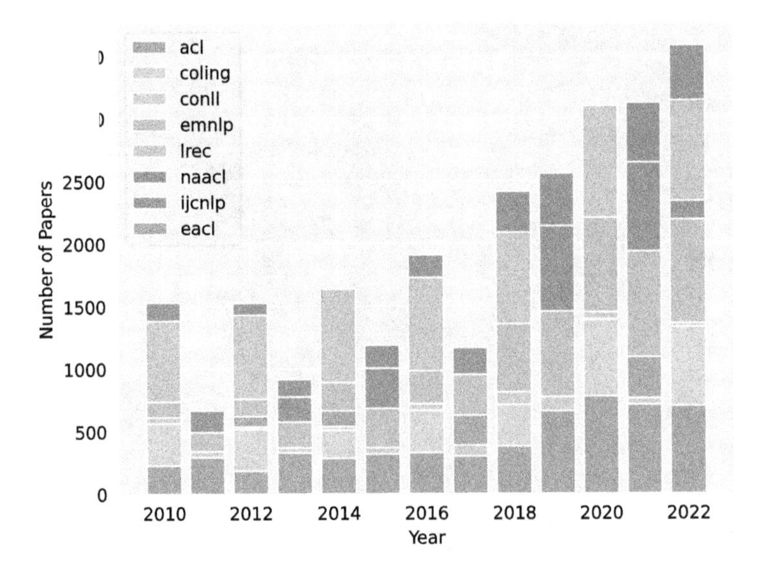

Fig. 7. Number of papers published in each conference per year

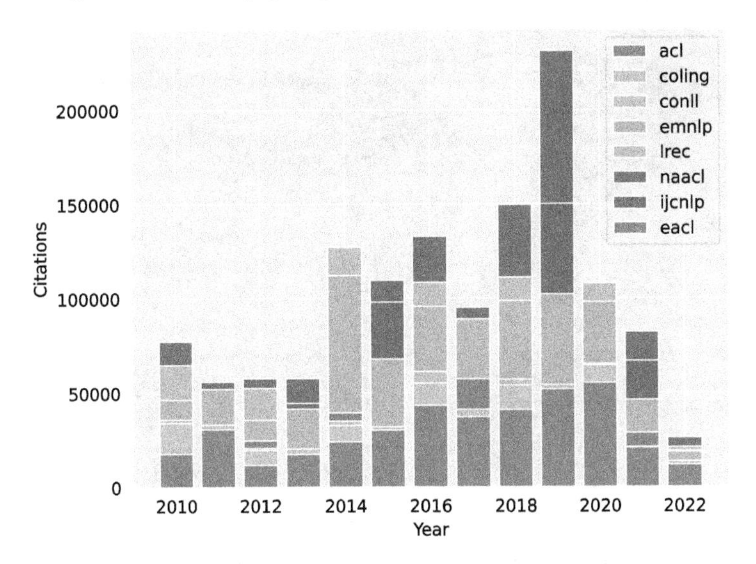

Fig. 8. Number of citations from each paper published in each conference per year

Following this, we analyzed the data on a per-country basis. To determine the country of each paper, extracted from the paper using the structured reader. Out of 25,559 papers, we were able to identify the country of origin for 13,365 papers. With this information, we plotted heatmaps showcasing the countries with the most papers published (Fig. 10), and the number of citations per country (Fig. 11). The countries with the most papers are China (3,141), the United

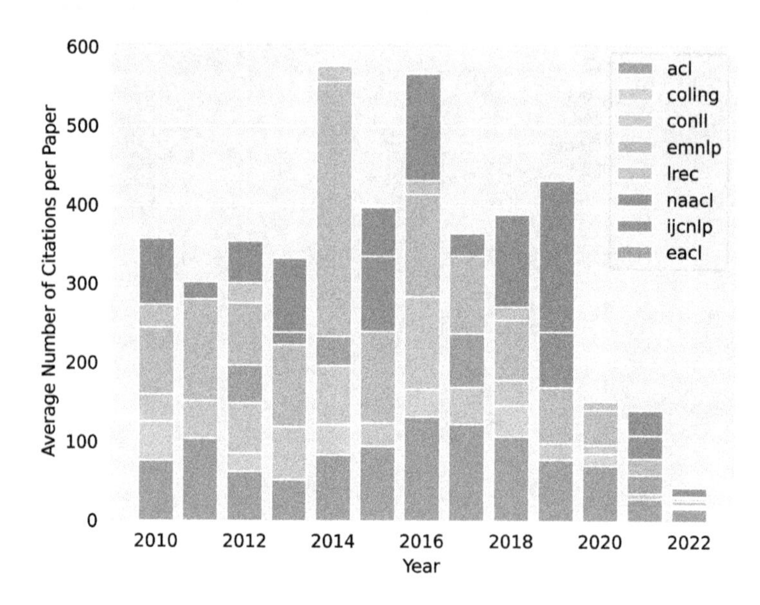

Fig. 9. Average number of citations per paper, per conference, per year

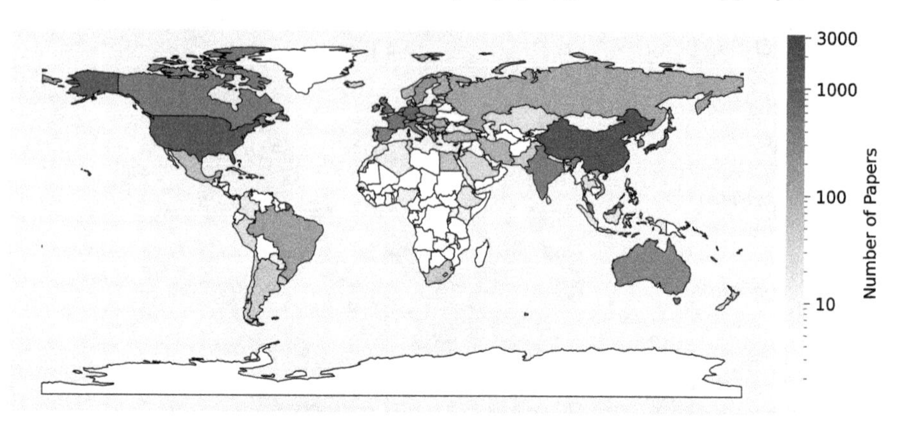

Fig. 10. Number of Papers published per country

States (2,538), and Germany (1,880). However, when comparing this value to the average number of citations per paper per country, the results differ, with the top countries being Mexico (101.66), Colombia (69.8), and the United States (67.72). This demonstrates that relevant research remains impact irrespective of the country of origin.

Turning to the environmental impact of NLP let's consider an example. In 2022, there were 557 mentions of the v100 GPU, which has a Thermal Design Power (TDP) of 300 W. TDP represents the maximum heat the GPU can generate under sustained workload conditions and is utilized here as an estimate of power draw. Assuming a model is trained, on average, for 1 week with

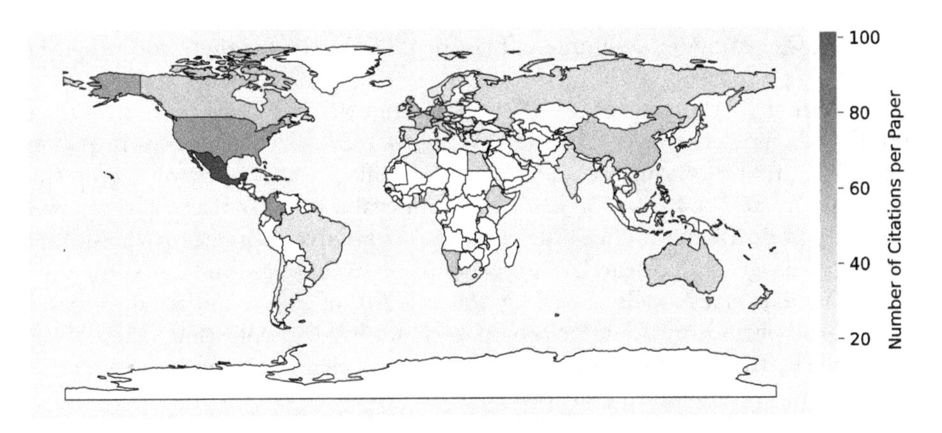

Fig. 11. Average number of citations per paper per country

Table 3. Average power consumption (kWh)

TDP	Hours	Average usage (kWh)			
		40%	50%	70%	100%
300 w	1	0.12	0.15	0.21	0.30
	1 * 24	2.88	3.60	5.04	7.20
	7 * 24	20.16	25.20	35.28	50.40
	30 * 24	86.40	108.00	151.20	216.00
600 w	1	0.24	0.30	0.42	0.60
	1 * 24	5.76	7.20	10.08	14.40
	7 * 24	40.32	50.40	70.56	100.80
	30 * 24	172.80	216.00	302.40	432.00
1200 w	1	0.48	0.60	0.84	1.20
	1 * 24	11.52	14.40	20.16	28.80
	7 * 24	80.64	100.80	141.12	201.60
	30 * 24	345.60	432.00	604.80	864.00
2400 w	1	0.96	1.20	1.68	2.40
	1 * 24	23.04	28.80	40.32	57.60
	7 * 24	161.28	201.60	282.24	403.20
	30 * 24	691.20	864.00	1209.60	1728.00

GPUs running at 70% usage, the power consumption, as per Table 3, would be $35.28 \times 557 = 19650.96$ kWh. This estimate is simplistic, covering only the GPU's power consumption, excluding the server's and data center's power usage. The latter is commonly estimated by the Power Usage Effectiveness (PUE) coefficient, estimated at 1.58 [17] (Google reported a PUE of 1.10 in 2023 [5]). Additionally, this does not account for models trained on multiple GPUs. For a more realistic estimate, assuming a TDP of 1200 W, the consumption would be $141.7 \times 557 = 78,926.9$ kWh.

Extending this to the total GPUs, assuming each GPU averages a TDP of 300 W with the server averaging 1200 W, with 1697 GPU references in papers from 2022, and prototyping/training done for 1 month at 50% power usage, this would imply $1697 \times 432 = 733,104$ kWh. This value slightly exceeds the power needed to train Llama 2 70B. Although this is a naive estimate, it shows that the energy usage of all research, barely compares to that used in large companies where foundational models are being trained. Although the impact of research pales in comparison to the larger companies, we need to note that many of the work done by researchers, utilize these models, and that often times these models are presented in conferences such as ACL.

4 Conclusion

In this study, we conducted an analysis of Natural Language Processing (NLP) trends from 2010 to 2022, focusing on ACL based conference papers. We utilize ChatGPT in order to perform unsupervised annotations to the dataset, before providing some in depth analysis. Our findings highlight an increasing entry cost to NLP, driven by the demand for expensive hardware. Despite uncertainties about hardware longevity, newer high VRAM options suggest potential stability. The environmental impact of NLP training appears relatively modest, with larger companies overshadowing individual researchers. The impact of the transformer architecture on **Research Trends** has driven increased output and exploration of complex topics. Anticipating continued impact, we foresee a rise in papers facilitated by tools like ChatGPT and other APIs, enabling research without dedicated hardware. Hugging Face and PyTorch currently dominate NLP software. In our analysis of **Geographic Trends**, we note that the primary determinant of impact is the paper's quality, and does not appear to be limited by country of origin. To conclude, our study provides insights into the evolving NLP landscape, briefly overviewing trends present in the last decade of research. Although this work may not offer any new insights, this work empirically confirms and presents many trends present in the field.

5 Limitations

The primary limitation of this work lies in the automatic extraction of GPUs used in papers. We acknowledge that this value may be underestimated since many papers do not include this information within the text. Despite this potential underestimation, we are confident that if a GPU was mentioned within an extracted context, it was almost always correctly identified, enhancing the reliability of this study.

As mentioned earlier, we desired to include information about the training time of algorithms to more accurately calculate the energy consumption of model training. However, we were unable to confidently extract this information for publication. While we naively estimate the power consumption of machine learning models, this value cannot be confidently estimated even with the training time of algorithms. We lack information about the time spent on prototyping beforehand, making any estimate regarding computing time inherently inaccurate.

In Figs. 5 and 6, the lists of tasks and algorithms may not be exhaustive. However, to the best of our ability, we ensured that all significant tasks and algorithms were included. Various other algorithms were tested but deemed non-significant and subsequently removed, including 'tf-idf', 'nn', 'random forest', 'knn', 'recurrent neural network', 'pca', 'rbf', and 'lda'.

Acknowledgments. This work was funded by the Foundation for Science and Technology (FCT) in the context of project UIDB/00127/2020 (https://doi.org/10.54499/UIDB/00127/2020). Richard A. A. Jonker is funded by the grant PRT/BD/154792/2023. Tiago Almeida is funded by the grant 2020.05784.BD (https://doi.org/10.54499/2020.05784.BD).

References

1. Barr, J.: Amazon EC2 update - Inf1 instances with AWS inferentia chips for high performance cost-effective inferencing. AWS News Blog (2019). Amazon Blog. https://aws.amazon.com/blogs/aws/amazon-ec2-update-inf1-instances-with-aws-inferentia-chips-for-high-performance-cost-effective-inferencing/
2. Brown, T., et al.: Language models are few-shot learners. In: Advances in Neural Information Processing Systems, vol. 33, pp. 1877–1901 (2020)
3. Devlin, J., Chang, M.W., Lee, K., Toutanova, K.: BERT: pre-training of deep bidirectional transformers for language understanding. In: Proceedings of the 2019 Conference of the North American Chapter of the Association for Computational Linguistics: Human Language Technologies, Volume 1 (Long and Short Papers), pp. 4171–4186. Association for Computational Linguistics, Minneapolis (2019). https://doi.org/10.18653/v1/N19-1423. https://aclanthology.org/N19-1423
4. Eloundou, T., Manning, S., Mishkin, P., Rock, D.: GPTs are GPTs: an early look at the labor market impact potential of large language models (2023)
5. Google: Efficiency - Data Centers - Google (2023). https://www.google.com/about/datacenters/efficiency/
6. StartUs Insights: 9 Natural Language Processing Trends in 2023 (2022). https://www.startus-insights.com/innovators-guide/natural-language-processing-trends/
7. Khurana, D., Koli, A., Khatter, K., Singh, S.: Natural language processing: state of the art, current trends and challenges. Multimedia Tools Appl. **82**(3), 3713–3744 (2023)
8. Leopold, G.: AWS to offer Nvidia's T4 GPUs for AI inferencing (2019). https://www.hpcwire.com/2019/03/19/aws-upgrades-its-gpu-backed-ai-inference-platform/
9. Mikolov, T., Chen, K., Corrado, G., Dean, J.: Efficient estimation of word representations in vector space. arXiv preprint arXiv:1301.3781 (2013)
10. Pandu, N.: Understanding searches better than ever before (2019). Google Blog. https://blog.google/products/search/search-language-understanding-bert/
11. Patterson, D., et al.: Carbon emissions and large neural network training. arXiv preprint arXiv:2104.10350 (2021)

12. Pennington, J., Socher, R., Manning, C.: GloVe: global vectors for word representation. In: Proceedings of the 2014 Conference on Empirical Methods in Natural Language Processing (EMNLP), pp. 1532–1543. Association for Computational Linguistics, Doha (2014). https://doi.org/10.3115/v1/D14-1162. https://aclanthology.org/D14-1162
13. Prabhakar, R.: How AI is powering a more helpful Google (2020). Google Blog. https://blog.google/products/search/search-on/
14. Radford, A., Narasimhan, K., Salimans, T., Sutskever, I., et al.: Improving language understanding by generative pre-training (2018)
15. Radford, A., Wu, J., Child, R., Luan, D., Amodei, D., Sutskever, I.: Language models are unsupervised multitask learners (2019)
16. Strubell, E., Ganesh, A., McCallum, A.: Energy and policy considerations for deep learning in NLP. In: Proceedings of the 57th Annual Meeting of the Association for Computational Linguistics, pp. 3645–3650. Association for Computational Linguistics, Florence (2019). https://doi.org/10.18653/v1/P19-1355. https://aclanthology.org/P19-1355
17. Taylor, P.: Data center average annual PUE worldwide 2023 (2023). https://www.statista.com/statistics/1229367/data-center-average-annual-pue-worldwide/
18. Touvron, H., et al.: Llama 2: open foundation and fine-tuned chat models (2023). https://doi.org/10.48550/arXiv.2307.09288. http://arxiv.org/abs/2307.09288. arXiv:2307.09288
19. Wolff, R.: 9 natural language processing (NLP) trends in 2022 (2020). https://monkeylearn.com/blog/nlp-trends/
20. Young, T., Hazarika, D., Poria, S., Cambria, E.: Recent trends in deep learning based natural language processing. IEEE Comput. Intell. Mag. **13**(3), 55–75 (2018)

Improving Serendipity for Collaborative Metric Learning Based on Mutual Proximity

Taichi Nakashima[1(✉)], Hanxiong Chen[1], and Kazutaka Furuse[2]

[1] University of Tsukuba, 1-1-1 Tennodai, Tsukuba-shi, Ibaraki 305-8577, Japan
s2420638@u.tsukuba.ac.jp, chx@cs.tsukuba.ac.jp

[2] Hakuoh University, 2-2-2 Ekihigashidori, Oyama-shi, Tochigi 323-8586, Japan
furuse@fc.hakuoh.ac.jp

Abstract. Today, in web space, where content is constantly expanding, recommendation systems that enable users to explore information passively have become essential technologies, and their accuracy is significantly improving. However, recent studies have focused not only on enhancing recommendation accuracy by suggesting items that match user preferences but also on increasing the appeal of recommendation systems by offering unexpected discoveries and recommendations that exceed expectations. Studies have been conducted to achieve serendipity, recommendations that are both beneficial to the user and marked by novelty and surprise. However, existing systems alter the embedding space compromising this flexibility. In this study, we propose a recommendation method called mutual proximity collaborative metric learning, which improves serendipity for users in an embedding-based recommendation method called collaborative metric learning. The proposed method improves existing techniques by refining the embedding space search algorithm, reducing the bias toward popular items in recommendations without altering the original embedding space, thereby enabling users to achieve serendipity. Furthermore, experimental results demonstrate that our approach outperforms existing methods across various evaluation metrics, such as unpopularity and serendipity. Our approach can thus achieve serendipity and preserve the applicability of the embedding space to diverse tasks.

Keywords: Serendipity · Mutual proximity · Metric learning

1 Introduction

In recent years, the web space has become inundated with content, rendering it increasingly challenging for users to actively seek out their favorite items. Hence, users are more commonly relying on recommendation systems to passively explore items. Many recommendation models are capable of suggesting items that are similar to user preferences and previous transactions, utilizing user

R. Wrembel et al. (Eds.): DaWaK 2024, LNCS 14912, pp. 177–191, 2024.
https://doi.org/10.1007/978-3-031-68323-7_14

and item attribute information [20] or past transaction data [15]. However, the recommendation of highly relevant items may often be anticipated by users, or the recommendations could lean toward highly popular items owing to inherent biases within the system [5]. This can lead to a lack of positive surprises, known as serendipity, for users of the recommendation system. Thus, the challenge for recommendation systems lies not only in suggesting highly relevant items but also in achieving serendipity [8].

For example, a recommendation model known as collaborative metric learning (CML) [11], embeds users and items in a common vector space to reflect their similarities and preferences. This model has been observed to recommend popular items, thus failing to enhance user serendipity [27]. In this paper, we propose an improvement to the search method within the embedding space of CML, aiming to enhance the serendipity of recommendations for users.

The structure of this paper is as follows. First, in Sect. 2, we explain the related research that forms the background of this study and the advantages of the proposed approach over those suggested in previous studies. Next, in Sect. 3, we describe the method developed in this research, and in Sect. 4, we present experiments conducted on the proposed method to verify its usefulness. Then, in Sect. 5, we discuss the results of the experiments, conclude this study, and present future works.

2 Background

2.1 Serendipity

The term "serendipity" refers to "good fortune resulting from unexpected coincidences," and within the context of recommendation systems, it denotes recommendations that are both beneficial to the user and marked by novelty and surprise. Various methods have been proposed for defining and evaluating serendipity. Kotkov et al. [16] suggested that serendipity comprises three components. The first is relevance, which indicates the degree to which the recommended item is likely to be of interest or use to users based on their past behavior, preferences, or explicit feedback. Traditional recommendation systems provide personalized recommendations by suggesting items highly related to those a user has previously engaged with through content-based filtering [20], or by recommending items that other users with similar preferences have enjoyed, via collaborative filtering [22]. The recommendations suggested by these traditional systems are based on the concept that items highly related to a user are beneficial to the user. In studies on recommendation systems, accuracy is often measured by relevance. The second component is novelty, which indicates the recommended items' unknown aspect from the perspective of the user; greater novelty is achieved as more unknown items are recommended. Various studies have explored ways to enhance novelty, suggesting that recommending less popular items among all available options can increase it [6,24] because the user is less likely to come across these items. The third component is unpredictability, which estimates the extent to which a recommendation was unexpected by the user. Recommendations are considered more unpredictable when they significantly differ from the established preferences of the user [1,17].

This study focuses on relevance and novelty in achieving serendipity in recommendations. We aim for recommendations that are both relevant and less popular items. However, a trade-off exists between relevance and unpopularity; suggesting items that are already known by the user could make the user lose interest in the items suggested, on the one hand, striving for a diverse range of recommendations that avoids a bias toward popular items can reduce relevance [8]. We demonstrate the trade-off between relevance and unpopularity in Sect. 4.3, and address this trade-off by evaluating the proposed method with a metric that comprehensively assesses both relevance and unpopularity.

2.2 Collaborative Metric Learning (CML)

CML is a method that combines collaborative filtering [22] and metric learning [25]. CML first learns from implicit feedback [12,21] by users on items and embeds both users and items into a common vector space. If a user prefers an item, the model embeds them such that the Euclidean distance between the two is close. Recommendations are then made to the user by performing a nearest neighbor search in the embedding space.

The embedding space of CML reflects not only the relationships between users and items but also the relationships among different users and different items. For example, users and items that are close to each other are considered to be more related and similar. This property is generated by the propagation of relevance through the principle of triangular inequality [11].

2.3 Mutual Proximity (MP)

Mutual proximity (MP) is a scaling method that reduces "hubness" [19] in vector spaces. Hubness refers to the tendency of some objects to be frequently chosen as the nearest neighbors of other objects in high-dimensional vector spaces; it represents a facet of the "curse of dimensionality." A "hub" is an object that is more likely to be selected as a nearest neighbor compared to others. Typically, hubs tend to emerge more frequently as the dimensionality increases, regardless of the data's domain or characteristics [19]. Moreover, the neighborhood relationship between a hub object and its nearest neighbors in a vector space is asymmetric [19]. For instance, consider two objects x, y in a vector space. If y is the nearest neighbor of x, and x is also the nearest neighbor of y, this neighborhood relationship is symmetric, indicating a high affinity between the two. However, if y is the nearest neighbor of x but x is not the nearest neighbor of y, the relationship is asymmetric, and the affinity between them is not necessarily high. MP aims to reduce hubness by scaling metrics such as Euclidean distance and cosine similarity to render asymmetric neighborhood relationships more symmetric.

In this study, we adopted the MP concept to enhance the recommendation of unpopular items. Given that a popular item in an embedding space can be characterized as an object frequently chosen as the nearest neighbor by many users, reducing hubness can also mitigate the bias toward popular items in recommendation systems. The details of the proposed method are elaborated in Sect. 3.

2.4 Advantages and Originality of the Proposed Method

Several studies have explored various methods for enhancing CML [18,27,28]. These methods universally aim to achieve their respective goals by modifying the loss function of CML in alignment with those goals. For example, Yoneda et al. [27], aimed to enhance CML and improve serendipity by learning specific embedding space. This is achieved using a specifically designed loss function, as expressed in Eq. (1).

$$L_{m_2}(d) = \sum_{(i,j)\in S} \alpha_j \sum_{(i,k)\notin S} w_{ij}[m + d(i,j)^2 - d(i,k)^2]_+ \tag{1}$$

Here, $[x]_+ = \max(0,x)$. α_j denotes the weight assigned to item j, and the loss increases as the amount of implicit feedback for j increases. It is defined by Eq. (2), using the number of implicit feedback X_j for item j across the dataset, along with the hyperparameters a and b.

$$\alpha_j = aX_k^{-b} \tag{2}$$

To achieve serendipity, the proposed method improves the loss function of CML and adjusts the embedding space. However, as mentioned in Sect. 2.2, the CML embedding space captures not only the relationships between users and items but also among different users and different items. This versatility enables the embedding space to be utilized for a wide range of tasks beyond top-k recommendations for users. Adjusting the embedding space solely for a specific purpose could compromise this flexibility. Therefore, our approach focuses on innovating the search method within the embedding space without altering the original CML embedding space, preserving its applicability to diverse tasks and achieving serendipity.

Furthermore, this research incorporates the concept of MP into the embedding space search method. Originally proposed to diminish the hubness in high-dimensional vector spaces, MP is not inherently designed to mitigate the bias toward popular items in recommendation systems. Although some studies have adapted MP to recommendation systems [7,14], to the best of our knowledge, this research is first in applying the hubness reduction technique specifically to reduce the bias toward popular items within such systems.

3 Methodology

In the following, when a user i provides implicit feedback on an item j, we refer to j as a positive item for i and the pair (i, j) as a positive pair. The symbols used in this study are outlined in Table 1.

Table 1. Symbols used in the proposed method

Symbol	Description
i, j, k, l	Indices for users or items
$\mathbf{u}_i, \mathbf{v}_j$	Embedding representation of user i, item j
S	Set of positive pairs
$d(i, j)$	Euclidean distance between $\mathbf{u}_i, \mathbf{v}_j$
$L_m(d)$	Loss function (4)
m	Hyperparameter for the loss function
w_{ij}	Weighted approximate rank pairwise loss [26]
$P(X)$	Distribution of distances from a certain user to all items
$P(Y)$	Distribution of distances from a certain item to all users
$\mathcal{F}_X, \mathcal{F}_Y$	Cumulative distribution functions for $P(X), P(Y)$
α	Hyperparameter for $MP_{i,j}$

3.1 Learning Embeddings

The proposed method initially embeds users and items from the dataset into a vector space, similar to CML. It adjusts the embedding positions starting from the user, aiming to bring positive items closer and distance nonpositive items. The representations of user i and item j in the embedding space are denoted as $\mathbf{u}_i$ and $\mathbf{v}_j$, respectively. The distance between $\mathbf{u}_i$ and $\mathbf{v}_j$ in the embedding space is defined by the Euclidean distance as expressed in Eq. (3). For a set of positive pairs denoted as S, the loss function is defined by Eq. (4).

$$d(i, j) = ||\mathbf{u}_i - \mathbf{v}_j|| \tag{3}$$

$$L_m(d) = \sum_{(i,j)\in S} \sum_{(i,k)\notin S} w_{ij}[m + d(i, j)^2 - d(i, k)^2]_+ \tag{4}$$

Here, $[x]_+ = \max(0, x)$. A loss is added if the distance between user i and positive item j plus a margin m is greater than the distance between that user and a nonpositive item k. Additionally, w_{ij} denotes the weight for positive pairs, utilizing the weighted approximate-rank pairwise loss [26]. w_{ij} is defined by the following Eq. (5):

$$w_{ij} = \log(\mathrm{rank}_d(i, j) + 1) \tag{5}$$

Here, $\text{rank}_d(i, j)$ denotes the rank of the distance between $\mathbf{u}_i$ and $\mathbf{v}_j$ among all positive items for user u, indicating the closeness of the item to the user compared to other positive items. In CML, the embedding space is obtained by minimizing Eq. (4) through learning. Model training is conducted using mini-batch stochastic gradient descent, and the learning rate is updated using Adam [13]. Figure 1 illustrates the process of computing the embedding space in CML [11].

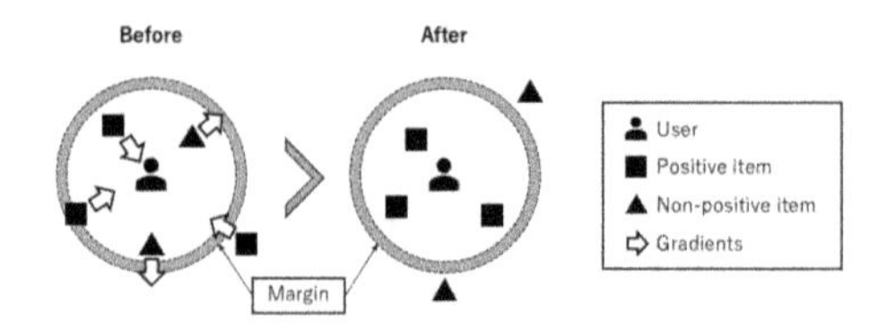

Fig. 1. Computing the embedding space in CML.

3.2 Searching Embedding Space and Recommending Items

In this study, we established a top-k recommendation task that aims to recommend K items that are most relevant to the user and yet least popular among all items. To recommend items that are highly relevant to the user, recommending items that are close to the user in the embedding space is effective. This approach is viable because the CML embedding space captures users' preferences for items; if a user prefers an item, the learning process ensures that the distance between the user and the item is minimized. To recommend less popular items, decreasing the recommendation of items that are nearest neighbors to many users in the embedding space is beneficial. To accomplish this, we introduced mutual proximity collaborative metric learning (MPCML), which draws inspiration from MP and generates recommendations through an exploration of the embedding space.

The proximity of objects within the embedding space is determined based on their Euclidean distance in that space. However, the proximity between two objects is relative, depending on their spatial relationships with other objects. For example, as illustrated in Fig. 2, even if the Euclidean distance is small, if there are many other objects nearby, the proximity is not necessarily high. Consequently, MPCML transforms the proximity between objects into a probabilistically interpretable value, akin to MP. The proximity of an item j from a user i is defined by Eq. (6), using the distance $d(i, j)$ between i and j, the distribution $P(X)$ of distances from i to all other items in the embedding space, and cumulative distribution functions for $P(X)$, $\mathcal{F}_X$. The same methodology is applied when considering the proximity of a user from the perspective of an item.

$$\begin{aligned} P(X > d(i,j)) &= 1 - P(X \leq d(i,j)) \\ &= 1 - \mathcal{F}_X(d(i,j)) \end{aligned} \tag{6}$$

For example, if $P(X)$ follows the distribution depicted in Fig. 3, then $P(X > d(i,j))$ corresponds to the area shaded in gray in the figure. As illustrated in this figure, the proximity $P(X > d(i,j))$ increases as the distance $d(i,j)$ between i and j becomes relatively small compared to other items.

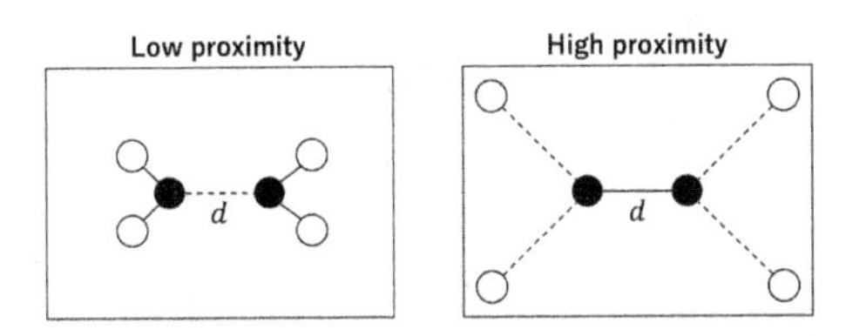

Fig. 2. In both figures, the distance between the black objects is equal. However, owing to their spatial relationships with the other white objects, their perceived proximity is different.

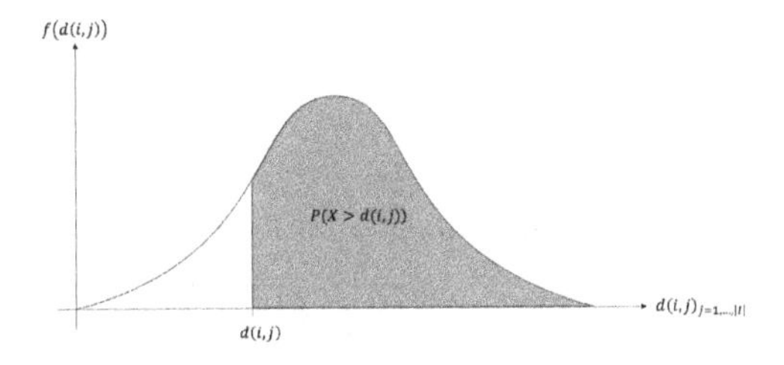

Fig. 3. Given the distance distribution $P(X)$ from i to all other items in the embedding space, the relationship between $d(i,j)$ and $P(X > d(i,j))$.

In addition, the proximity between a user and an item is not necessarily symmetrical; what may be highly relevant from one perspective could be less relevant from the other. This asymmetric proximity is more likely to occur with popular items, as a large number of users tend to be clustered around these items. MPCML addresses this asymmetry by prioritizing recommendations that are highly relevant to both parties, avoiding popular items. The proximity of user i and item j in MPCML is defined by the distance $d(i,j)$, the distribution of distances $P(X)$ from user i to all other items, the distribution of distances $P(Y)$ from item j to all other users, cumulative distribution functions $\mathcal{F}_X$, $\mathcal{F}_Y$, and the hyperparameter α, as expressed in Eq. (7).

$$\begin{aligned} MP_{i,j}(d(i,j)) &= P(X > d(i,j))^{\alpha} \cdot P(Y > d(i,j))^{(1-\alpha)} \\ &= (1 - \mathcal{F}_X(d(i,j)))^{\alpha} \cdot (1 - \mathcal{F}_Y(d(i,j)))^{(1-\alpha)} \end{aligned} \tag{7}$$

In MPCML, given the embedding space by CML, the recommendation is conducted using the following algorithm. Notably, the sets of users and items are denoted as U and I, respectively, and the distance distributions $P(X)$ and $P(Y)$ are assumed to follow gamma distributions $\Gamma_x(\hat{k}_x, \theta_x)$ and $\Gamma_y(\hat{k}_y, \theta_y)$, respectively, adopting the method proposed in previous research [3,4].

1. Calculate the distance $d(i, j)$ from user i to each item j in the embedding space for all $j \in I$.
2. Approximate the parameters of the distance distribution $P(X)$ from user i to all items by computing Eq. (8) with the values of $d(i, j)$ obtained in step 1.
3. For each j, sample T users and define the set U'_j.
4. For each j, calculate the distance $d(l, j)$ from item j to each user $l \in U'_j$.
5. Approximate the parameters of the distance distribution $P_j(Y)$ from item j to all users by computing Eq. (9) for the values of $d(l, i)$ obtained in step 4.
6. Calculate $MP_{i,j}$ from user i to any item j in the embedding space by Eq. (7), with the values of $d(i, j)$ obtained in step 1, $P(X)$ estimated in step 2, and $P_j(Y)$ estimated in step 5.
7. Recommend the top K items with the highest $MP_{i,j}$ values.

$$\mu_x = \frac{1}{|I|}\sum_{j\in I} d(i,j), \quad \sigma_x^2 = \frac{1}{|I|}\sum_{j\in I}(d(i,j)-\mu_x)^2, \quad \hat{k}_x = \frac{\mu_x^2}{\sigma_x^2}, \quad \theta_x = \frac{\sigma_x^2}{\mu_x} \tag{8}$$

$$\mu_y = \frac{1}{T}\sum_{l\in U'_j} d(l,j), \quad \sigma_y^2 = \frac{1}{T}\sum_{l\in U'_j}(d(l,j)-\mu_y)^2, \quad \hat{k}_y = \frac{\mu_y^2}{\sigma_y^2}, \quad \theta_y = \frac{\sigma_y^2}{\mu_y} \tag{9}$$

As shown in Sect. 4.3, the number of samples for parameter approximation T is sufficient at approximately 100, because it does not significantly affect the execution time. The time complexity for performing the recommendation task is $O(|U| + |I|)$, enabling linear-time recommendation for the number of objects in the embedding space.

4 Experiments

4.1 Datasets

We conducted experiments using MovieLens dataset [10]. This dataset contains users' rating histories for movies and is extensively used for evaluating the performance of recommendation systems. It includes results of users' rating histories for movies (hereinafter referred to as "items") on a five-point scale from 1 to 5. Given that CML is a model predicated on implicit feedback, similar to the experiments conducted by Hsieh et al. [11], our experiments considered only ratings of 4 and 5 as "positive feedback;" we extracted records with such feedback. Moreover, considering the division of the dataset into training and test data at a ratio of 3:1, the experiments were performed with only those users who provided positive feedback for four or more items.

To examine the impact of dataset size on performance, we conducted tests with two versions of the MovieLens dataset: the MovieLens 100K dataset (hereafter "MovieLens 100K") and the MovieLens 20M dataset (hereafter "MovieLens 20M"). After the aforementioned preprocessing, the MovieLens 100K dataset comprised 940 users, 1,447 items, and 55,369 instances of positive feedback, whereas the MovieLens 20M dataset included 137,330 users, 20,720 items, and 9,993,294 instances of positive feedback.

4.2 Metrics

To quantitatively evaluate the proposed method, we used the following offline evaluation metrics:

Recall. Recall is a metric that measures the extent to which a set of recommended items covers the set of items for which a user provided positive feedback. Widely used to assess the relevance of recommendations, it is defined as follows:

$$\text{Recall} = \frac{TP}{TP + FN} \tag{10}$$

TP denotes the number of items for which the user provided positive feedback and the recommendation system predicted that the user would provide positive feedback. FN denotes the number of items for which the user provided positive feedback but the recommendation system did not predict that the user would give positive feedback.

Unpopularity. This metric evaluates how well the recommendation system can recommend unpopular items. As a preliminary step, we first define the popularity of an item i as follows:

$$\text{Popularity}(i) = \frac{|U_i|}{|U|} \tag{11}$$

Here, $|U_i|$ denotes the number of users who provided positive feedback to item i, and $|U|$ denotes the number of all users. On this basis, we define the following metric for the unpopularity of the recommendation system.

$$\text{Unpopularity}(R) = \sum_{i \in R} \frac{1}{\text{Popularity}(i)} \tag{12}$$

where R denotes the set of items recommended to the user by the recommendation system.

Serendipity. As discussed in Sect. 2.1, achieving serendipity in a recommendation system requires both high relevance and unpopularity. However, as indicated in Sect. 4.3, these two factors are in a trade-off relationship. Therefore, to comprehensively evaluate serendipity from these two perspectives, the following metric is employed:

$$\text{Serendipity}(R) = \sum_{i \in R} \frac{1}{\text{Popularity}(i)} \times 1_A(i) \tag{13}$$

$$1_A(i) = \begin{cases} 1 & \text{if } i \in A, \\ 0 & \text{if } i \notin A. \end{cases} \tag{14}$$

Here, A denotes the set of items for which the user provided positive feedback.

Long-Tail Rate. We evaluate the recommendation system's ability to recommend long-tail items [2] using the long-tail rate [9] metric. The long-tail rate is defined as the proportion of long-tail items within the recommendation list. In this context, long-tail items are identified as those in the bottom 80% of feedback counts among all items in the recommendation system.

4.3 Results

Popularity Basis of Items. In the MovieLens dataset, we investigated the presence of bias in item popularity. We counted the number of users who provided positive feedback to each item across the dataset. Figures 4 and 5 present bar graphs that sort items by the number of feedbacks received, with feedback counts plotted on the vertical axis. The dark areas of the graphs represent the feedback counts for the top 20% of items in terms of feedback volume (Head). For the MovieLens 20M dataset (Fig. 5), owing to the significant difference in feedback counts between popular and less popular items, we used a logarithmic scale for better visualization. These experimental results reveal a substantial bias between popular and unpopular items in the MovieLens data.

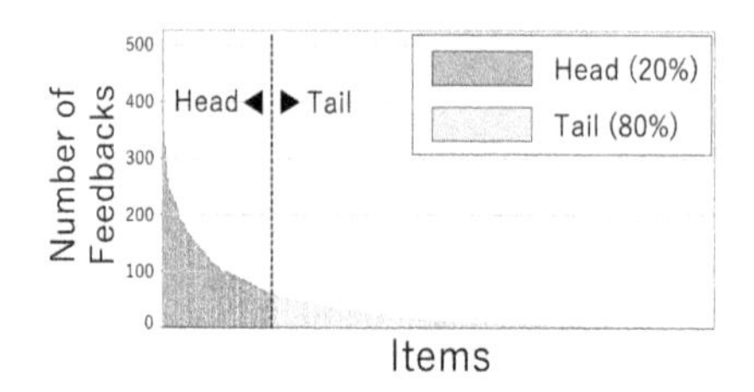

Fig. 4. MovieLens 100K

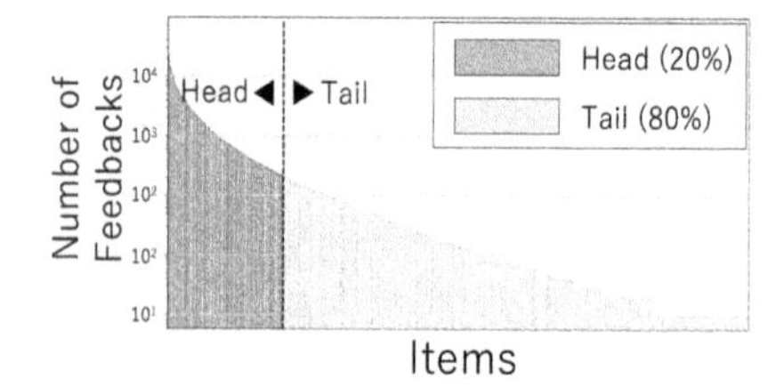

Fig. 5. MovieLens 20M (Log scale)

Performance Evaluation. We compared the performance of MPCML with that of the original CML using the evaluation metrics described in Sect. 4.2. The scores for each metric were measured by varying the value of the hyperparameter α within the range of 0 to 1 in increments of 0.1, as per Eq. (7). For all metrics, the number of recommended items k was set to $k = 10$. The data was preprocessed as described in Sect. 4.1, splitting the dataset into training and test data at a ratio of 3:1. Furthermore, records of items that did not receive positive feedback from the user were randomly added to the test data, ensuring the ratio of items with positive feedback to those without was 1:10. The value of T, the number of samples for determining the parameters of the probability distribution, mentioned in Sect. 3.2, was set to $T = 110$ based on the experimental results presented in the last part of this section. For each α, the score of each evaluation metric was calculated 10 times, and the average was considered the score. The implementation was based on the source code from Ryo [23], extended to test the proposed method.

The experimental results, as shown in Figs. 6 and 7, indicate that recall and unpopularity are in a trade-off relationship. As the hyperparameter α approaches 1, recall increases, whereas when it approaches 0, unpopularity increases. However, when $\alpha = 1$, MPCML inherently performs the same search as the original CML. We note that the original CML is equivalent to MPCML when $\alpha = 1$. For unpopularity, serendipity, and long-tail rate, MPCML significantly outperforms the original CML. These results suggest that MPCML can recommend less popular items while maintaining a certain level of relevance to the user, in comparison to the original method. Additionally, the results demonstrate that the balance between relevance and popularity can be adjusted through the hyperparameters.

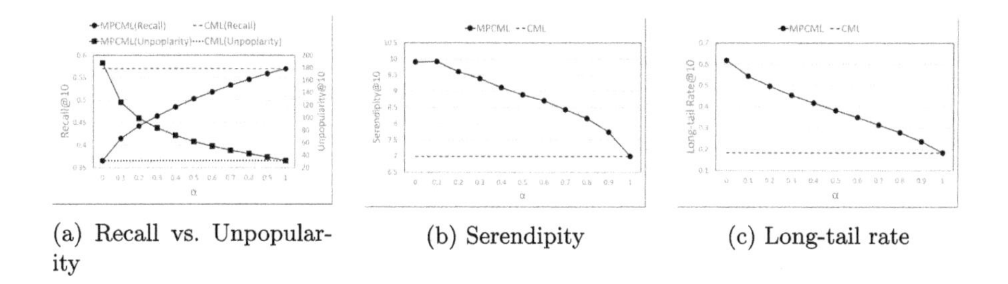

(a) Recall vs. Unpopularity (b) Serendipity (c) Long-tail rate

Fig. 6. MovieLens 100K.

Number of Samples for Distance Distribution Approximation. We validated the value of the number of samples T for estimating the parameters of the distance distribution $P_i(Y)$ from an item i to all users, as described in Sect. 3.2. We measured the score of each evaluation metric by varying the value of T in increments of 10 from 10 to 150. At each T, the score of each evaluation metric was calculated 10 times, and their average was used as the final score. The hyperparameter α in Eq. (7) was set to $\alpha = 0.5$ or 0.8, with other experimental conditions being the same as those described in Sect. 4.3.

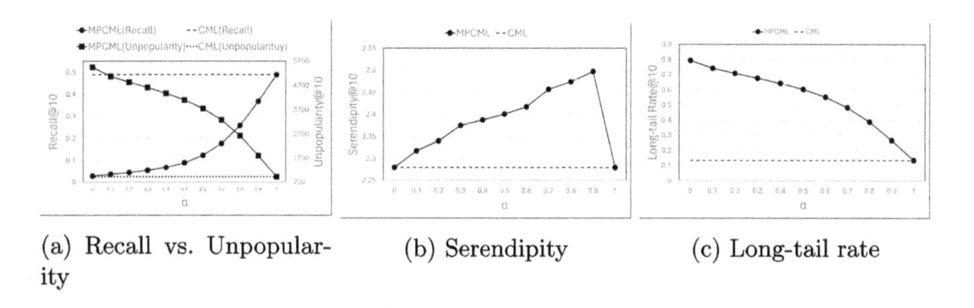

(a) Recall vs. Unpopularity (b) Serendipity (c) Long-tail rate

Fig. 7. MovieLens 20M.

As indicated in Figs. 8a to 9d, for all evaluation metrics except Serendipity for MovieLens20M, a T value of approximately 50 is sufficient to produce stable scores; increasing it further does not significantly improve the scores. For the Serendipity score of MovieLens20M, a value of $T = 110$ is also adequate; the score does not significantly increase with larger values. This suggests that MPCML

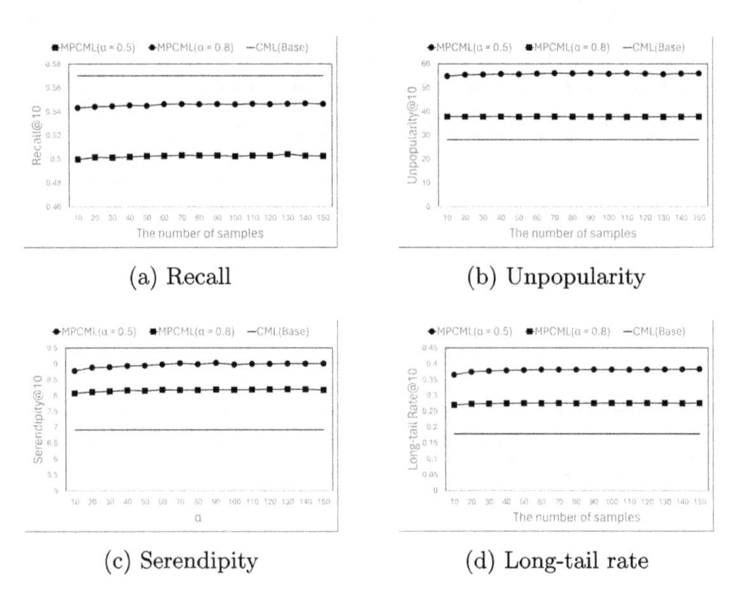

(a) Recall (b) Unpopularity

(c) Serendipity (d) Long-tail rate

Fig. 8. MovieLens 100K.

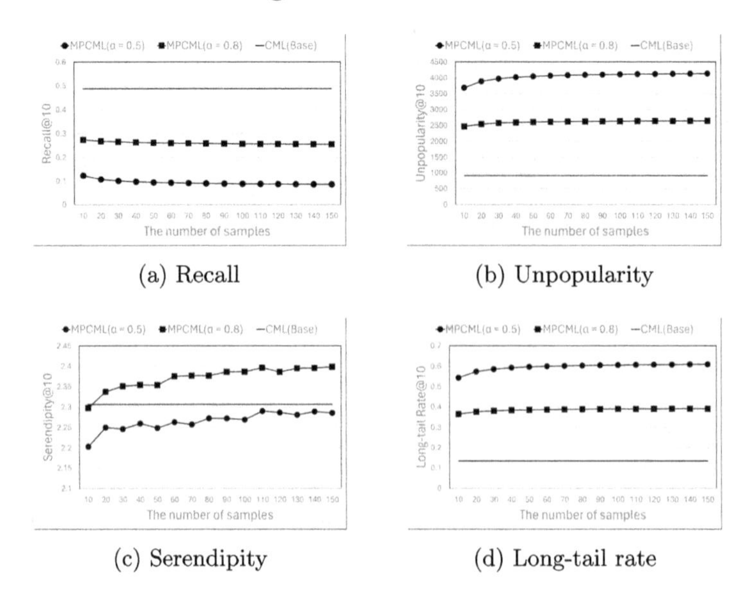

(a) Recall (b) Unpopularity

(c) Serendipity (d) Long-tail rate

Fig. 9. MovieLens 20M.

is capable of performing calculations in linear time complexity relative to the dataset size of $O(|U| + |I|)$ by assuming a gamma distribution for the distance distribution.

5 Conclusions and Discussion

In this study, we proposed MPCML, a method that applies a search approach based on the MP concept to the embedding space in CML. This approach mitigates the bias toward popular items and facilitates the recommendation of items that are both highly relevant to users and less popular, enhancing the likelihood of generating serendipity for users. In contrast to existing methods, MPCML does not alter the embedding space of CML to enhance serendipity, enabling it to achieve its objectives without impairing the potential applicability of CML to other tasks. In addition, this study implies that CML has the potential to offer a variety of recommendations by innovating its search method.

Future research should investigate whether MPCML can maintain consistent performance across different data sizes and domains. Given the observed differences in evaluation results between the MovieLens 100K and 20M datasets, the scalability of MPCML across various dataset sizes remains an open question. Additionally, the effectiveness of MPCML beyond movie rating data has not been tested, indicating the need for further validation across a broader spectrum of data. Furthermore, the evaluation methodology poses challenges. Although the metrics of unpopularity and serendipity used in this study are based on intuitive formulas, their effectiveness has not been empirically verified. Furthermore, this study relied solely on offline evaluations; conducting online evaluations is deemed necessary for a more accurate assessment of utility. Specifically, the generation of serendipity by recommendations is determined by individual user perceptions, which vary among users, rendering fully validating effectiveness through offline evaluation alone difficult.

References

1. Adamopoulos, P., Tuzhilin, A.: On unexpectedness in recommender systems: or how to better expect the unexpected. ACM Trans. Intell. Syst. Technol. **5**(4) (2014). https://doi.org/10.1145/2559952
2. Anderson, C.: The long tail: why the future of business is selling less of more. Hyperion (2008). https://cir.nii.ac.jp/crid/1130000794762659200
3. Cai, R., Zhang, C., Zhang, L., Ma, W.Y.: Scalable music recommendation by search, pp. 1065–1074 (2007). https://doi.org/10.1145/1291233.1291466
4. Casey, M., Rhodes, C., Slaney, M.: Analysis of minimum distances in high-dimensional musical spaces. IEEE Trans. Audio Speech Lang. Process. **16**(5), 1015–1028 (2008). https://doi.org/10.1109/TASL.2008.925883
5. Celma, Ò., Cano, P.: From hits to niches?: or how popular artists can bias music recommendation and discovery. In: NETFLIX 2008 (2008). https://api.semanticscholar.org/CorpusID:6701551

6. Celma Herrada, Ò.: Music recommendation and discovery in the long tail. Ph.D. thesis, Universitat Pompeu Fabra (2009)
7. Flexer, A., Stevens, J.: Mutual proximity graphs for improved reachability in music recommendation. J. New Music Res. **47**(1), 17–28 (2018). https://doi.org/10.1080/09298215.2017.1354891. pMID: 29348779
8. Ge, M., Delgado-Battenfeld, C., Jannach, D.: Beyond accuracy: evaluating recommender systems by coverage and serendipity. In: Proceedings of the Fourth ACM Conference on Recommender Systems, RecSys 2010, pp. 257–260. Association for Computing Machinery, New York (2010). https://doi.org/10.1145/1864708.1864761
9. Ge, Y., et al.: Explainable fairness in recommendation. In: Proceedings of the 45th International ACM SIGIR Conference on Research and Development in Information Retrieval, SIGIR 2022. ACM (2022). https://doi.org/10.1145/3477495.3531973
10. Harper, F.M., Konstan, J.A.: The movielens datasets: history and context. ACM Trans. Interact. Intell. Syst. **5**(4) (2015). https://doi.org/10.1145/2827872
11. Hsieh, C.K., Yang, L., Cui, Y., Lin, T.Y., Belongie, S., Estrin, D.: Collaborative metric learning. In: Proceedings of the 26th International Conference on World Wide Web, WWW 2017, pp. 193–201. International World Wide Web Conferences Steering Committee, Republic and Canton of Geneva (2017). https://doi.org/10.1145/3038912.3052639
12. Hu, Y., Koren, Y., Volinsky, C.: Collaborative filtering for implicit feedback datasets. In: 2008 Eighth IEEE International Conference on Data Mining, pp. 263–272 (2008). https://doi.org/10.1109/ICDM.2008.22
13. Kingma, D.P., Ba, J.: Adam: a method for stochastic optimization (2017)
14. Knees, P., Schnitzer, D., Flexer, A.: Improving neighborhood-based collaborative filtering by reducing hubness. In: Proceedings of International Conference on Multimedia Retrieval, ICMR 2014, pp. 161–168. Association for Computing Machinery, New York (2014). https://doi.org/10.1145/2578726.2578747
15. Koren, Y., Bell, R., Volinsky, C.: Matrix factorization techniques for recommender systems. Computer **42**(8), 30–37 (2009). https://doi.org/10.1109/MC.2009.263
16. Kotkov, D., Wang, S., Veijalainen, J.: A survey of serendipity in recommender systems. Knowl.-Based Syst. **111**, 180–192 (2016). https://doi.org/10.1016/j.knosys.2016.08.014. https://www.sciencedirect.com/science/article/pii/S0950705116302763
17. Murakami, T., Mori, K., Orihara, R.: Metrics for evaluating the serendipity of recommendation lists. In: Satoh, K., Inokuchi, A., Nagao, K., Kawamura, T. (eds.) JSAI 2007. LNCS (LNAI), vol. 4914, pp. 40–46. Springer, Heidelberg (2008). https://doi.org/10.1007/978-3-540-78197-4_5
18. Park, C., Kim, D., Xie, X., Yu, H.: Collaborative translational metric learning. In: 2018 IEEE International Conference on Data Mining (ICDM), pp. 367–376 (2018). https://doi.org/10.1109/ICDM.2018.00052
19. Radovanović, M., Nanopoulos, A., Ivanović, M.: Hubs in space: popular nearest neighbors in high-dimensional data. J. Mach. Learn. Res. **11**(86), 2487–2531 (2010). http://jmlr.org/papers/v11/radovanovic10a.html
20. Rendle, S.: Factorization machines. In: 2010 IEEE International Conference on Data Mining, pp. 995–1000 (2010). https://doi.org/10.1109/ICDM.2010.127
21. Rendle, S., Freudenthaler, C., Gantner, Z., Schmidt-Thieme, L.: BPR: Bayesian personalized ranking from implicit feedback. CoRR abs/1205.2618 (2012). http://arxiv.org/abs/1205.2618

22. Resnick, P., Iacovou, N., Suchak, M., Bergstrom, P., Riedl, J.: GroupLens: an open architecture for collaborative filtering of netnews. In: Proceedings of the 1994 ACM Conference on Computer Supported Cooperative Work, CSCW 1994, pp. 175–186. Association for Computing Machinery, New York (1994). https://doi.org/10.1145/192844.192905
23. Ryo, M.: Pytorchcml (2021). https://github.com/hand10ryo/PyTorchCML
24. Vargas, S., Castells, P.: Rank and relevance in novelty and diversity metrics for recommender systems. In: Proceedings of the Fifth ACM Conference on Recommender Systems, RecSys 2011, pp. 109–116. Association for Computing Machinery, New York (2011). https://doi.org/10.1145/2043932.2043955
25. Weinberger, K.Q., Saul, L.K.: Distance metric learning for large margin nearest neighbor classification. J. Mach. Learn. Res. **10**, 207–244 (2009)
26. Weston, J., Bengio, S., Usunier, N.: Large scale image annotation: learning to rank with joint word-image embeddings. Mach. Learn. **81**, 21–35 (2010). https://doi.org/10.1007/s10994-010-5198-3
27. Yoneda, A., Matsunae, R., Yamashita, H., Goto, M.: A method to improve serendipity of recommendation lists based on collaborative metric learning. Total Qual. Sci. **9**(2), 62–73 (2023). https://doi.org/10.17929/tqs.9.62
28. Zhou, X., Liu, D., Lian, J., Xie, X.: Collaborative metric learning with memory network for multi-relational recommender systems. CoRR abs/1906.09882 (2019). http://arxiv.org/abs/1906.09882

Ada2vec: Adaptive Representation Learning for Large-Scale Dynamic Heterogeneous Networks

Ranran Bian[1,2](✉) and R. Willem Vervoort[1,2]

[1] The University of Sydney, Camperdown, NSW 2006, Australia
ranran.bian@sydney.edu.au

[2] Australian Research Council Training Centre in Data Analytics for Resources and Environments (DARE), Eveleigh, Australia

Abstract. Representation learning generates the embedding vector of an object based on its relationships with others in a network. The generated vectors are inputs to various downstream machine learning tasks, such as classification, clustering and similarity search. The research area has attracted great interest and effort in recent years. However, due to its complexity, few of the existing representation learning methods have been developed for dynamic heterogeneous networks. Comparing with a static homogeneous network (graph), which contains single-typed objects (nodes) and relationships (edges) and remains unchanged over time, a dynamic heterogeneous network contains multiple-typed objects and relationships and evolves with time. We develop a novel adaptive representation learning algorithm, named Ada2vec, to address the challenges of embedding learning in large-scale dynamic heterogeneous networks. The key challenge is how to efficiently and effectively handle the network dynamics and heterogeneity features simultaneously. Ada2vec employs a statistical bound to capture network dynamics, and metapath-guided random walks to capture network heterogeneity. To the best of our knowledge, Ada2vec is the first approach that leverages the Hoeffding bound for modeling changes in dynamic heterogeneous network embedding. Extensive experiments demonstrate that Ada2vec significantly outperforms the state-of-the-art benchmarks in terms of embedding learning accuracy and efficiency. Compared to other dynamic heterogeneous network representation learning models, the size of the experiment datasets that we use are orders of magnitude larger (millions instead of tens of thousands).

1 Introduction

Network representation learning or network embedding, projects objects in a network into a low-dimensional space while preserving network structure and properties [9,39,44]. The popularity of network embedding arises due to its usefulness in various downstream machine learning tasks and application domains.

In the past decade, researchers have focused on network representation learning in homogeneous networks. A homogeneous **network (graph)** consists of only

R. Wrembel et al. (Eds.): DaWaK 2024, LNCS 14912, pp. 192–207, 2024.
https://doi.org/10.1007/978-3-031-68323-7_15

one type of **objects** (**nodes**) and **relationships** (**edges**). For instance, the homogeneous network in Fig. 1(a) has author-typed objects and co-authorship-typed relationships (denoted by the black solid lines). Real-world networks are usually heterogeneous, containing multiple types of objects and relationships. Figure 1(b) shows a heterogeneous network with objects of different types (author, paper and publication venue). Relationships in the heterogeneous network are co-authorship, authorship (denoted by the red dashed lines) and published at (denoted by the green dotted lines). Homogeneous network embedding methods' inability of distinguishing object and relationship types leads to low embedding accuracy [3,5,10]. In addition, a real-world network is often dynamic, where the numbers of objects and relationships evolve over time. Few studies have been conducted on dynamic heterogeneous network embedding [5,9,39,44]. This research gap limits the feasibility and applicability of network representation learning for real-life problem domains.

To address the challenges of dynamic heterogeneous network embedding, we develop the **ada**ptive representation learning model, Ada2vec. Our model treats a dynamic heterogeneous network as snapshots of static heterogeneous networks with consecutive time stamps t (t = 1, 2, 3, ..., i). For example, the static heterogeneous networks in Fig. 1(b) and (c) are the snapshots of a dynamic heterogeneous network at time stamps t and t + 1 respectively. For the embedding accuracy challenge created from the heterogeneity characteristic, Ada2vec employs metapath-guided random walks to model the heterogeneous k-hop neighbourhood, which effectively incorporates the local and higher-order proximity relationships. To maintain efficiency for large-scale dynamic networks, Ada2vec identifies *structurally-changed* objects at each time stamp t ($t \geq 2$) as **coreset**. It then maps from the coreset instead of all the nodes that exist in the network to produce metapath-guided random walks. Afterwards, the coreset based metapath-guided random walks are used as inputs to generate up-to-date embedding vectors of all the nodes in the network at the specific time stamp t. Since networks usually evolve smoothly instead of volatile restructures between each time step [49], the intuition behind Ada2vec is that lower storage requirements and faster run times can be achieved by mapping from the changed objects in the coreset, which is a significantly smaller subnet of the overarching network.

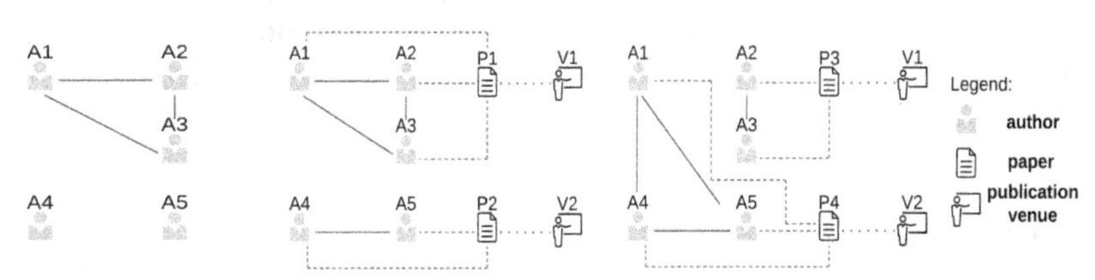

(a) Static homogeneous network (b) Static heterogeneous network at time stamp **t** (c) Static heterogeneous network at time stamp **t+1**

Fig. 1. Different types of networks. (Color figure online)

2 Related Work

We review the related research for homogeneous network embedding, heterogeneous network embedding, and change detection.

Homogeneous Network Embedding. With the advancements in natural language processing, DeepWalk [27], LINE [35] and node2ve [15] adopt the skip-gram models in word representation learning [24] for static homogeneous network embedding. DynamicTriad [49] was developed for dynamic homogeneous network embedding. Empirical results demonstrate its superior performance over other dynamic homogeneous network embedding methods, such as Temporal Network Embedding (TNE) [50]. More recent representation learning methods developed for dynamic homogeneous networks include [14,17,23,25]. **Heterogeneous Network Embedding.** Existing heterogeneous network embedding methods can be categorized into four categories: (1) Methods based on metapath-guided random walks [10,12,28]. Metapath2vec++ [10] is a state-of-the-art representation learning model for static heterogeneous networks. The framework employs metapath [11,31,33] guided random walks to model the node neighborhood and leverages a heterogeneous skip-gram architecture for embedding learning. (2) Methods based on neural networks [6,30,40,47]. (3) Specific task-oriented (e.g., prediction and recommendation) methods [7,16,38,48]. (4) Methods that decompose a heterogeneous network into simple networks for representation learning [29,34,43]. **Change Detection** or concept drift detection, aims at finding changes in the underlying data distribution. A common practice for drift detection methods is to employ well-established statistical bounds (e.g., Chernoff [8] and Bernstein Inequality [2]) to detect a difference in the data distribution. Due to its generality, Hoeffding Inequality [18] has been widely used to monitor concept drifts [13]. Bifet et al. [4] developed ADWIN2 which adapts the Hoeffding bound to signal concept drifts when learning from data sequences that may vary with time.

Our Method. Majority of the aforementioned representation learning methods target static homogeneous networks, dynamic homogeneous networks and static heterogeneous networks [22,45]. However, real-world networks are usually dynamic and heterogeneous. There has been an increasing trend in dynamic heterogeneous network embedding research with new models [3,19,21,41,42,46] being developed since 2019. However, few of them were evaluated on large-scale dynamic heterogeneous networks.

3 Problem Definition

Definition 1. *A **Dynamic Heterogeneous Network** is an undirected and unweighted graph $G = (V, E, T)$ where each node v and each edge e is associated with the mapping functions $\phi(v) : V \rightarrow T_V$ and $\varphi(e) : E \rightarrow T_E$ respectively. T_V represents the object types ($|T_V| > 1$). T_E represents the relationship types ($|T_E| > 1$). $|V^t| \neq |V^{t+1}|$ and/or $|E^t| \neq |E^{t+1}|$ between two consecutive time* *stamps t and t + 1.* $|V^t|$ *and* $|V^{t+1}|$ *denote the number of objects at time stamps t and t + 1.* $|E^t|$ *and* $|E^{t+1}|$ *denote the number of relationships at time stamps t and t + 1.* T_V *and* T_E *are considered as static across all time stamps.*

Problem 1. **Dynamic Heterogeneous Network Embedding.** Given a dynamic heterogeneous network G, which consists of snapshots of the static heterogeneous networks G^t with different time steps t, the task is to learn the dynamic d-dimensional embedding vectors $X^t \in \mathbb{R}^{|V^t| \times d}$, $d \ll |V^t|$ that are able to approximate the structural relationships among the objects in G^t.

4 The Ada2vec Framework

4.1 Part 1 Dynamic

Between two consecutive time stamps t and t + 1 (t = 1, 2, 3, ..., i), Ada2vec models two types of nodes: (1) New nodes at time stamp t + 1:

$$V_{new} = \{v : v \notin V^t, v \in V^{t+1}\} \quad (1)$$

(2) Existing nodes from time stamp t:

$$V_{existing} = \{v : v \in V^t, v \in V^{t+1}\} \quad (2)$$

As the network G evolves from time stamp t to t + 1, if a structural change of an existing node is identified, then the node is categorised as a changed existing node and placed in the coreset. At the initial time stamp t = 1, metapath-guided random walks are produced based on all the nodes that exist in G^1. Those random walks are then used as inputs by the metapath2vec++ algorithm [10] to generate the embedding vectors X^1 of all the nodes in G^1. From time stamps t $\geq$ 2, the input of metapath2vec++ becomes the metapath-guided random walks that are produced based on new and changed nodes only. Networks generally evolve smoothly over time [49], therefore, the number of new and changed nodes should only be a fraction of all nodes in G^t, which justifies the storage and run time efficiency of Ada2vec for networks with smooth transitions. Ada2vec outputs the up-to-date embedding vectors X^t of all the nodes in G^t (t = 1, 2, 3, ..., i).

4.2 Part 2 Heterogeneity

At each time stamp t (t = 1, 2, 3, ..., i), Ada2vec leverages the metapath2vec++ algorithm to produce up-to-date embedding vectors of all the nodes in the network, where metapath-guided random walks are used as inputs by the algorithm's heterogeneous skip-gram architecture. Ada2vec also employs metapath-guided random walks to model the node heterogeneous k-hop neighbour connectivity. The value of k measures the number of hops that one node needs to take to reach another node. For instance, we consider node v_j as node v_i's 1-hop neighbour if an edge exists between v_j and v_i. Node v_j becomes v_i's 2-hop neighbour when there exists no edge between them but both of them are connected to a third node v_l. Based on the random walks starting from an existing node v_i, Ada2vec constructs a $(|V_{neighbour}| \times 1)$ matrix for v_i to model the connectivity between v_i and v_i's k-hop neighbour nodes. $|V_{neighbour}|$ represents the number of *distinct* k-hop neighbours of v_i visited in the random walks.

The entries in the matrix represent the edge existence between v_i and its k-hop neighbours, v_j, in G, where the value of row q is 0 if and only if there exists no edge between them. Each value of $(q, 1)$ is defined as follows:

$$f[q,1] = \begin{cases} 1 & \text{if } i = j \\ 1 & \text{if } i \neq j \text{ and } (i,j) \in E^t \\ 0 & \text{if } i \neq j \text{ and } (i,j) \notin E^t \end{cases} \tag{3}$$

We illustrate the heterogeneous k-hop neighbour connectivity scheme with the example in Fig. 2. At time stamp t, following the metapath "AVA" (A denotes *Author* and V denotes *Publication Venue*), a random walk of "$A_1V_1A_2V_2A_3$" is generated for the author node A_1. At time stamps t and t + 1, an edge exists between A_1 and each visited node (excluding A_1 itself) in the random walk. Therefore, two identical (4×1) matrices of 1 s were derived for the two time stamps respectively. The semantic meaning of the two random walk-based matrices is that author A_1 has co-authored paper(s) with A_2 and A_3 and has published papers in V_1 and V_2. During the time between t + 1 and t + 2, author A_1 stopped collaborating with A_2 and A_3 and publishing papers in V_1 and V_2, which resulted in a (4×1) matrix of 0 s in the most left column of time stamp t + 2. If A_1 only stopped collaborating with A_2 and publishing papers in V_1 during the time span between t + 1 and t + 2, then the resulted (4×1) matrix would have entries of two 0 s and two 1 s.

Compared to representation learning methods that only model 1-hop neighbour relationship (first-order proximity), the k-hop neighbour connectivity (k-order proximity) scheme in Ada2vec approximates the complex network semantic and structural properties more effectively. We analyze the random walk based k-hop neighbour nodes instead of full k-hop neighbour nodes for higher efficiency as we do not need to consider all node pairs. Node pairs that co-occur on random walks are expressive with flexible stochastic definition of node similarity which incorporates both local and higher-order neighbourhood information.

4.3 Part 3 Change

Ada2vec produces the $(|V_{neighbour}| \times 1)$ matrix for each existing node v_i at time stamp t (t = 1, 2, 3, ..., i). The metapath-guided random walks captured in the matrix at time stamp t + 1 are carried over from time stamp t. However, entries in the t + 1 matrix are determined by the connectivity between v_i and each matrix-forming object at t + 1 (instead of t). Ada2vec employs the Hoeffding bound to detect change between the two sets of binary entries in the matrices of time stamps t and t + 1. In the scenario where no change is detected between the two sets of entries, the metapath-guided random walks from t are carried over again to time stamp t + 2 for matrix construction. If a change is identified, new metapath-guided random walks for node v_i are produced and followed to construct a second matrix which describes the connectivity between v_i and its k-hop neighbour nodes appeared in the new random walks at t + 1. Notice that $|V_{neighbour}|$ is the number of rows of each constructed matrix, which is determined by the number of *distinct* objects visited in the corresponding random walks. To illustrate the design, we consider the author node A_1 in Fig. 2 as an existing node for four time stamps t, t + 1, t + 2 and t + 3. At time stamp t, a metapath-guided random walk of "$A_1V_1A_2V_2A_3$" is generated for A_1, based on which a (4×1) matrix is produced. The matrix has entries of 1 s, which means that there exists an edge between A_1 and each matrix-forming object. These edges remain as G evolves from t to t + 1, therefore no change is detected and the random walk is carried over from t + 1 to t + 2 for matrix construction. At t + 2, A_1 loses its connectivity with the four objects ($V_1A_2V_2A_3$) visited in the random walk generated at t, which causes a structural change of A_1 being identified. To capture the up-to-date heterogeneous structural property of A_1, a new metapath-guided random walk of "$A_1V_3A_4V_4A_5$" is produced at t + 2. The new random walk is followed to construct a second matrix at t + 2 and carried over to t + 3 for the matrix based change detection.

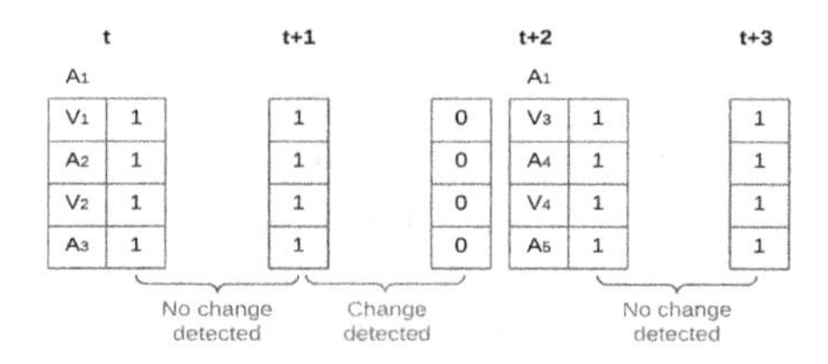

Fig. 2. Illustration of change detection.

Hoeffding Inequality provides an upper bound on the probability that the sum of bounded independent random variables deviates from its expected value by more than a certain amount [18]. Ada2vec leverages the Hoeffding bound to identify connectivity and structural changes for each existing node v_i when the network G evolves from time stamp t to t + 1. The connectivity and structural status of v_i at a particular time stamp is represented by the binary entries in the matrices of the corresponding time stamp. We produce the Hoeffding bound using:

$$m = \frac{1}{1/|M^t| + 1/|M^{t+1}|} \text{ (harmonic mean of } M^t \text{ and } M^{t+1}) \tag{4}$$

$$\epsilon = \sqrt{\frac{2}{m} \cdot \sigma^2 \cdot \ln\frac{2}{\delta}} + \frac{2}{3m} \cdot \ln\frac{2}{\delta} \tag{5}$$

where $|M^t|$ and $|M^{t+1}|$ are the row numbers of the two matrices (M^t and M^{t+1}) produced for the node v_i at time stamps t and t + 1. Because the two matrices are constructed based on the same metapath-guided random walks of v_i, $|M^t|$ and $|M^{t+1}|$ are equal to each other. σ^2 denotes the variance of the two sets of binary entries in M^t and M^{t+1}.

Let μ_{M^t} and $\mu_{M^{t+1}}$ be the averages of the binary entries in M^t and M^{t+1} respectively. From time stamps t ≥ 2, for each existing node v_i from time stamp t $-$ 1, Ada2vec tests the null hypothesis H_0 that $\mu_{M^t} = \mu_{M^{t+1}}$, against the alternate hypothesis H_1 that entries in M^t and M^{t+1} are from different distributions with $\mu_{M^t} \neq \mu_{M^{t+1}}$. The hypothesis test is as follows: we accept hypothesis H_1 whenever $Pr(|\mu_{M^t} - \mu_{M^{t+1}}| \geq \epsilon) > \delta$, where δ lies in the range [0, 1] and is a parameter that controls the maximum allowable false positive rate, while ϵ is a function of δ and the change threshold used to model the difference between the binary entries in M^t and M^{t+1}. We use the Hoeffding Inequality to calculate the change detection threshold ϵ. Ada2vec checks whether the difference of the observed averages in the two matrices exceeds the change threshold ϵ. We monitor the underlying distribution of the entries within M^t against M^{t+1}. If there is a significant difference in distributions then a real concept drift is detected. Intuitively, when the entries in M^t have a similar distribution compared to those in M^{t+1} ($|\mu_{M^t} - \mu_{M^{t+1}}| < \epsilon$), then there are no changes (concept drifts) within the v_i's k-hop neighbour connectivity. When the distributions of the entries in the two matrices are different ($|\mu_{M^t} - \mu_{M^{t+1}}| \geq \epsilon$) then there are changes (concept drifts) within the v_i's k-hop neighbour connectivity.

Hoeffding bound provides a rigorous performance guarantee that a change is signaled with probability at most δ (a user defined confidence value). The change detection threshold ϵ and the confidence value δ are negatively correlated with each other. The higher the confidence value is, the lower the change threshold, which results in tighter and more sensitive bound for signaling concept drifts.

Usually, the objective of representation learning methods is to maximize a network probability in consideration of node neighbourhood [15,24,27,35]. Ada2vec follows this common objective and maximizes the probability of the network $G = (V, E, T)$ at each different time stamp t (t = 1, 2, 3, ..., i) as below:

$$\underset{\theta}{\operatorname{argmax}} \sum_{v^t \in V^t} \sum_{t_v \in T_V} \sum_{c^t_{t_v} \in N^t_{t_v}(v^t)} \log p(c^t_{t_v} | v^t; \theta) \tag{6}$$

where $N^t_{t_v}(v^t)$ represents the t^{th}_v typed neighbour nodes of the node v^t in the network G at time step t, $p(c^t_{t_v}|v^t;\theta)$ denotes the conditional probability of having a context node $c^t_{t_v}$ of the t^{th}_v type given a node v^t at t and is commonly defined as a softmax function [1,24], that is:

$$p(c^t_{t_v} | v^t; \theta) = \frac{e^{X^t_{c_{t_v}} \cdot X^t_v}}{\sum_{u^t_{t_v} \in V^t_{t_v}} e^{X^t_{u_{t_v}} \cdot X^t_v}} \tag{7}$$

where X^t_v is the vth row of X^t, denoting the vector representation for node v at time stamp t, $V^t_{t_v}$ is the node set of type t_v in the network at t. The softmax function is adjusted to the node type t_v of the neighbour c_{t_v} as well as the specific time stamp t. To illustrate this, for the heterogeneous bibliographical network in Fig. 1(b), at time stamp t, the neighbours of the author node A_1 can be topologically close to other authors (A_2, A_3), papers (P_1) and venues (V_1). Ada2vec employs the stochastic gradient descent algorithm to optimize the objective. In addition, the framework leverages the heterogeneous negative sampling [10] for higher efficiency.

5 Experimental Evaluations

We first introduce the details of datasets and benchmark techniques. We use classification and clustering as proxy tasks to evaluate the accuracy of Ada2vec's learned representations. We discuss the scalability, efficiency, parameter sensitivity and limitations of Ada2vec. Our solution was implemented in Python, C and C++. The replication package including source code, processed datasets and labeled dynamic ground truth is publicly available[1].

5.1 Data

To evaluate the performance of Ada2vec, we use the Digital Bibliographic Library Project (DBLP) datasets with three time stamps (2011, 2013 and 2015). The DBLP datasets were extracted from the AMiner[2] Computer Science (CS) dataset [36] to construct a dynamic heterogeneous network, G, which consists of three types of nodes: authors, papers and publication venues. The edges in G represent different types of relationships: co-authorship between two authors, authorship between authors and papers, publication relationships between papers and venues. G^{2011} (the network at the year 2011 time stamp), G^{2013} and G^{2015} contain papers that published till 2011, 2013 and 2015 respectively (two year time spans). To capture the dynamics of node connectivity within the same time span, G contains no papers that was published earlier than the year 2009. The statistics of the three processed datasets is summarized in Table 1. To model meaningful semantic correlations in G, we only consider the binary entries for author to author and author to venue connectivity.

Table 1. Statistics of datasets.

Dataset Year	Sequence	# Authors	# Venues	# Papers
2011	1st	349551	2615	288514
2013	2nd	712140	4475	785497
2015	3rd	1149954	7376	1566389

5.2 Benchmarks

We compare Ada2vec with four recent dynamic heterogeneous network embedding methods (Dhne [46], Dyhne [41], Lime [26] and Thine [19]). We also select a static heterogeneous network embedding method (metapath2vec++ [10]) and a dynamic homogeneous network embedding method (DynamicTriad [49]) as

[1] https://github.com/Ada2vec/ada2veccode.git.
[2] https://www.aminer.cn/citation.

benchmarks due to their empirically proven superior performance over other state-of-the-art embedding techniques, e.g., DeepWalk [27], LINE [35], PTE [34], node2vec [15] and TNE [50].

We use the public code of benchmark techniques and apply the following parameters (if applicable) for fair comparisons: (1) the embedding dimension: 128; (2) the number of random walks per node: 50; (3) the random walk length: 20; (4) the neighborhood size: 7; (5) the size of negative samples: 5. In addition, the metapaths were set to "APVPA" and "VPAPV", where A denotes *Author*, P represents *Paper* and V denotes *Publication Venue*. The hyper-parameters for different benchmarks are all optimized according to the original papers and we report their best performance.

Table 2. Venue node classification results in AMiner data.

Metric	Year	Method	5%	10%	20%	30%	40%	50%	60%	70%	80%	90%
Macro-F1	2011	DynamicTriad	0.0618	0.0996	0.1952	0.1988	0.2458	0.2897	0.2911	0.3370	0.3512	0.3646
		Metapath2vec++	**0.1999**	**0.3434**	**0.5836**	**0.7468**	**0.8307**	**0.8731**	**0.8800**	**0.9038**	**0.9078**	**0.8872**
		Ada2vec	**0.1999**	**0.3434**	**0.5836**	**0.7468**	**0.8307**	**0.8731**	**0.8800**	**0.9038**	**0.9078**	**0.8872**
		Dhne	0.0876	0.1365	0.2830	0.2975	0.3568	0.4510	0.4673	0.5002	0.5214	0.5039
		Dyhne	0.1108	0.2579	0.3864	0.3918	0.4592	0.5797	0.6211	0.6544	0.6681	0.6692
		Lime	0.1026	0.2144	0.3199	0.3645	0.4128	0.5320	0.5418	0.5836	0.5811	0.5862
		Thine	0.1033	0.2216	0.3026	0.3620	0.4267	0.5217	0.5366	0.5625	0.5910	0.5926
	2013	DynamicTriad	0.0469	0.0938	0.2211	0.2419	0.2390	0.2941	0.3273	0.3088	0.3344	0.3550
		Metapath2vec++	0.1898	0.3476	0.6145	0.7354	0.8210	0.8832	**0.8901**	0.8983	0.8830	0.8728
		Ada2vec	**0.2095**	**0.3506**	**0.6315**	**0.7751**	**0.8686**	**0.8960**	0.8871	**0.9081**	**0.9066**	**0.8861**
		Dhne	0.1201	0.2070	0.3043	0.4570	0.4840	0.4838	0.5216	0.5200	0.5349	0.5488
		Dyhne	0.1498	0.2546	0.3690	0.5577	0.5776	0.5991	0.6418	0.6533	0.6774	0.6920
		Lime	0.1333	0.2461	0.3578	0.5291	0.5332	0.5380	0.6021	0.6085	0.6197	0.6112
		Thine	0.1406	0.2395	0.3699	0.5116	0.5498	0.5399	0.6077	0.6129	0.6255	0.6342
	2015	DynamicTriad	0.1988	0.1274	0.2560	0.2837	0.3377	0.3615	0.3699	0.3916	0.4027	0.4003
		Metapath2vec++	**0.3475**	**0.5197**	**0.7350**	**0.8550**	**0.9203**	**0.9346**	**0.9505**	**0.9570**	**0.9525**	**0.9515**
		Ada2vec	0.3084	0.4624	0.7014	0.8500	0.9135	0.9325	0.9449	0.9363	0.9346	0.9330
		Dhne	0.2210	0.3549	0.4066	0.5199	0.6534	0.6815	0.7180	0.7241	0.7356	0.7301
		Dyhne	0.2490	0.4216	0.5001	0.6243	0.7273	0.7670	0.7885	0.8024	0.8210	0.8227
		Lime	0.2371	0.4012	0.4544	0.5705	0.6611	0.7026	0.7399	0.7544	0.7491	0.7561
		Thine	0.2365	0.4096	0.4628	0.5881	0.6590	0.7222	0.7326	0.7616	0.7760	0.7614
Micro-F1	2011	DynamicTriad	0.1160	0.1702	0.2248	0.2569	0.2880	0.3103	0.3388	0.3451	0.3865	0.4040
		Metapath2vec++	**0.3004**	**0.4157**	**0.6227**	**0.7546**	**0.8290**	**0.8712**	**0.8815**	**0.9097**	**0.9215**	**0.9240**
		Ada2vec	**0.3004**	**0.4157**	**0.6227**	**0.7546**	**0.8290**	**0.8712**	**0.8815**	**0.9097**	**0.9215**	**0.9240**
		Dhne	0.1518	0.2261	0.3366	0.3696	0.4545	0.5512	0.5652	0.6004	0.6298	0.6237
		Dyhne	0.2031	0.2570	0.4517	0.4805	0.5276	0.6003	0.6566	0.7033	0.7216	0.7471
		Lime	0.1886	0.2435	0.4022	0.4447	0.5093	0.5898	0.6042	0.6822	0.6730	0.6998
		Thine	0.1747	0.2579	0.4130	0.4351	0.5172	0.5971	0.6012	0.6976	0.7237	0.7429
	2013	DynamicTriad	0.1079	0.1863	0.2390	0.2664	0.2899	0.3371	0.3547	0.3690	0.3894	0.4118
		Metapath2vec++	0.2889	0.4137	0.6490	0.7464	0.8219	0.8814	0.8883	0.9027	0.9010	0.9070
		Ada2vec	**0.3086**	**0.4182**	**0.6615**	**0.7800**	**0.8641**	**0.8900**	**0.9023**	**0.9184**	**0.9213**	**0.9227**
		Dhne	0.1627	0.2076	0.2660	0.3667	0.4013	0.5058	0.5249	0.5551	0.6344	0.6351
		Dyhne	0.2142	0.2881	0.3533	0.4779	0.4990	0.5666	0.6118	0.6337	0.7079	0.7120
		Lime	0.2006	0.2671	0.3386	0.4445	0.4788	0.5503	0.6026	0.6237	0.7001	0.7041
		Thine	0.2028	0.2526	0.3390	0.4528	0.4869	0.5412	0.6007	0.6119	0.6992	0.7195
	2015	DynamicTriad	0.1688	0.1927	0.2673	0.2614	0.2776	0.2939	0.3482	0.3679	0.4104	0.4387
		Metapath2vec++	**0.4381**	**0.5804**	**0.7610**	**0.8616**	**0.9206**	**0.9352**	**0.9509**	**0.9605**	**0.9600**	**0.9669**
		Ada2vec	0.3891	0.5285	0.7317	0.8543	0.9110	0.9306	0.9451	0.9397	0.9446	0.9507
		Dhne	0.1595	0.2138	0.2719	0.3571	0.4244	0.5102	0.5377	0.5656	0.6395	0.6588
		Dyhne	0.2319	0.2997	0.3644	0.4849	0.4877	0.5670	0.6266	0.6489	0.7147	0.7273
		Lime	0.2178	0.2641	0.3434	0.4526	0.4612	0.5423	0.6012	0.6210	0.7052	0.7129
		Thine	0.2196	0.2566	0.3482	0.4661	0.4800	0.5438	0.6125	0.6200	0.7101	0.7163

5.3 Classification

Similar to the evaluation strategy in [10], we employ third-party labels to determine the class of each node. The first step is to generate venue ground truth, where we match the eight categories[3] of venues in Google Scholar[4] with those in the network G. Author ground truth is generated in the second step, for each author who published in the matched venues, they are assigned to the category with highest number of publications. When a tie exists, the author is randomly assigned to a candidate category. The two steps are carried out at each time stamp of G to produce dynamic ground truth. For G^{2011}, among all of the Google Scholar 160 venues (20 per category $\times$ 8 categories), 98 of them were successfully matched and labeled, and 47285 authors were labeled with research categories. For G^{2013}, 107 venues and 84558 authors were successfully labeled with research categories. For G^{2015}, 126 venues and 132831 authors were successfully labeled.

The node embeddings were learned from the full network at each time stamp. The vectors of above labeled nodes were directly input to a logistic regression classifier [20]. The training data sets were set between 5% and 90% with the rest used for testing. Each experiment is repeated thirty times and the average performance is reported in Macro-F1 and Micro-F1 scores. We consistently observed statistically insignificant standard deviations.

Results. Tables 2 and 3 provide the eight-class venue and author classification results (the confidence value used for Ada2vec is 0.3). Both Macro-F1 and Micro-F1 scores are in the range of [0, 1] with 1 indicating that the result is identical to the ground truth. Results in the two tables show that metapath2vec++ consistently outperform the four dynamic heterogeneous network embedding methods (Dhne, Dyhne, Lime and Thine) and the dynamic homogeneous network embedding method (DynamicTriad) in terms of embedding accuracy. In most cases, metapath2vec++ learns more accurate node embeddings than Ada2vec. For G^{2013} and G^{2015}, Ada2vec anchors from the new and changed existing nodes only, which is a subset of the entire nodes that are mapped by metapath2vec++. It is expected that the different anchoring targets would result in the embedding accuracy loss of Ada2vec. However, the accuracy losses observed were consistently insignificant (0.0021 to 0.0573). Meanwhile, Ada2vec gains more than 60% efficiency improvements (discussed in details in the Efficiency Section) by only targeting subsets of nodes. Overall, the results demonstrate that Ada2vec achieves a fair tradeoff between embedding accuracy (relatively 0.22% to 11% loss) and efficiency (relatively >60% gain).

---

[3] 1. Computational Linguistics, 2. Computer Graphics, 3. Computer Networks & Wireless Communication, 4. Computer Vision & Pattern Recognition, 5. Computing Systems, 6. Databases & Information Systems, 7. Human Computer Interaction, and 8. Theoretical Computer Science.

[4] https://scholar.google.com/citations?view_op=top_venues&hl=en&vq=eng Accessed on November, 2023.

Table 3. Author node classification results in AMiner data.

Metric	Year	Method	5%	10%	20%	30%	40%	50%	60%	70%	80%	90%
Macro-F1	2011	DynamicTriad	0.4405	0.4453	0.4471	0.4466	0.4490	0.4491	0.4517	0.4538	0.4549	0.4552
		Metapath2vec++	**0.9506**	**0.9545**	**0.9581**	**0.9601**	**0.9616**	**0.9629**	**0.9636**	**0.9642**	**0.9647**	**0.9651**
		Ada2vec	**0.9506**	**0.9545**	**0.9581**	**0.9601**	**0.9616**	**0.9629**	**0.9636**	**0.9642**	**0.9647**	**0.9651**
		Dhne	0.7543	0.7566	0.7591	0.7611	0.7627	0.7637	0.7645	0.7659	0.7661	0.7670
		Dyhne	0.8297	0.8337	0.8396	0.8443	0.8438	0.8459	0.8461	0.8466	0.8472	0.8480
		Lime	0.8049	0.8059	0.8088	0.8261	0.8264	0.8265	0.8265	0.8278	0.8282	0.8295
		Thine	0.8180	0.8245	0.8294	0.8311	0.8320	0.8326	0.8339	0.8346	0.8343	0.8355
	2013	DynamicTriad	0.4033	0.4029	0.4058	0.4077	0.4116	0.4135	0.4160	0.4166	0.4198	0.4204
		Metapath2vec++	**0.9271**	**0.9332**	**0.9380**	**0.9405**	**0.9419**	**0.9429**	**0.9435**	**0.9439**	**0.9444**	**0.9445**
		Ada2vec	0.9163	0.9227	0.9289	0.9319	0.9333	0.9341	0.9350	0.9355	0.9365	0.9367
		Dhne	0.7110	0.7151	0.7186	0.7213	0.7255	0.7260	0.7275	0.7305	0.7309	0.7314
		Dyhne	0.7924	0.7943	0.7967	0.7998	0.8008	0.8015	0.8082	0.8085	0.8090	0.8092
		Lime	0.7633	0.7651	0.7702	0.7882	0.7894	0.7899	0.7899	0.7911	0.7912	0.7926
		Thine	0.7777	0.7781	0.7785	0.7899	0.7911	0.7922	0.7938	0.7951	0.7950	0.7954
	2015	DynamicTriad	0.3826	0.3844	0.3868	0.3889	0.3953	0.3961	0.3963	0.3963	0.3979	0.3983
		Metapath2vec++	**0.9077**	**0.9149**	**0.9195**	**0.9214**	**0.9226**	**0.9229**	**0.9235**	**0.9236**	**0.9241**	**0.9236**
		Ada2vec	0.8970	0.9049	0.9102	0.9122	0.9135	0.9139	0.9142	0.9148	0.9149	0.9152
		Dhne	0.6742	0.6786	0.6789	0.6788	0.6816	0.6827	0.6884	0.6901	0.6911	0.6912
		Dyhne	0.7480	0.7477	0.7492	0.7536	0.7542	0.7547	0.7552	0.7643	0.7644	0.7658
		Lime	0.7266	0.7271	0.7284	0.7289	0.7290	0.7328	0.7356	0.7360	0.7379	0.7388
		Thine	0.7250	0.7283	0.7291	0.7299	0.7345	0.7370	0.7429	0.7436	0.7439	0.7442
Micro-F1	2011	DynamicTriad	0.4630	0.4631	0.4644	0.4669	0.4671	0.4691	0.4727	0.4727	0.4736	0.4738
		Metapath2vec++	**0.9527**	**0.9563**	**0.9598**	**0.9618**	**0.9632**	**0.9643**	**0.9650**	**0.9656**	**0.9660**	**0.9664**
		Ada2vec	**0.9527**	**0.9563**	**0.9598**	**0.9618**	**0.9632**	**0.9643**	**0.9650**	**0.9656**	**0.9660**	**0.9664**
		Dhne	0.7552	0.7567	0.7576	0.7582	0.7640	0.7669	0.7683	0.7689	0.7711	0.7726
		Dyhne	0.8310	0.8321	0.8328	0.8339	0.8367	0.8446	0.8449	0.8453	0.8457	0.8466
		Lime	0.8081	0.8085	0.8089	0.8111	0.8125	0.8129	0.8133	0.8146	0.8157	0.8189
		Thine	0.8162	0.8169	0.8274	0.8277	0.8341	0.8349	0.8371	0.8382	0.8386	0.8389
	2013	DynamicTriad	0.4418	0.4419	0.4432	0.4451	0.4453	0.4466	0.4468	0.4520	0.4529	0.4528
		Metapath2vec++	**0.9321**	**0.9380**	**0.9426**	**0.9447**	**0.9459**	**0.9468**	**0.9474**	**0.9476**	**0.9480**	**0.9481**
		Ada2vec	0.9220	0.9282	0.9338	0.9366	0.9377	0.9386	0.9393	0.9396	0.9404	0.9406
		Dhne	0.7202	0.7215	0.7235	0.7234	0.7330	0.7352	0.7355	0.7374	0.7376	0.7380
		Dyhne	0.7943	0.7969	0.7991	0.8043	0.8062	0.8078	0.8110	0.8112	0.8156	0.8176
		Lime	0.7642	0.7647	0.7698	0.7703	0.7746	0.7781	0.7830	0.7846	0.7828	0.7866
		Thine	0.7810	0.7817	0.7859	0.7937	0.7952	0.7978	0.7988	0.7991	0.7995	0.7999
	2015	DynamicTriad	0.4168	0.4169	0.4173	0.4186	0.4193	0.4245	0.4243	0.4286	0.4288	0.4291
		Metapath2vec++	**0.9133**	**0.9199**	**0.9239**	**0.9257**	**0.9266**	**0.9270**	**0.9274**	**0.9274**	**0.9277**	**0.9277**
		Ada2vec	0.9037	0.9109	0.9156	0.9174	0.9184	0.9188	0.9192	0.9196	0.9196	0.9197
		Dhne	0.7140	0.7145	0.7174	0.7220	0.7230	0.7242	0.7275	0.7278	0.7279	0.7300
		Dyhne	0.7821	0.7866	0.7865	0.7891	0.7895	0.7926	0.7929	0.8001	0.8036	0.8043
		Lime	0.7402	0.7430	0.7436	0.7460	0.7521	0.7584	0.7585	0.7593	0.7597	0.7598
		Thine	0.7540	0.7549	0.7558	0.7576	0.7642	0.7643	0.7689	0.7688	0.7685	0.7697

Table 4. Clustering performance comparison in AMiner data.

	Venue-NMI			Author-NMI			Venue-ARI			Author-ARI		
	2011	2013	2015	2011	2013	2015	2011	2013	2015	2011	2013	2015
DynamicTriad	0.2949	0.3188	0.3462	0.2459	0.2173	0.2004	0.2418	0.2407	0.2643	0.2207	0.2332	0.2196
Metapath2vec++	**0.8011**	**0.8154**	**0.8640**	**0.8208**	**0.7641**	**0.7048**	**0.6583**	**0.6776**	**0.7817**	**0.8367**	**0.7886**	**0.7266**
Ada2vec	**0.8011**	0.7950	0.8557	**0.8208**	0.7384	0.6798	**0.6583**	0.6576	0.7589	**0.8367**	0.7506	0.6923
Dhne	0.4664	0.4540	0.4588	0.4237	0.4224	0.4349	0.4180	0.4141	0.4278	0.4544	0.4688	0.4766
Dyhne	0.6073	0.6140	0.6092	0.6100	0.5924	0.5955	0.5443	0.5377	0.6112	0.6233	0.6195	0.6148
Lime	0.5421	0.5298	0.5518	0.5266	0.5366	0.5400	0.5023	0.5156	0.5579	0.5322	0.5364	0.5427
Thine	0.5782	0.5204	0.5473	0.5340	0.5218	0.5536	0.5168	0.5264	0.5479	0.5331	0.5422	0.5571

5.4 Clustering

For the clustering task, we employ the same eight categories (clusters) venue and author dynamic ground truth used in the classification task. The learned representations of labeled nodes are inputs to the k-means algorithm [33] to cluster the venue and author separately. Clustering results are evaluated in terms of Normalized Mutual Information (NMI) [32] and Adjusted Rand Index (ARI) [37] scores. Each experiment is conducted thirty times and the average performance is reported. We consistently observed statistically insignificant standard deviations.

Results. Table 4 presents the eight-class venue and author clustering results (the confidence value used for Ada2vec is 0.3). Both NMI and ARI scores are in the range of [0, 1] with higher value indicating that the result is closer to the ground truth. Among the seven network embedding methods, metapath2vec++ achieves the best clustering performance with regards to the two evaluation metrics. Overall, the accuracy losses of Ada2vec over metapath2vec++ are insignificant again (relatively 0.96% to 4.82%). The constant gain achieved by Ada2vec is around 42% to 62% over DynamicTriad.

5.5 Performance Analysis

Scalability. Using the same mechanism in word2vec [24] and node2vec [15], Ada2vec can be parallelized on multiple threads. Our experiments are performed in a computing cloud server with 48 core 2.4 GHz Intel Xeon CPUs E5-2673 v3 Operating System Ubuntu 18.04-LTS. We conduct experiments on the AMiner CS 2015 network (G^{2015}) using different number of threads, i.e., 1, 2, 4, 8, 16, 32, 48, each of them utilizing one CPU core. With 48 cores, Ada2vec's embedding learning process takes 18.2 min for G^{2015}, which consists of over 1 million authors with 1.5 million papers published in more than 7300 venues. The empirical results show that Ada2vec is scalable for large-scale dynamic heterogeneous networks with millions of nodes.

Efficiency. From time stamp t + 1 (t = 1, 2, 3, ..., i), Ada2vec produces metapath-guided random walks *only* for the new and changed existing nodes, the embeddings are then learned from the random walks. This mechanism equips Ada2vec with higher storage and computational efficiency from t + 1. Our experimental results show that Ada2vec can reduce the storage and computational costs of metapath2vec++ by more than 60%. For the AMiner 2015 data, the file size of the random walks generated for all the venue and author nodes by metapath2vec++ is 33.5G. As Ada2vec produces random walks for a subset of all the nodes only, the file size decreases to 13.2G. Meanwhile, run time costs of the learning processes with 48 threads decrease from 47.3 min (metapath2vec++) to 18.2 min (Ada2vec). Overall, DynamicTriad takes 30% to 40% less learning time than Ada2vec. However, its high computational efficiency is achieved by being agnostic to object and relationship types, which results in the significant loss of embedding accuracy as shown in the previous sections. The runtime for the remaining heterogeneous network embedding methods were a couple of days longer than Ada2vec.

Parameter Sensitivity. Ada2vec introduces the confidence value δ as a new parameter. Following the practice in [4,13], the values of δ are set as 0.3, 0.2, 0.1 and 0.05. We run Ada2vec on the dynamic heterogeneous network G to produce the embeddings of venue and author nodes. Table 5 shows the percentages of changed existing venue and author nodes with different values of δ as G evolves over time. Figure 3(a) shows the classification results as a function of various values of δ in G^{2013} (the network at the year 2013 time stamp). By measuring the clustering performance, we show the sensitivity analysis of Ada2vec to different values of δ in Fig. 3(b). Overall, we find that Ada2vec is not sensitive to the confidence value parameter and is able to achieve high classification and clustering performance with a sensible confidence value.

Table 5. Percentage of changed existing nodes.

Time Span	$\delta = 0.3$	$\delta = 0.2$	$\delta = 0.1$	$\delta = 0.05$
2011–2013	5.5014%	3.2098%	1.2897%	0.6525%
2013–2015	4.6851%	2.9614%	1.4041%	0.7858%

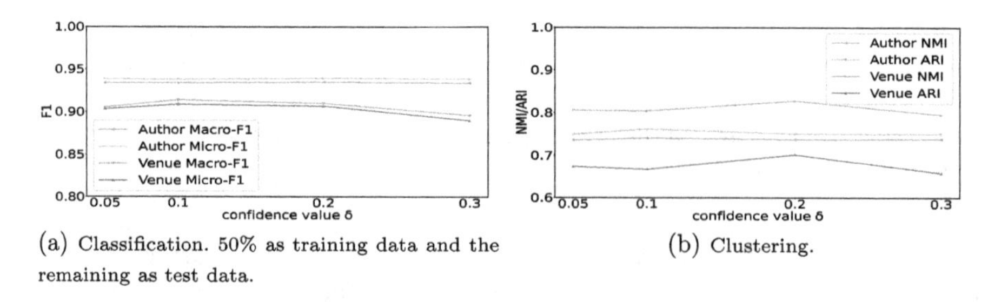

(a) Classification. 50% as training data and the remaining as test data.

(b) Clustering.

Fig. 3. Parameter sensitivity in G^{2013}.

Limitations. Ada2vec gains efficiency by mapping from a subset of the entire nodes in a network. This performance advantage is justified in a dynamic network with smooth transitions due to the relatively small subset sizes. For example, a bibliographic network (object types: authors, papers, venues) with one year as the time span; an open source software network (object types: developers, projects, APIs) with one month as the time span.

6 Conclusion and Future Work

We developed a novel representation learning approach, Ada2vec, for large-scale dynamic heterogeneous networks. Ada2vec leverages a statistical bound as a guidance to model object structural changes in a dynamic environment. Meanwhile, metapath-guided random walks are employed for monitoring a object's heterogeneous k-order proximity. Our experimental results demonstrate the superiority of Ada2vec over the state-of-the-art benchmarks. For future work, we plan to apply Ada2vec to non-bibliographic networks such as health, groundwater, natural resource management related dynamic heterogeneous networks to investigate the interpretability of the learned representations in different application domains.

References

1. Bengio, Y., Courville, A.: Representation learning: a review and new perspectives. IEEE Trans. Pattern Anal. Mach. Intell. **35**(8), 1798–1828 (2013)
2. Bernstein, S.: On the Order of the Best Approximation of Functions Continuous by Polynômes of Degree é Given é, vol. 4. Hayez, Printer of Royal Academies (1912)
3. Bian, R., Koh, Y.S., Dobbie, G., Divoli, A.: Network embedding and change modeling in dynamic heterogeneous networks. In: Proceedings of the 42nd ACM SIGIR (2019)
4. Bifet, A., Gavaldà, R.: Learning from time-changing data with adaptive windowing. In: Proceedings of the 2007 SIAM International Conference on Data Mining (2007)
5. Cen, Y., Zou, X., Zhang, J., Yang, H.: Representation learning for attributed multiplex heterogeneous network. In: Proceedings of the 25th ACM SIGKDD (2019)
6. Chang, S., Han, W., Tang, J., Qi, G.J.: Heterogeneous network embedding via deep architectures. In: Proceedings of the 21st ACM SIGKDD (2015)
7. Chen, H., Yin, H., Wang, W.: PME: projected metric embedding on heterogeneous networks for link prediction. In: Proceedings of the 24th ACM SIGKDD (2018)
8. Chernoff, H.: On the distribution of the likelihood ratio. Ann. Math. Stat. **25**(3), 573–578 (1954)
9. Cui, P., Wang, X., Pei, J., Zhu, W.: A survey on network embedding. IEEE Trans. Knowl. Data Eng. **31**(5), 833–852 (2019)
10. Dong, Y., Chawla, N.V., Swami, A.: Metapath2vec: scalable representation learning for heterogeneous networks. In: Proceedings of the 23rd ACM SIGKDD (2017)
11. Dong, Y., Zhang, J., Tang, J., Chawla, N.V., Wang, B.: CoupledLP: link prediction in coupled networks. In: Proceedings of the 21st ACM SIGKDD (2015)
12. Fu, T., Lee, W.C.: HIN2Vec: explore meta-paths in heterogeneous information networks for representation learning. In: Proceedings of the 2017 ACM CIKM (2017)
13. Gama, J., Žliobaite, I., Bifet, A., Pechenizkiy, M., Bouchachia, A.: A survey on concept drift adaptation. ACM Comput. Surv. **46**(4), 1–37 (2014)
14. Goyal, P., Chhetri, S.R., Canedo, A.: Dyngraph2vec: capturing network dynamics using dynamic graph representation learning. Knowl.-Based Syst. **187**, 104816 (2020)
15. Grover, A., Leskovec, J.: Node2vec: scalable feature learning for networks. In: Proceedings of the 22nd ACM SIGKDD, pp. 855–864 (2016)

16. Han, X., Shi, C., Wang, S., Philip, S.Y., Song, L.: Aspect-level deep collaborative filtering via heterogeneous information networks. In: Proceedings of the 2018 IJCAI, pp. 3393–3399 (2018)
17. Heidari, F., Papagelis, M.: Evolving network representation learning based on random walks. Appl. Netw. Sci. **5**(1), 1–38 (2020)
18. Hoeffding, W.: Probability inequalities for sums of bounded random variables. In: Fisher, N.I., Sen, P.K. (eds.) The Collected Works of Wassily Hoeffding, pp. 409–426. Springer, New York (1994). https://doi.org/10.1007/978-1-4612-0865-5_26
19. Huang, H., Shi, R., Zhou, W., Wang, X., Jin, H., Fu, X.: Temporal heterogeneous information network embedding. In: Proceedings of the 30th IJCAI, pp. 1470–1476 (2021)
20. Ji, M., Han, J., Danilevsky, M.: Ranking-based classification of heterogeneous information networks. In: Proceedings of the 17th ACM SIGKDD, p. 1298–1306 (2011)
21. Ji, Y., Jia, T., Fang, Y., Shi, C.: Dynamic heterogeneous graph embedding via heterogeneous Hawkes process. In: Oliver, N., Pérez-Cruz, F., Kramer, S., Read, J., Lozano, J.A. (eds.) ECML PKDD 2021. LNCS (LNAI), vol. 12975, pp. 388–403. Springer, Cham (2021). https://doi.org/10.1007/978-3-030-86486-6_24
22. Kazemi, S.M., et al.: Representation learning for dynamic graphs: a survey. J. Mach. Learn. Res. **21**(70), 1–73 (2020)
23. Kumar, S., Zhang, X., Leskovec, J.: Predicting dynamic embedding trajectory in temporal interaction networks. In: Proceedings of the 25th ACM SIGKDD, pp. 1269–1278 (2019)
24. Mikolov, T., Sutskever, I., Chen, K., Corrado, G.S., Dean, J.: Distributed representations of words and phrases and their compositionality. In: Advances in Neural Information Processing Systems, vol. 26, pp. 3111–3119 (2013)
25. Pareja, A., et al.: EvolveGCN: evolving graph convolutional networks for dynamic graphs. In: Proceedings of the 2020 AAAI, pp. 5363–5370 (2020)
26. Peng, H., et al.: Lime: low-cost incremental learning for dynamic heterogeneous information networks. IEEE Trans. Comput. **71**(3), 628–642 (2021)
27. Perozzi, B., Al-Rfou, R., Skiena, S.: DeepWalk: online learning of social representations. In: Proceedings of the 20th ACM SIGKDD, pp. 701–710 (2014)
28. Shang, J., Qu, M., Liu, J., Kaplan, L.M., Han, J., Peng, J.: Meta-path guided embedding for similarity search in large-scale heterogeneous information networks. arXiv preprint arXiv:1610.09769 (2016)
29. Shi, C., Hu, B., Zhao, W.X., Yu, P.S.: Heterogeneous information network embedding for recommendation. IEEE Trans. Knowl. Data Eng. **31**(2), 357–370 (2019)
30. Shi, Y., Zhu, Q., Guo, F., Zhang, C., Han, J.: Easing embedding learning by comprehensive transcription of heterogeneous information networks. In: Proceedings of the 24th ACM SIGKDD, pp. 2190–2199 (2018)
31. Sun, Y., Han, J., Yan, X., Yu, P.S., Wu, T.: PathSim: meta path-based top-k similarity search in heterogeneous information networks. Proc. VLDB Endow. **4**, 992–1003 (2011)
32. Sun, Y., Han, J., Zhao, P., Yin, Z., Cheng, H., Wu, T.: RankClus: integrating clustering with ranking for heterogeneous information network analysis. In: Proceedings of the 12th EDBT, pp. 565–576 (2009)
33. Sun, Y., Norick, B., Han, J., Yan, X., Yu, P.S., Yu, X.: PathSelClus: integrating meta-path selection with user-guided object clustering in heterogeneous information networks. ACM Trans. Knowl. Discov. Data (TKDD) **7**, 1–23 (2013)
34. Tang, J., Qu, M., Mei, Q.: PTE: predictive text embedding through large-scale heterogeneous text networks. In: Proceedings of the 21st ACM SIGKDD, pp. 1165–1174 (2015)

35. Tang, J., Qu, M., Wang, M., Zhang, M., Yan, J., Mei, Q.: LINE: large-scale information network embedding. In: Proceedings of the 24th WWW, pp. 1067–1077 (2015)
36. Tang, J., Zhang, J., Yao, L., Li, J., Zhang, L., Su, Z.: ArnetMiner: extraction and mining of academic social networks. In: Proceedings of the 14th ACM SIGKDD, pp. 990–998 (2008)
37. Wagner, S., Wagner, D.: Comparing clusterings: an overview (2007)
38. Wang, H., Zhang, F., Hou, M., Xie, X., Guo, M., Liu, Q.: SHINE: signed heterogeneous information network embedding for sentiment link prediction. In: Proceedings of the 11th ACM WSDM, pp. 592–600 (2018)
39. Wang, X., Bo, D., Shi, C., Fan, S., Ye, Y., Yu, P.S.: A survey on heterogeneous graph embedding: methods, techniques, applications and sources. arXiv preprint arXiv:2011.14867 (2020)
40. Wang, X., et al.: Heterogeneous graph attention network. In: Proceedings of the 2019 WWW, pp. 2022–2032 (2019)
41. Wang, X., Lu, Y., Shi, C., Wang, R., Cui, P., Mou, S.: Dynamic heterogeneous information network embedding with meta-path based proximity. IEEE Trans. Knowl. Data Eng. **34**(3), 1117–1132 (2020)
42. Xie, Y., et al.: Learning and updating node embedding on dynamic heterogeneous information network. In: Proceedings of the 14th ACM WSDM, pp. 184–192 (2021)
43. Xu, L., Wei, X., Cao, J., Yu, P.S.: Embedding of embedding (EOE): Joint embedding for coupled heterogeneous networks. In: Proceedings of the 10th ACM WSDM, pp. 741–749 (2017)
44. Xue, G., Zhong, M., Li, J., Chen, J., Zhai, C., Kong, R.: Dynamic network embedding survey. arXiv preprint arXiv:2103.15447 (2021)
45. Yang, C., Xiao, Y., Zhang, Y., Sun, Y., Han, J.: Heterogeneous network representation learning: a unified framework with survey and benchmark. IEEE Trans. Knowl. Data Eng. **34**(10), 4854–4873 (2020)
46. Yin, Y., Ji, L.X., Zhang, J.P., Pei, Y.L.: DHNE: network representation learning method for dynamic heterogeneous networks. IEEE Access **7**, 134782–134792 (2019)
47. Zhang, J., Xia, C., Zhang, C., Cui, L., Fu, Y., Yu, P.S.: BL-MNE: emerging heterogeneous social network embedding through broad learning with aligned autoencoder. In: Proceedings of the 2017 ICDM, pp. 605–614 (2017)
48. Zhao, K., et al.: Deep adversarial completion for sparse heterogeneous information network embedding. In: Proceedings of the 2020 WWW, pp. 508–518 (2020)
49. Zhou, L., Yang, Y., Ren, X., Wu, F., Zhuang, Y.: Dynamic network embedding by modeling triadic closure process. In: Proceedings of the 32nd AAAI (2018)
50. Zhu, L., Guo, D., Yin, J., Steeg, G.V., Galstyan, A.: Scalable temporal latent space inference for link prediction in dynamic social networks. IEEE Trans. Knowl. Data Eng. **28**(10), 2765–2777 (2016)

Differentially-Private Neural Network Training with Private Features and Public Labels

Islam A. Monir(✉) and Gabriel Ghinita

Hamad Bin Khalifa University, Ar-Rayyan, Qatar
{ismo58166,gghinita}@hbku.edu.qa

Abstract. Training neural networks (NN) with differential privacy (DP) protection has been extensively studied in the past decade, with the DP-SGD (stochastic gradient descent) mechanism representing the benchmark approach. Conventional DP-SGD assumes that both the features and the labels of training samples must be protected. A recent variation of DP-SGD considers training when the input sample features are non-private, and only labels must be protected, which improves accuracy by reducing the amount of noise injected by DP. We argue that in some scenarios, the converse holds, namely the labels may be publicly known, while the features themselves are sensitive. We provide a customized technique for this setting, we identify several design trade-offs, and we show how one can factor in such trade-offs to revise the architecture of the NN in order to improve accuracy. Extensive experiments on real data show that our approach significantly outperforms the DP-SGD baseline.

Keywords: Differential Privacy · Machine Learning · Neural Networks

1 Introduction

Neural networks (NN) trained on large datasets consisting of individual data witnessed an explosive growth in the past decade. Significant privacy concerns arise, as an adversary may use the resulting models to derive sensitive individual details [16]. Differential privacy (DP) [6], a robust semantic protection model, established itself as the de-facto standard for private publication of individuals' data. The benchmark supervised learning method is stochastic gradient descent (SGD - see Sect. 2.2 for details) which measures the difference between predicted and true labels in the training data and *back-propagates* into the network a parameter change to correct the model.

In conventional SGD [17], both features and labels of training samples are public. The DP-SGD technique [1], a DP-compliant version of SGD illustrated in Fig. 1b, performs two steps: (1) *clipping* the gradient to bound its *sensitivity* (a key DP concept detailed in Sect. 2.1 which determines how much noise is needed); and (2) adding noise to the clipped gradient.

Figure 1a provides a visual representation of the SGD and DP-SGD approaches with respect to the sensitivity of the features and labels in the

R. Wrembel et al. (Eds.): DaWaK 2024, LNCS 14912, pp. 208–222, 2024.
https://doi.org/10.1007/978-3-031-68323-7_16

training data. These approaches are the polar extremes of the privacy spectrum: SGD offers no privacy, whereas DP-SGD protects both features and labels privacy. Recently, a third cell of the quadrant has been explored, where the features are public whereas the labels are private [7]. In this setting, it is assumed that while the feature values of a sample may not be sensitive (i.e., publicly observable features of individuals that an adversary may have access to), the association between the label and those features needs to be protected. Therefore, it is sufficient if one protects the label only, thus reducing the amount of noise required by DP, and improving accuracy.

While private-label training is valuable, there is an equally significant but so far overlooked scenario: when labels are public, but the actual features remain private. Consider an online community where users, using pseudonyms, rate various products. In this case, the public labels are the product ratings, but users want to keep their personal features (e.g., age, ethnicity, gender, income, or education levels) private. As another example, in a dataset about the effectiveness of treatments for a medical condition, participants share public labels representing treatment outcomes such as improved blood sugar levels or overall well-being. However, participants are cautious about the privacy of their health information, including genetic markers and lifestyle details.

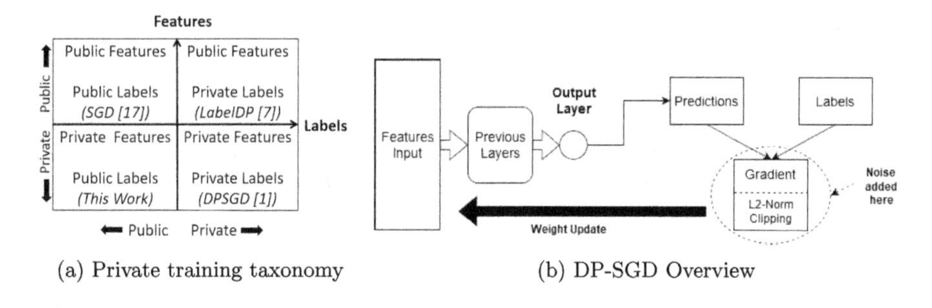

(a) Private training taxonomy (b) DP-SGD Overview

Fig. 1. Private NN Learning Taxonomy and DP-SGD

To address the private features-public label NN training problem, we propose a customized technique that takes a significantly different approach than DP-SGD. When DP-SGD adds noise to the gradient, which is a choke-point in the processing pipeline, it may destroy lots of useful information. Instead, we add noise to an intermediate NN layer we call *sanitization layer*, allowing more information to be carried forward. We investigate the emerging trade-off between the position and width of the sanitization layer in the NN pipeline, which controls the balance between privacy and accuracy. Studying this trade-off allows us to explore several choices of how to adapt the NN architecture to improve private training accuracy.

Our specific contributions are:

- We formulate a new version of the private neural network training problem, where the labels are public and the features are private. This completes the quadrant of the private/public features/labels space in NN training.

- We design an alternative approach to DP-SGD, whereby DP-compliant noise is added to the outputs of a designated sanitization layer, rather than gradients.
- We explore the effects of NN architecture choices on model accuracy (i.e., placement and width of the sanitization layer).
- We perform an extensive experimental evaluation on real data which shows that our approach significantly outperforms benchmarks.

We introduce preliminary concepts and definitions in Sect. 2. Section 3 reviews related work. We present the proposed approach in Sect. 4. We evaluate empirically our technique in Sect. 5 in comparison with benchmarks, and conclude with directions for future work in Sect. 6.

2 Background

2.1 Differential Privacy

Differential privacy is a powerful semantic privacy framework that prevents adversaries from learning whether the data of any single individual has been included or not in the input of a computation process. This is achieved by introducing controlled random noise into the computation, and thus bounding the probability of an adversary to discern the contribution of any specific individual to the data.

Definition 1. *A randomized mechanism M achieves (ϵ, δ)-differential privacy if for any two neighboring datasets D and D' (differing by one single record) and for any output measurement S, the following inequality holds:*

$$Pr[M(D) \in S] \leq e^{\epsilon} Pr[M(D') \in S] + \delta \tag{1}$$

This condition ensures that the probability of observing a specific output S from the mechanism M on dataset D is not significantly different from the probability of observing the same output on a neighboring (or sibling) dataset D', up to a factor of e^{ϵ} and a small additive term δ [1,12]. The sibling relationship between datasets is denoted as $D \sim D'$.

The parameter ϵ is commonly known as *privacy budget*. It serves as an adjustable parameter which allows fine-tuning of the amount of privacy a mechanism provides. A smaller ϵ values ensures higher protection, but at the same time requires the addition of more noise, hence decreasing data utility [6].

A key concept in DP is that of *sensitivity*, which quantifies the maximal amount by which the output of a computation can change following the addition or removal of a single record from the input.

Definition 2. *ℓ_2-Sensitivity: Consider a vector-valued function $q : \mathcal{D} \rightarrow \mathbb{R}^d$ defined over the data domain. The ℓ_2-sensitivity of q, denoted as $\Delta_2(q)$, or simply $\Delta(q)$, is defined as:*

$$\Delta_2(q) = \sup_{D \sim D'} \|q(D) - q(D')\|_2. \tag{2}$$

A prominent DP protection mechanism used by DP-SGD is the Gaussian mechanism [1].

Definition 3. *(Gaussian Mechanism): Suppose an algorithm* $f : \mathcal{D} \rightarrow \mathbb{R}^d$ *has* ℓ_2 *sensitivity* Δ_f*, i.e., for any pair of neighboring datasets* D, D', $\|f(D) - f(D')\| \leq \Delta_f$*. Then, for any* $\epsilon > 0, \delta \leq 1$*, the mechanism* $M(x) = f(x) + w$ *satisfies* (ϵ, δ)*-differential privacy (DP), where* $w \sim \mathcal{N}(0, \sigma^2 I_d)$*, and*

$$\sigma = \frac{\Delta_f \sqrt{2 \ln(1.25/\delta)}}{\epsilon} \tag{3}$$

2.2 DP-SGD

Stochastic Gradient Descent (SGD) is a widely adopted optimization algorithm in training machine learning models. SGD repeatedly updates model parameters, guided by the gradients of a loss function computed on a random sample of the training data, referred to as a batch. The specific update rule governing the adjustment of model parameters is [9,17]

$$\theta_{t+1} = \theta_t - \eta \nabla J(\theta_t; x^{(i)}, y^{(i)})$$

where θ_t represents the model parameters at iteration t, η is the learning rate, and $J(\theta_t; x^{(i)}, y^{(i)})$ is the loss function for the current batch $(x^{(i)}, y^{(i)})$.

Differentially Private Stochastic Gradient Descent (DP-SGD) [1] incorporates differential privacy into the optimization process. Sanitization is performed at the gradient, which represents a chokepoint in the training pipeline. The gradient is first clipped, to bound sensitivity, and then Gaussian noise is added according to Eq. (3), at the location illustrated in Fig. 1b.

One important aspect in DP-SGD is the privacy budget usage: due to the iterative nature of SGD, the budget consumption can grow very large, if one uses the conventional theorem of *sequential composition* [8], which prescribes that the budget of the entire learning process equals the sum of the budgets used in each iteration. The seminal work in [1] provides a tight bound on privacy budget consumption, called *moments accountant (MA)*, which shows that in the presence of sampling, the budget consumption can be significantly reduced. MA analyzes the composition of Poisson sampling and Gaussian noise addition. Budget consumption of Gaussian noise addition along training epochs is calculated using Eq. (4), where subscript i refers to the current iteration, and k is the total number of iterations. Poisson sampling introduces randomness in the selection of training batches, ensuring that the inclusion or exclusion of individual data records is duly accounted in the training process.

$$\epsilon_{training} \approx \sqrt{2 \sum_{i=1}^{k} \frac{\ln(1.25/\delta)}{\sigma_i^2}}, \tag{4}$$

3 Related Work

The seminal work in [1] proposed the DP-SGD approach and the moments' accountant privacy budgeting technique. Numerous follow-up works introduced various optimizations to further improve accuracy. In [20] and [11], the focus lies on adaptive strategies for privacy budget allocation, whereby different noise magnitudes are used at distinct iterations. In [20], decaying Gaussian noise is added to gradients during training, while [11] dynamically adapts the learning rate in SGD based on estimated errors. The work in [3] proposed a private backtracking line search algorithm for noisy gradients and loss function error estimates. It adjusts the privacy budget based on the reliability of the noisy gradient, avoiding the use of large privacy budgets. The DiceSGD approach [19] investigates the bias introduced by gradient clipping and chooses clipping thresholds using an error-feedback mechanism to eliminate the bias.

The impact of normalization layers on accuracy is studied in [5], where batch normalization is integrated with DP-SGD to mitigate accuracy loss. The work in [13] investigates Batch Clipping (BC) and Adaptive Layerwise Clipping (ALC) methods to further enhance accuracy and accelerate convergence.

The work in [14] addresses hyperparameter tuning in DP-SGD, and proposes three algorithms: evolutionary, Bayesian, and reinforcement learning. Through extensive experiments, the authors demonstrate the significant boost in accuracy of these algorithms over conventional grid search baselines. The study showcases that evolutionary and Bayesian optimization achieve the lowest privacy leakage while concurrently maintaining high accuracy.

The work in [4] introduces a framework that automates search for model architectures suitable for private learning. It integrates private learning with architecture search, introducing a DP-aware method for training candidate models during the search process. The work in [15] develops the SmoothNet architecture tailored for deep learning with DP. It identifies components with strong individual performance and demonstrates increased robustness, outperforming standard architectures on benchmark datasets.

Closer to our work, [7] proposed label differential privacy (label-DP), where only the labels need to be protected for privacy. The focus is on training models with label-DP guarantees by leveraging unsupervised and semi-supervised learning. Novel techniques are presented to add less noise while obtaining the same privacy, leading to a better privacy-utility trade-off. Our work addresses the converse problem, learning on data with private features and public labels, and investigates the effect of NN model architecture on accuracy.

4 Proposed Approach

We introduce our proposed approach for private learning on data with *private* features and *public* labels. Our technique addresses several key questions:

1. *Where is the best place in the NN pipeline to add DP-compliant noise?* In DP-SGD, noise is added to the gradient, because it represents a choke point

in the pipeline, and it is thus easy to control sensitivity. While this may be suitable when labels are private, it introduces unnecessary constraints in the case of public labels. We introduce a *sanitization layer* in the NN pipeline, which allows us to flexibly control the accuracy loss due to DP noise.

2. *How to perform clipping?* DP-SGD performs ℓ_2-norm clipping to bound sensitivity, which means that it effectively bounds the magnitude of the gradient norm. By doing so, it allows a single outlier value in the gradient vector to disproportionately affect all other gradient elements, compromising accuracy. In contrast, we perform element-wise clipping, which prevents an outlier from excessively affecting the output, but the resulting sensitivity is higher, and hence an important trade-off emerges (details in Sect. 4.2).
3. *How to adapt the NN architecture for a good privacy-accuracy trade-off?* The flexibility in choice of location where to add noise and width of sanitization layer lead to important architecture choices that can influence the privacy-accuracy trade-off, which we investigate in-depth.

4.1 Sanitization Layer

As discussed in Sect. 2.2, DP-SGD applies DP at the gradient, which is a choke point within the NN pipeline. Specifically, it clips and adds noise to the average gradient for a batch. This approach is convenient and generic, since the gradient captures the contribution of both features and labels, and by adding noise to the gradient, one can enforce the DP indistinguishability constraint (Eq. (1)) with relative ease. Furthermore, DP-SGD performs ℓ_2-norm clipping, where each element of the gradient is clipped by the same factor, in order to cap the value of the resulting ℓ_2 norm after clipping. This comes at a cost in terms of adaptability to training data characteristics, leading to subpar accuracy.

Consider the case where a single element in the gradient vector is an outlier, having a value much larger than all other elements. This element has a disproportionately large effect on the norm, and if one performs ℓ_2-norm clipping, the values of all other gradient elements are obliterated by clipping, since all elements are scaled down by the same amount, equal to the ratio between the clipping threshold C and the original gradient norm [1].

Our approach, illustrated in Fig. 2, takes a different approach. First, it performs clipping and noise addition at a designated *sanitization layer*, *before* gradient computation. This way, a set of sanitized values equal to the width of the sanitization layer are propagated forward, conveying more information towards gradient calculation. Furthermore, we perform *element-wise clipping*, hence each one of these values is clipped and added noise to independently, preventing a single outlier value from disproportionately affecting the clipping. This can lead to more robust accuracy and faster convergence.

An immediate question that arises is: at which layer is it viable to perform clipping and noise addition? Doing so early in the pipeline (i.e., closer to the input layer) is likely to have a negative effect on accuracy, due to the compounding effect of noise in subsequent calculations (i.e., at intermediate layers). We choose to perform sanitization on the output of the penultimate layer in the

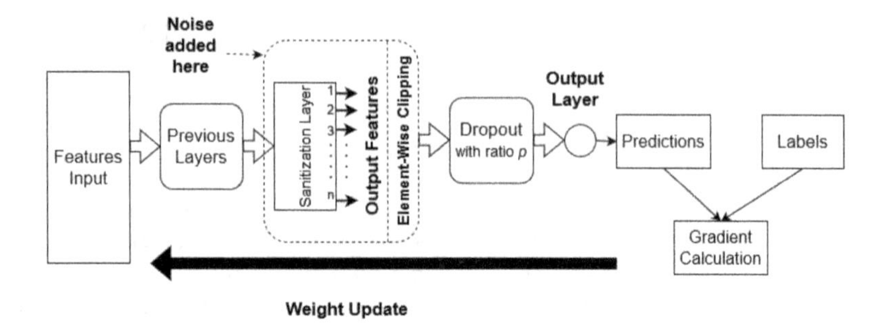

Fig. 2. Proposed Sanitization Layer Approach

neural network. By targeting the penultimate layer, we mitigate the downstream propagation of noise, thus preserving the integrity of the training process and achieving the least possible disruption to the overall model performance. At the same time, the penultimate layer is wide enough to prevent outlier values from significantly altering gradient computation. We discuss in more detail the impact of layer width on noise magnitude and accuracy in Sect. 4.3.

4.2 Bounding Sensitivity and Adding Noise

As discussed in Sect. 2.2, training proceeds in epochs, where each epoch takes as input a data sample and updates the model according to the difference between actual and predicted labels for each batch. DP-compliant noise must be added in each epoch, to protect the input sample used in each iteration.

The sensitivity (Sect. 2.1) of the neural network training process is difficult to quantify, and the preferred approach in literature is to impose a clipping constant that artificially bounds the change in algorithm output that any single input may have. Specifically, to limit the sensitivity of the sanitized output of the layer, denoted by Δf, we set a clipping threshold C and clip the values of *each element in the output of the layer to this threshold.* Note that, this is substantially different than what conventional DP-SGD does, where the gradient is clipped using l_2-norm clipping. If we denote the width of the sanitization layer by n, then by applying element-wise clipping, the l_2-sensitivity of that output becomes $C\sqrt{n}$. After clipping, our algorithm adds Gaussian noise drawn from $\sim \mathcal{N}(0, C^2 * \sigma^2 * n)$ to the layer output to ensure (ϵ, δ)-differential privacy.

Our choice of element-wise clipping and adding noise at the sanitization layer instead of the l_2-norm gradient clipping has two advantages:

1. Propagating multiple, independent sanitization layer output values towards the gradient reduces the risk of information loss due to large noise drawn from the Gaussian distribution. In the single invocation of the random generator that DP-SGD performs to perturb the gradient, the effect of a large noise drawn can destroy the utility of the gradient. In contrast, we add noise to sanitization layer output values, and even if a subset of those is affected by a

large-drawn noise, the averaging effect of gradient computation reduces the negative impact of noise, enhancing the learning dynamics of the model.

2. By performing element-wise clipping at the sanitization layer output, our technique explicitly limits the effect of outliers in the model, as each element is independently clipped and noised, and thus the effect of a small subset of outliers is offset. In contrast, outliers in DP-SGD are directly influencing the gradient clipping factor, and thus can negatively impact the gradient accuracy.

4.3 Design Choices and Tradeoffs

We investigate privacy-accuracy trade-offs that emerge in our technique, and we explore how to adapt the architecture of the NN to boost accuracy.

Choosing Sanitization Layer Width. Choosing an appropriate setting for the sanitization layer width n is an important decision: an initial increase in n will lead to better accuracy, as outliers are clipped and more feature information flows into the gradient calculation. However, increasing n too much will translate in larger sensitivity (proportional to $\sqrt{n}$, since we perform element-wise clipping and not ℓ_2-norm clipping), and consequently, increased noise magnitude. Adding large amounts of noise will likely lead to overfitting, and cause the model to miss new patterns, thus harming accuracy.

We observe that the sanitization layer width needs to be large-enough to produce a suitable amount of features with sufficient information for the training process to converge fast. At the same time, we need to control sensitivity, in order to limit the impact of the increased noise magnitude when n increases. To address this problem, we add more layers in the NN pipeline to create a funnel-shaped structure. In order to maintain this shape and to fix the number of outputs of the sanitization layer to be n, we add fully connected layers having widths equal to $8n$, $4n$, $2n$ and n, respectively. The width of the output layer is dictated by the number of classes in the dataset. Increasing the number of fully connected layers in the proposed funnel-shape structure, illustrated in Fig. 3, helps get more focus on specific features. As we delve deeper in the pipeline, the noise is added to fewer, but more concentrated number of features, without loss of important information.

Choosing Clipping Threshold. Selecting an appropriate clipping threshold also involves a nuanced analysis. Opting for a low clipping threshold may be advantageous in terms of facilitating a low noise magnitude suitable for model training. However, this approach has drawbacks: a harsh clipping threshold on sanitization layer outputs will result in the loss of crucial information, negatively impacting the NN model accuracy. Conversely, opting for a high clipping threshold preserves more information, but increases sensitivity, and consequently the magnitude of added noise.

Table 1 illustrates two cases of low and high clipping threshold choices, respectively. When a low clipping threshold is used (1.0), it leads to aggressive clipping

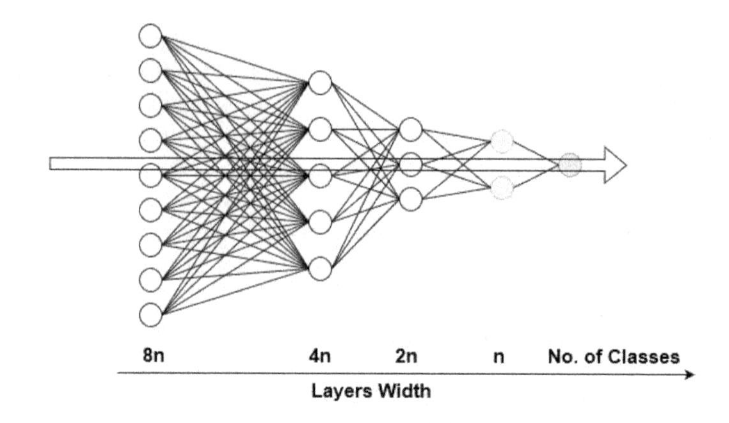

Fig. 3. NN Architecture Customization with Funnel-shape Structure of Layers

of the sanitization layer output, but sensitivity is reduced, and the added noise is smaller, leading to a less perturbed result. Conversely, with a high clipping threshold (5.0), the sanitization layer output may remain largely unclipped, but sensitivity is larger, hence the added noise has a higher magnitude.

Table 1. Clipping Threshold Magnitude Example

Scenario	Original	Clipped	Noise	Noisy Output
Low Threshold (1.0)	[3.0, 1.5, 2.5]	[1.0, 1.0, 1.0]	0.5	[1.38, 1.39, 2.22]
High Threshold (5.0)	[3.0, 1.5, 2.5]	[3.0, 1.5, 2.5]	2.5	[1.06, 4.62, 2.13]

Regularization with Dropout. Adding a moderate amount of noise during training can act as a regularization technique. It introduces a level of uncertainty and prevents the model from becoming overly-sensitive to the training data, potentially improving its ability to generalize to new, unseen data. On the other hand, if the noise injected is excessive, it may lead to overfitting. The model might start to memorize the noise itself, treating it as if it were a meaningful signal. In such cases, the model may perform well on the training data but struggle to generalize to new data because it has learned to rely on the injected noise.

To fight against overfitting, we apply after the sanitization layer a dropout layer with ratio p during each training epoch. Dropout is a regularization technique used to mitigate overfitting in neural networks. It involves randomly "dropping out" (setting to zero) a fraction of the units (neurons) in a layer during training. This process occurs independently for each example in the training set and helps prevent the network from relying too much on any particular set of features.

Previously, we scaled up the sensitivity by $\sqrt{n}$ to account for the application of element-wise clipping on the n outputs of the sanitization layer. However, if high overfitting occurs during model training, dropping out some outputs with ratio p may alleviate this problem. If dropout is applied, then sensitivity is decreased, and the noise magnitude must only be scaled by $\sqrt{n * (1-p)}$. Hence, the magnitude of the Gaussian noise that is required to guarantee DP becomes $\sim \mathcal{N}(0, C^2 * \sigma^2 * n * (1-p))$.

Algorithm 1 presents the pseudocode that integrates all the steps in our approach.

Algorithm 1. Private Learning with Public Labels and Private Features

Require: Sampling ratio q, Dropout ratio p
  Sanitization layer width n, Clipping threshold C, Noise multiplier σ
Require: Training set X
  Sample X into mini-batches B_i for $i = 1, 2, ..., m$ (number of batches)
  for batch B_i **do**
    Feed forward the batch till the target layer output.
    for each output x in B_i **do**
      $x \leftarrow \text{clip}(\Delta f, C)$
      $x \leftarrow x + \sim \mathcal{N}(0, C^2\sigma^2 n(1-p))$
    end for
    mask $\leftarrow$ random binary mask with probability p
    $B_i \leftarrow B_i \odot$ mask {Element-wise multiplication for applying Dropout}
    Propagate noisy B_i forward in the network
  end for

5 Experimental Evaluation

5.1 Experimental Settings

We evaluated our approach in comparison with several prominent benchmarks [1,3,4,15,18,20]. Following most prior work in the area, we employ two datasets, MNIST and CIFAR-10, and consider two distinct NN architectures: a Convolutional Neural Network (CNN) and a Residual Network (ResNet-18). We briefly review the characteristics of each network.

CNNs, a general and customizable model architecture, are commonly used for image classification on MNIST and CIFAR-10 datasets. The basic model we implemented consists of three convolutional layers and three max pooling layers, followed by four fully connected layers in a funnel-like shape (see Sect. 4.3) and one output layer. As we discussed earlier, we add DP-compliant noise to the penultimate layer, deemed sanitization layer. During training, a dropout layer was added before the sanitization layer to mitigate the effect of added noise and its impact on overfitting. Dropout ratio p is defined as a hyper parameter to balance between controlling sensitivity (and thus required noise magnitude) on the one hand, and improving accuracy (i.e., avoiding overfitting) on the other hand.

The ResNet-18 architecture (Residual Network with 18 layers) is a popular deep neural network introduced by He et al. [10] and consists of a series of convolutional and residual blocks. In the case of ResNet-18, we also perform clipping and noise addition at the penultimate layer output, which in this case is a Global Average Pooling (GAP) layer: a 1D tensor with 512 elements. In ResNet, this tensor typically represents high-level semantic features extracted from the input image. Each element in the tensor can be viewed as a summary or an aggregated value that captures the presence of specific patterns, textures, or object parts in the image.

In the context of image classification tasks (which ResNet-18 is often used for), these features are meant to be discriminative enough to distinguish between different classes. The 512-element tensor serves as a compressed representation of the most relevant information extracted from the entire image. We apply our proposed sanitization mechanism directly to this 1D tensor and we set $n = 512$.

5.2 Results

First, we focus on the privacy strength achieved by our approach, which follows the (ϵ, δ)-differential privacy model, and is subject to the moments accountant privacy budget accounting technique. The moments accountant technique takes as input the noise multiplier (σ), the size of training data and number of training steps (i.e., epochs), and calculates the consumed privacy budget. We considered a broad range of noise multiplier values from 1 to 5, with different values for δ (a typical suggested setting in literature for δ is one over dataset cardinality). For instance, by setting noise multiplier to 1 and $\delta = 10^{-5}$ for MNIST, our method consumed privacy budget of $\epsilon = 1.0$. Figures 4a and 4b illustrate the privacy budget consumption for MNIST and CIFAR-10 using the CNN architecture after 100 epochs of private training. Note that, in most cases, our approach requires a privacy budget of less than 0.5, which is far better than benchmarks, which in some cases use a budget well in excess of 1 (and, in some cases, up to 10). Our method incurred a privacy consumption of 1.4 in the worst case, for the least amount of noise magnitude and highest δ.

At the same time, the validation accuracy of our approach is high. Figures 5a and 5b show the accuracy results for MNIST and CIFAR-10 datasets, respectively, using the CNN architecture. Recall from Sect. 4 that the accuracy of our approaches is heavily influenced by the width of the sanitization layer n: while a larger value preserves more information, it also increases the sensitivity in the case of element-wise clipping, thus requiring a higher noise magnitude for the same privacy bound ϵ. The results confirm this behavior: an initial increase in n from 10 to 60 increases accuracy (sensitivity grows proportionally to $\sqrt{n}$), but subsequently an excessive growth in n leads to higher noise, and the validation accuracy drops. Our results show that a sanitization layer width in the range n = 40–60 is the best choice for both MNIST and CIFAR-10. In terms of absolute validation accuracy values, our approach obtained 98% accuracy on MNIST data and close to 60% on CIFAR-10, clearly outperforming conventional DP-SGD (side-by-side benchmark results are shown in Table 2).

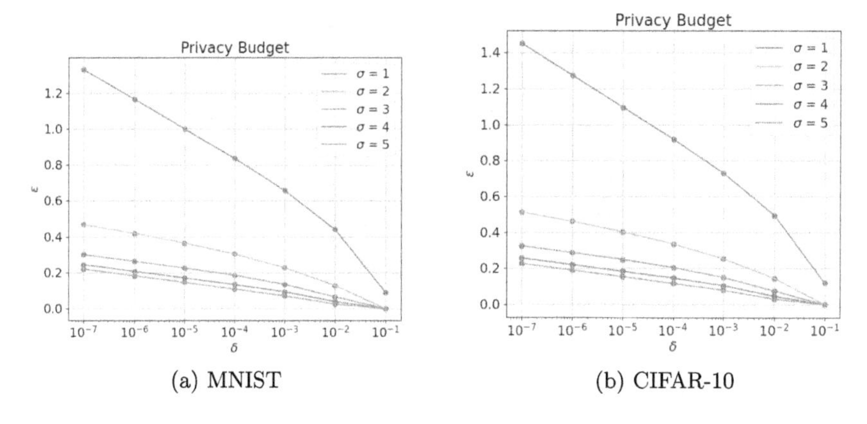

Fig. 4. Privacy Budget Consumption for Proposed Approach

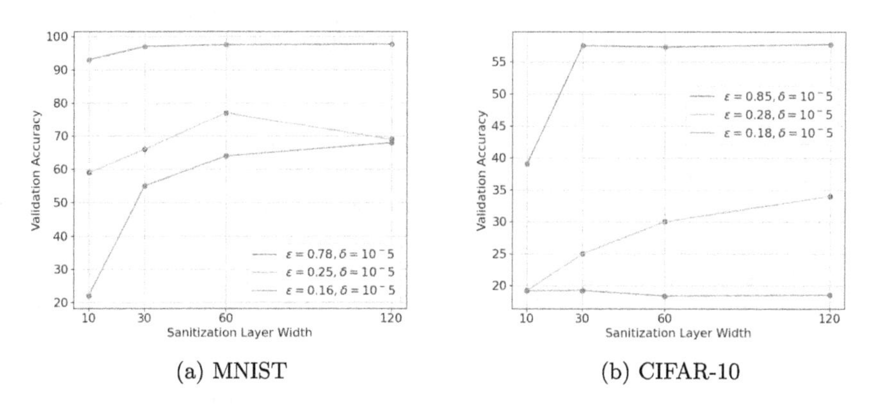

Fig. 5. Accuracy with Varying Sanitization Layer Width

Next, we focus on the training convergence of our approach for both CNN and ResNet architectures. As shown in Fig. 6, training converges well in all cases. For RESNET-18 trained on CIFAR-10, which is the most challenging of the two datasets considered, we fixed $\sigma = 1$ and clipping threshold $C = 0.5$ with a batch size of 250. The obtained validation accuracy was 78.37% satisfying $(4, 10^{-}5)) - DP$ after 82 training epochs and 79% satisfying $(4.46, 10^{-}5)) - DP$ after 100 training epochs.

Comparison with Benchmarks. Tables 2 and 3 show side-by-side accuracy results for similar privacy bounds of our approach compared to most prominent existing benchmarks. We provide results for both MNIST and CIFAR-10, and both considered architectures, namely CNN and ResNet-18.

Conventional DP-SGD [1] lags behind our approach, underperforming by 18% on MNIST and by 15% on CIFAR-10. The work in [3], which introduced adaptivity in the privacy budget being used by the DP-SGD algorithm in each epoch, achieved 90% and 45% accuracy for MNIST and CIFAR-10, respectively.

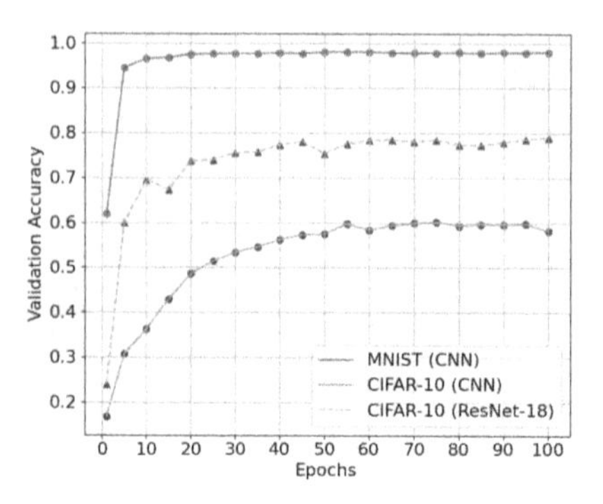

Fig. 6. Convergence on CNN and ResNet-18 Architectures

Having privacy guarantees of $\epsilon = 1$ and $\delta = 10^{-5}$, the DPNASNET [4] approach was able to achieve an accuracy of 97.22% and 52.95% on MNIST and CIFAR-10 respectively. For the same privacy guarantees, our approach outperforms DPNASNET by 1% on MNIST and 10% on CIFAR-10. Another recent adaptive privacy preserving deep learning framework proposed in [20] introduced a way of decaying the noise magnitude on the gradient along time, trying to mitigate the effect of the added noise. Under privacy bounds of $\epsilon = 1.19$ and $\delta = 10^{-}5$, they achieved an accuracy of 97.7%.

The work in [2] measured the effect of different activation functions on the performance of DP-SGD, and they found that the use of bounded *relu* or *tanh* functions is the best option, giving accuracy levels of 96.02% and 44.42% for MNIST and CIFAR-10, respectively, with $\epsilon = 2.0$.

We also tested the ResNet-18 model architecture modified with our methodology and compared its performance against other proposed benchmarks. In the experiments conducted by Remerscheid et al. [15], differentially private training was performed on the ResNet-18 architecture for CIFAR-10, alongside their novel model, SmoothNets, designed specifically to address privacy. The reported accuracy for ResNet-18 was 59.8%, while SmoothNets achieved a higher accuracy of 73.5% with an excessive $\epsilon = 7$ for both models. These results further underscore the superior performance of our algorithm, outperforming all existing models with significantly enhanced privacy assurances. Finally, the work by Xiao et al. [18] proposed a solution to the challenges encountered in DP-SGD by focusing on two key operations: iterative perturbation and gradient clipping. Their approach, called ModelMix, offers an optimization framework that incorporates random aggregation of intermediate model states. They tested their approach with CIFAR-10 on RESNET-20 model architecture and reported accuracy as high as 79.1% with ($\epsilon = 6.1, \delta = 10^{-}5$). Our approach gave an almost similar accuracy with tighter privacy guarantees.

Table 2. Comparison with Benchmarks on CNN Architecture

Dataset	MNIST		CIFAR-10	
Study	(ϵ, δ)-DP	Accuracy (%)	(ϵ, δ)-DP	Accuracy (%)
Chen et al. [3]	ADAPTIVE	90%	ADAPTIVE	45%
DPNASNet [4]	(1, 10^-5)-DP	97.22%	(1, 10^-5)-DP	52.95%
Ayoub et al. [2]	(1.43, 10^-5)-DP	96.02%	(2.20, 10^-5)-DP	44.42%
DP-SGD	(1, 10^-5)-DP	80%	(1.096, 10^-5)-DP	45%
Our Study	**(1, 10^-5)-DP**	**98%**	**(1.096, 10^-5)-DP**	**60%**

Table 3. Comparison with Benchmarks on ResNet-18 Architecture (CIFAR-10)

Study	(ϵ, δ)-DP	Accuracy	Model
Xiao et al. [18]	(6.1, 10^-5)-DP	79.1%	ResNet-20
Remerscheid et al. [15]	($\epsilon = 7$)-DP	59.8%	ResNet-18
Remerscheid et al. [15]	($\epsilon = 7$)-DP	73.5%	SmoothNets
Our Study	**(4, 10^-5)-DP**	**78.37%**	ResNet-18
	(4.46, 10^-5)-DP	**79%**	

6 Conclusion

We proposed an innovative approach to perform differentially-private training of neural networks when input data have private features and public labels. We investigated alternative techniques to perform clipping and noise addition, markedly different than the existing DP-SGD benchmark. Specifically, our approach is more flexible, and is able to better insulate the effect of outliers by performing element-wise clipping and injecting noise at a dedicated sanitization layer, as opposed to ℓ_2-norm clipping and noise addition at the gradient. We investigated emerging trade-offs, and identified NN architecture modifications that lead to better accuracy. Extensive experimental results show that our accuracy is superior to benchmarks on multiple datasets and architectures. In the future, we plan to investigate avenues for additional accuracy gains, through a fine-tuned setting of clipping thresholds and non-uniform budget allocation across epochs. Furthermore, we will research private strategies to stop early the training process when an accuracy plateau is reached, in order to save privacy budget.

References

1. Abadi, M., et al.: Deep learning with differential privacy. In: Proceedings of ACM SIGSAC Conference on Computer and Communications Security, pp. 308–318 (2016)
2. Arous, A., Guesmi, A., Hanif, M., Alouani, I., Shafique, M.: Exploring machine learning privacy/utility trade-off from a hyperparameters lens. In: International Joint Conference on Neural Networks (IJCNN) (2023)

3. Chen, C., Lee, J.: Stochastic adaptive line search for differentially private optimization. In: 2020 IEEE International Conference on Big Data (Big Data), pp. 1011–1020 (2020)
4. Cheng, A., Wang, J., Zhang, X.S., Chen, Q., Wang, P., Cheng, J.: DPNAS: neural architecture search for deep learning with differential privacy. In: Proceedings of the AAAI Conference on Artificial Intelligence, vol. 36, no. 6, pp. 6358–6366 (2022)
5. Davody, A., Adelani, D.I., Kleinbauer, T., Klakow, D.: On the effect of normalization layers on differentially private training of deep neural networks. arXiv preprint arXiv:2006.10919 (2020)
6. Dwork, C.: Differential privacy. In: Bugliesi, M., Preneel, B., Sassone, V., Wegener, I. (eds.) ICALP 2006. LNCS, vol. 4052, pp. 1–12. Springer, Heidelberg (2006). https://doi.org/10.1007/11787006_1
7. Ghazi, B., Golowich, N., Kumar, R., Manurangsi, P., Zhang, C.: Deep learning with label differential privacy. In: Neural Information Processing Systems (NeurIPS), pp. 27131–27145 (2021)
8. Gong, M., Xie, Y., Pan, K., Feng, K., Qin, A.K.: A survey on differentially private machine learning. IEEE Comput. Intell. Mag. **15**(2), 49–64 (2020)
9. Gower, R.M., Loizou, N., Qian, X., Sailanbayev, A., Shulgin, E., Richtárik, P.: SGD: general analysis and improved rates. In: Chaudhuri, K., Salakhutdinov, R. (eds.) International Conference on Machine Learning (ICML), vol. 97, pp. 5200–5209 (2019)
10. He, K., Zhang, X., Ren, S., Sun, J.: Deep residual learning for image recognition. In: IEEE Conference on Computer Vision and Pattern Recognition, pp. 770–778 (2016)
11. Koskela, A., Honkela, A.: Learning rate adaptation for differentially private learning. In: International Conference on Artificial Intelligence and Statistics, pp. 2465–2475 (2020)
12. Mironov, I.: Rényi differential privacy. In: 2017 IEEE 30th Computer Security Foundations Symposium (CSF), pp. 263–275 (2017)
13. Nguyen, T.N., Nguyen, P.H., Nguyen, L.M., Van Dijk, M.: Batch clipping and adaptive layerwise clipping for differential private stochastic gradient descent. arXiv preprint arXiv:2307.11939 (2023)
14. Priyanshu, A., Naidu, R., Mireshghallah, F., Malekzadeh, M.: Poster: efficient hyperparameter optimization for differentially private deep learning (2021)
15. Remerscheid, N.W., Ziller, A., Rueckert, D., Kaissis, G.: SmoothNets: optimizing CNN architecture design for differentially private deep learning. arXiv preprint arXiv:2205.04095 (2022)
16. Shokri, R., Stronati, M., Song, C., Shmatikov, V.: Membership inference attacks against machine learning models. In: 2017 IEEE Symposium on Security and Privacy (SP), pp. 3–18 (2016)
17. Tian, Y., Zhang, Y., Zhang, H.: Recent advances in stochastic gradient descent in deep learning. Mathematics **11**, 682 (2023)
18. Xiao, H., Wan, J., Devadas, S.: Differentially private deep learning with modelmix. arXiv arXiv:abs/2210.03843 (2022)
19. Zhang, X., Bu, Z., Wu, S., Hong, M.: Differentially private SGD without clipping bias: an error-feedback approach. In: The Twelfth International Conference on Learning Representations (2024)
20. Zhang, X., Ding, J., Wu, M., Wong, S.T., Van Nguyen, H., Pan, M.: Adaptive privacy preserving deep learning algorithms for medical data. In: IEEE/CVF Winter Conference on Applications of Computer Vision, pp. 1169–1178 (2021)

Time Series

Series2Graph++: Distributed Detection of Correlation Anomalies in Multivariate Time Series

Phillip Wenig[1(✉)] and Thorsten Papenbrock[2]

[1] Hasso Plattner Institute, University of Potsdam, Potsdam, Germany
phillip.wenig@hpi.de
[2] Philipps University of Marburg, Marburg, Germany
papenbrock@informatik.uni-marburg.de

Abstract. Multivariate time series are a form of real-valued sequence data that simultaneously record different time-dependent variables. They originate mostly from multi-sensor setups and serve a variety of important analytical purposes, including the detection of normal and abnormal behavior. To efficiently detect both single-channel and correlation anomalies in time series of real-world size, we propose Series2Graph++ (S2G++), an unsupervised, distributable anomaly detection algorithm for wide and long multivariate time series. S2G++ extends the univariate S2G algorithm and its distributed variant DADS by adding support for multidimensional time series and, hence, correlation anomalies. For this, we translate S2G's graph-based anomaly detection approach into multidimensional spaces. We additionally propose a root cause feature that serves to relate the detected anomalies to the anomalous channel(s). Our experiments demonstrate that S2G++ is significantly faster than related algorithms and still competes for the best quality results.

1 Introduction

Time Series (TS) analytics on multi-sensor systems is becoming increasingly popular [1–3]. As a result, more and more TS data is being produced. Much of this TS data stems from multiple sensors that simultaneously record different variables that are time related and, hence, feed into the same TS. The effect of TSs becoming increasingly multidimensional is accompanied by the fact that recorded TSs also become ever longer. For TS analytical algorithms, keeping up with this development is a challenge [4]. To tackle issues such as performance and root cause identification in high-dimensional TS anomaly detection, we propose the novel TS anomaly detector Series2Graph++, which is an unsupervised multivariate anomaly detection algorithm. S2G++' anomaly detection is based on a clever graph encoding of the TS [5] and can be distributed on a computer cluster for improved scalability [6]. Our contributions in S2G++ enable the algorithm to detect anomalies in *multivariate* and *particularly long* TS; they also allow S2G++ to indicate the channels with the highest influence on a reported anomaly.

The remainder of this work is organized as follows. Section 2 first gives a brief overview of current anomaly detection approaches. In Sect. 3, we introduce the S2G++

R. Wrembel et al. (Eds.): DaWaK 2024, LNCS 14912, pp. 225–230, 2024.
https://doi.org/10.1007/978-3-031-68323-7_17

algorithm and report on our experiments in Sect. 4. Section 5 concludes this paper with a critical summary.

2 Related Work

Series2Graph++ is an unsupervised and distributable Time Series Anomaly Detection (TS-AD) algorithm that can find different types of anomalies in TS with multiple channels. The algorithm extends the existing univariate anomaly detection algorithm Series2Graph (S2G) [5] and its distributed adaptation called Distributed Anomaly Detection System (DADS) [6]. For a general overview of univariate (and multivariate) anomaly detection algorithms, we refer to comprehensive surveys [7–9] and an exhaustive evaluation [4].

The Series2Graph (S2G) [5] algorithm is an unsupervised anomaly detector that transforms a univariate input TS into a weighted graph representation that represents anomalous sub-sequences as low-weighted paths in the graph. The algorithm reduces each sub-sequence to a three dimensional vector with the PCA [10] algorithm. S2G, then, rotates the reduced sub-sequences around the origin of a 2-dimensional space. S2G creates a directed graph by down-sampling the 2-dimensional space. The edges represent transitions of sub-sequences in the input TS and receive an anomaly score proportional to the number of transitions; edges that are often involved in transitions of sub-sequences have smaller anomaly scores than rarely involved ones. S2G++ works in a very similar way, but is able to process *multivariate* input TS by projecting every TS channel separately and combining the 2-dimensional spaces into one space. DADS [6] is an extension of S2G that enables the algorithm to run in parallel and distributedly on a computer cluster while guaranteeing the same results as the single machine version. S2G++'s distribution follows a similar strategy, but works also on multivariate TS.

Some algorithms for TS-AD are able to relate the discovered anomalies to their causes, i.e., positions and channels in the input TS [11–14]. However, they are neural network approaches that require a significant amount of training time and data. S2G++ does not require training data and explains the importance of every channel for every single anomalous sub-sequence in the input TS.

3 Series2Graph++

Series2Graph++ (S2G++) is a multivariate extension to the S2G algorithm that can also indicate which channels in a TS contribute to a detected anomaly. S2G++ can be run in a parallelized and distributed way with a distribution strategy similar to DADS [6] that yields the same results independent of the degree of parallelization. For univariate TS inputs, S2G++ yields the same results as S2G.

Through sliding windows, S2G++ creates for every channel in the TS a so-called *phase space* matrix, as S2G does. We stack these matrices into a phase space tensor. For every channel of the TS (and similar to S2G), S2G++ calculates a two-dimensional rotated embedding by reducing the sub-sequences with the PCA [10] algorithm. Because the embeddings are time-wise related, we can stack them in a single matrix, in which all corresponding, two-dimensional embedded points are combined into one high-dimensional point. Similar multidimensional subsequences still appear close to each other in the rotated and stacked space. The embedding from the previous step represents a multidimensional TS that is rotated around the origin of a high-dimensional space. Then, S2G++ draws high-dimensional hyperplanes that equally span in the space of the rotated TS embedding. For every transition in the TS, S2G++ determines the high-dimensional intersection points between the transition embeddings and the hyperplanes. S2G++ clusters the intersection points of every hyperplane into groups that represent graph nodes. The clustering uses a proximity-based approach that automatically finds a suitable number of clusters. S2G and DADS both use KDE [15] clustering for this purpose. KDE can technically estimate densities in multidimensional data as well, but scanning a high-dimensional grid does not scale well with a growing number of dimensions. To circumvent this issue, S2G++ uses a custom, multidimensional and linearly scaling version of KDE clustering that performs KDE independently on every channel, combines the dimensions to high-dimensional peaks, and only keeps those peaks that are close to actual embeddings. S2G++ connects the nodes via edges to a graph representation of the input TS. To this goal, S2G++ iterates over all transitions following their rotation order and generates an edge between each two contiguous graph nodes that are associated with the current iteration step. The final step in S2G++ consolidates the nodes and edges into a single graph and aggregates all edges between same source and target nodes into one edge that receives the number of aggregated edges as score. The more edges are aggregated into one, the more normal the corresponding sub-sequences are. The final anomaly score indicates the degree of normality for every point in the TS depending on the normality of corresponding sub-sequences and edges.

S2G++ returns an anomaly score that indicates at what point in time anomalies occur. Additionally, it offers a feature that scores the TS channels by their contribution to the overall anomaly score. We recall the node estimation step clustered all intersections of a hyperplane (X) to estimate the cluster centers ($\bar{X}$), which are then used to build graph nodes. We refer to a single value of one intersection x in dimension c as x_c and to all intersection values x_c of one hyperplane in a single dimension $0 \leq c < d$ as X_i. The formula $X_c(\bar{x}) = \{x_c | x_c \in X_c \wedge x_c \ contributes \ to \ density \ \bar{x}_c\}$ returns all intersection values of X_i that belong to the cluster of $\bar{x}$. We count these assigned intersections and normalize this count by the number of all intersections with the current hyperplane to calculate the relevance of that dimension for the graph node $\bar{x}$. More specifically, the *anomaly contribution score* of dimension c for a graph node $\bar{x}$ is: $e_c(\bar{x}) = 1 - \frac{|X_c(\bar{x})|}{|X_c|}$ The more intersections belong to the same cluster coordinate $\bar{x}$, the lower the score gets. An anomalous cluster center has, therefore, fewer assignees in the anomalous dimension, which corresponds to a TS channel.

4 Experiments

We assess the quality with the ROC curve (AUROC) score [17]. S2G++ is a Python-wrapped Rust program[1] that is based on the actor model library *Actix* and its distribution extension *Actix-Telepathy* [18]. We use the real-world *Exathlon* data collec-

[1] https://github.com/HPI-Information-Systems/S2Gpp.

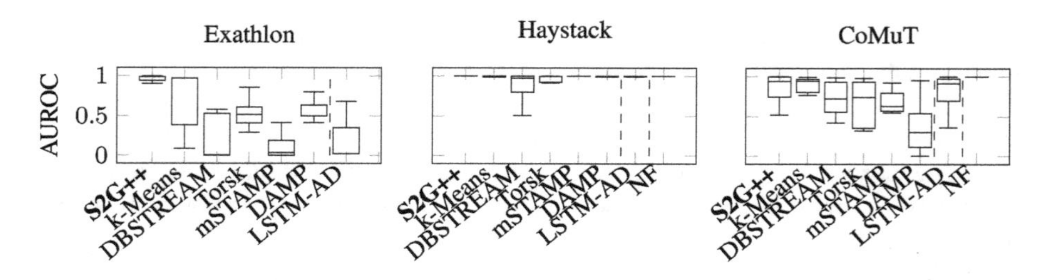

Fig. 1. AUROC results of unsupervised, semi-supervised, and supervised anomaly detectors on the Exathlon [16], Haystack, and COMUT dataset collections.

tion [16] (average length 47 530; average channel count 45) and generated[2] multivariate datasets with the *GutenTAG* tool [19]: *haystack* (3 to 20 channels; single channel and combined anomalies; length of 10 000), *COMUT* (2 to 4 channels; only correlation anomalies; length of 10 000), and *scalability* (1 and 100 channels; lengths between 10 000 and 5 120 000). We compare the performance of S2G++ with k-Means [20], Torsk [21], MSTAMP [22], DAMP [23] and DBSTREAM [24], the semi-supervised LSTM-AD [25], and the supervised NF [26].

As reported in Fig. 1, S2G++ competes well on all datasets; a slightly better approach sometimes exists, but the scores are reliably good and the runtime usually by far best. On the real-world dataset collection Exathlon [16], S2G++ achieved a mean and median AUROC of 0.93 in $\sim$ 4 min while k-Means achieved a mean AUROC of only 0.72 and a median AUROC of 0.96 in $\sim$ 29 min. For many datasets, Normalizing Flow (NF) and LSTM-AD exceed the time limits. On the *COMUT* TS collection that contains correlation anomalies – a particular difficult type of anomaly that only multivariate algorithms can detect [9], no unsupervised algorithm yielded a perfect result. S2G++ and k-Means still perform comparatively very well.

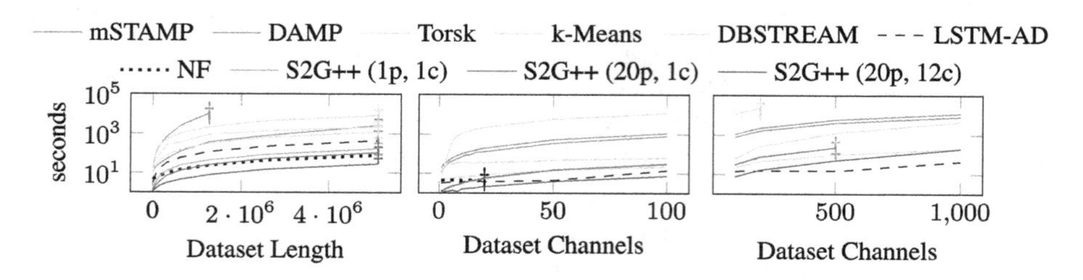

Fig. 2. Execution times of eight anomaly detectors on multivariate time series with increasing lengths (left) and increasing numbers of channels (right). OOMs (‡) and TOs (†) are marked at the last successful experiment of the corresponding algorithms.

The scalability results w.r.t. the TS lengths (Fig. 2 left) show that S2G++ is clearly the fastest algorithm in the experiment: The algorithm processes 5 million points in $\sim$ 3

[2] https://github.com/HPI-Information-Systems/s2gpp_experiments.

min (single threaded), $\sim$ 2 min (parallel) and $\sim$ 30 s (distributed), respectively. Only the (semi-)supervised algorithms LSTM-AD and NF can compete with these runtimes, because we do not count their training times. In the two right charts of Fig. 2, we see the algorithms' runtimes for a TS with 10 000 points and an increasing number of channels. The NF algorithm stops relatively early, because its training time takes too long. The k-Means algorithm scales very well and surpasses the non-distributed S2G++ after 50 channels. The distributed S2G++ still manages to outperform k-Means by taking less then 8 s for 100 channels. If we do not consider training times, LSTM-AD is the overall fastest algorithm. For TSs with a practically realistic number of channels (usually well below 100), S2G++ is the fastest approach in our experiments.

On the *haystack* collection, we also evaluate S2G++'s root cause feature that scores the TS channels by their impact on the anomaly score. We found that, out of the 24 TSs, S2G++ correctly indicated the anomalous channel in all TSs.

5 Conclusion

Our S2G++ algorithm is an accurate, reliable and fast algorithm for anomaly detection in multivariate TS and root-cause analysis. It surpasses many established algorithms in quality, while being much faster in many scenarios. The high efficiency of S2G++ allows data scientists to run the algorithm on relatively large TS; the efficiency also supports exploratory and interactive anomaly detection sessions. As future work, we aim to investigate automatic self-tuning strategies for S2G++ and investigate dimensionality reduction methods that rotate the entire multidimensional space at once.

Acknowledgment. The work was funded by the German government as part of the LuFo VI call I program (Luftfahrtforschungsprogramm) under the grant number 20D1915. The management of Rolls-Royce Deutschland Ltd. & Co. KG is gratefully acknowledged for supporting the work and permitting the presentation of results.

References

1. Liu, D., Kim, H., Kim, S.-H., Kim, T., Lee, D., Xie, Y.: Distribution-free multivariate time-series monitoring with analytically determined control limits. Int. J. Prod. Res. **61**(20), 6960–6977 (2023)
2. Sprint, G., Cook, D., Weeks, D., Dahmen, J., La Fleur, A.: Analyzing sensor-based time series data to track changes in physical activity during inpatient rehabilitation. Sensors **17**(10), 2219 (2017)
3. Derse, C., El Baghdadi, M., Hegazy, O., Sensoz, U., Gezer, H., Nil, M.: An anomaly detection study on automotive sensor data time series for vehicle applications. In: 2021 Sixteenth International Conference on Ecological Vehicles and Renewable Energies (EVER) (2021)
4. Schmidl, S., Wenig, P., Papenbrock, T.: Anomaly detection in time series: a comprehensive evaluation. Proc. VLDB Endow. **15**(9), 1779–1797 (2022)
5. Boniol, P., Palpanas, T.: Series2Graph: graph-based subsequence anomaly detection for time series. Proc. VLDB Endow. **13**(12), 1821–1834 (2020)
6. Schneider, J., Wenig, P., Papenbrock, T.: Distributed detection of sequential anomalies in univariate time series. VLDB J. **30**(4), 579–602 (2021)

7. Braei, M., Wagner, S.: Anomaly detection in univariate time-series: a survey on the state-of-the-art. arXiv preprint arXiv:2004.00433 (2020)
8. Blázquez-García, A., Conde, A., Mori, U., Lozano, J.A.: A review on outlier/anomaly detection in time series data. arXiv preprint arXiv:2002.04236 (2020)
9. Wenig, P., Schmidl, S., Papenbrock, T.: Anomaly detectors for multivariate time series: the proof of the pudding is in the eating. In: Proceedings of the 1st International Workshop on Multivariate Time Series Analytics (2024)
10. Hotelling, H.: Analysis of a complex of statistical variables into principal components. J. Educ. Psychol. **24**(6), 417 (1933)
11. Assaf, R., Giurgiu, I., Pfefferle, J., Monney, S., Pozidis, H., Schumann, A.: An anomaly detection and explainability framework using convolutional autoencoders for data storage systems. In: Proceedings of the Twenty-Ninth International Joint Conference on Artificial Intelligence (2021)
12. Deng, A., Hooi, B.: Graph neural network-based anomaly detection in multivariate time series. In: AAAI (2021)
13. Siddiqui, S.A., Mercier, D., Munir, M., Dengel, A., Ahmed, S.: TSViz: demystification of deep learning models for time-series analysis. IEEE Access **7**, 67027–67040 (2019)
14. Munir, M., Siddiqui, S.A., Küsters, F., Mercier, D., Dengel, A., Ahmed, S.: TSXplain: demystification of DNN decisions for time-series using natural language and statistical features. In: Tetko, I.V., Kůrková, V., Karpov, P., Theis, F. (eds.) Artificial Neural Networks and Machine Learning – ICANN 2019: Workshop and Special Sessions (2019)
15. Scott, D.W.: Multivariate density estimation and visualization. In: Gentle, J., Härdle, W., Mori, Y. (eds.) Handbook of Computational Statistics. SHCS, pp. 549–569. Springer, Heidelberg (2012). https://doi.org/10.1007/978-3-642-21551-3_19
16. Jacob, V., Song, F., Stiegler, A., Rad, B., Diao, Y., Tatbul, N.: Exathlon: a benchmark for explainable anomaly detection over time series. Proc. VLDB Endow. **14**(11), 2613–2626 (2021)
17. Hanley, J.A., McNeil, B.J.: The meaning and use of the area under a receiver operating characteristic (ROC) curve. Radiology **143**(1), 29–36 (1982)
18. Wenig, P., Papenbrock, T.: Actix-telepathy. In: Proceedings of the 10th ACM SIGPLAN International Workshop on Reactive and Event-Based Languages and Systems (2023)
19. Wenig, P., Schmidl, S., Papenbrock, T.: TimeEval: a benchmarking toolkit for time series anomaly detection algorithms. Proc. VLDB Endow. **15**(12), 3678–3681 (2022)
20. Yairi, T., Kato, Y., Hori, K.: Fault detection by mining association rules from house-keeping data. In: Proceedings of the 6th International Symposium on Artificial Intelligence, Robotics and Automation in Space (2001)
21. Heim, N., Avery, J.E.: Adaptive anomaly detection in chaotic time series with a spatially aware echo state network. arXiv preprint arXiv:1909.01709 (2019)
22. Yeh, C.-C.M., Kavantzas, N., Keogh, E.: Matrix profile VI: meaningful multidimensional motif discovery. In: 2017 IEEE International Conference on Data Mining (ICDM) (2017)
23. Lu, Y., Wu, R., Mueen, A., Zuluaga, M.A., Keogh, E.: Matrix profile XXIV: scaling time series anomaly detection to trillions of datapoints and ultra-fast arriving data streams. In: Proceedings of the ACM SIGKDD Conference on Knowledge Discovery and Data Mining (SIGKDD) (2022)
24. Hahsler, M., Bolaños, M.: Clustering data streams based on shared density between micro-clusters. IEEE Trans. Knowl. Data Eng. **28**(6), 1449–1461 (2016)
25. Malhotra, P., Vig, L., Shroff, G., Agarwal, P., et al.: Long short term memory networks for anomaly detection in time series. In: Proceedings of the European Symposium on Artificial Neural Networks, Computational Intelligence and Machine Learning (ESANN) (2015)
26. Ryzhikov, A., Borisyak, M., Ustyuzhanin, A., Derkach, D.: Normalizing flows for deep anomaly detection. arXiv preprint arXiv:1912.09323 (2019)

Anomaly Detection from Time Series Under Uncertainty

Paul Wiessner[1(✉)], Grigor Bezirganyan[2], Sana Sellami[2], Richard Chbeir[3], and Hans-Joachim Bungartz[1]

[1] Department of Informatics, Technische Universität München, Garching, Germany
wiessner@in.tum.de, bungartz@cit.tum.de
[2] Aix Marseille Univ, CNRS, LIS, Marseille, France
grigor.bezirganyan@etu.univ-amu.fr, sana.sellami@univ-amu.fr
[3] Univ Pau & Pays Adour, E2S-UPPA, LIUPPA, EA3000, Anglet, France
richard.chbeir@univ-pau.fr

Abstract. Anomalies in data can cause potential issues in downstream tasks, making their detection critical. Data collection processes for continuous data are often defective and imprecise. For example, sensors are resource-constrained devices, raising questions about their reliability. This imprecision in measurements can be characterized as noise. In machine learning, noise is referred to as data (aleatoric) uncertainty. Additionally, the model itself introduces a second layer of uncertainty, known as model (epistemic) uncertainty. In this paper, we propose an LSTM Autoencoder that quantifies both data and model uncertainty, enabling a deeper understanding of noise recognition. Our experimental results across different real-world datasets show that consideration of uncertainty effectively increases the robustness to noise and point outliers, making predictions more reliable for longer periodic sequential data.

Keywords: Anomaly detection · Uncertainty quantification · Time series · Deep Neural Networks · Bayesian network

1 Introduction

Anomaly detection in time series [18] is a challenging task spreading over various domains. Presence of uncertainty further complicates the anomaly detection in several ways [6], as it obscures the distinction between normal and anomalous behavior and thus affects the reliability of model predictions. Indeed, uncertainty [1] encompasses ambiguity or lack of information about the data, distinguishing it from anomalies, which represent deviations from expected patterns. Traditional anomaly detection techniques [2] struggle to adapt to these uncertainties, leading to either missed anomalies or a high rate of false positives.

The goal of this work is to find anomalies while taking uncertainty, such as measurement impreciseness, i.e. noise, into consideration. We can differ uncertainty into aleatoric and epistemic uncertainty. The first describes the inherent

R. Wrembel et al. (Eds.): DaWaK 2024, LNCS 14912, pp. 231–238, 2024.
https://doi.org/10.1007/978-3-031-68323-7_18

uncertainty in data. The measure of aleatoric uncertainty assigns this impreciseness a value describing the inherent randomness in data. Then, we leverage an algorithm to find anomalies in data. Normally, the output of an algorithm is a guess for the likelihood that a given observation is an anomaly. What remains behind the scenes is the question: How certain is the model with this guess? This uncertainty introduced by the model is referred to as the epistemic uncertainty. Considering both uncertainties is important. On the one side, it can make the model more robust to noise. On the other, it acknowledges and alleviates shortcomings of a decision model.

In this work, we explore a way to consider both aleatoric and epistemic uncertainties for anomaly detection in time series data. Based on existing works [11,19], our approach constructs an LSTM (Long Short-Term Memory) [9] Autoencoder. Further, we implement a solution for separate aleatoric and epistemic uncertainty quantification. We argue that involving uncertainty supports the anomaly detection process to achieve more precise predictions. Consequently, we aim to have a closer look on uncertainty and how it behaves in presence of anomalies. The rest of this paper is structured as follows. Section 2 describes work related to this study. Section 3 describes our approach on anomaly detection under uncertainty. Section 4 presents experimental results. Finally, we draw a conclusion in Sect. 5.

2 Related Work

With the rise of deep neural networks various methods have been developed to approach anomaly detection in time series [7,21]. [3] uses a Long Short-Term Memory (LSTM) network to predict healthy electrocardiogram signals. A further variation of LSTMs has been developed by [15] where LSTM layers are stacked. It is supposed to be trained on normal data and uses multiple predictions in the future to find anomaly based on prediction error.

Autoencoders [5] try to compress the information into a lower dimensional (latent) space, to extract the most relevant features. [17] uses this approach to detect anomalies from satellite telemetry and artificial data.

Finally, there are approaches that combine several presented techniques. [14] presents a combination of an LSTM with encoder and decoder for multi-sensor time series data.

Uncertainty is often modeled through probabilistic approaches. However, probabilistic approaches alone fail to properly distinguish between *aleatoric* and *epistemic* uncertainties [10].

Bayesian deep learning provides tools for uncertainty quantification of deep learning models [11]. It was also applied to autoencoders. [13] shows in an empirical study how integration of uncertainty in the score function is likely to improve anomaly detection for general datasets.

[4] uses a LSTM and turns it into Bayesian network to predict anomalies in satellite telemetry data comparing various kinds of posterior approximations. [19] employs an LSTM autoencoder with point-wise prediction including model uncertainty calculation.

Most approaches focus on quantifying epistemic uncertainty, neglecting the inherent data uncertainty. [11] proposed to use an additional network output to quantify aleatoric uncertainty in computer vision. A method for calculating the predictive uncertainty, which can be separated into model and data uncertainties, is provided using negative log likelihood (NLL) [12,22].

In comparison, we focus in this work on especially separately involving data uncertainty as well as model uncertainty into the anomaly detection process accounting for inherent randomness in data.

3 Proposed Approach

In this section, we describe the approach we developed, called LSTMAE-UQ[1] (Long Short-Term Memory Autoencoder with Aleatoric and Epistemic Uncertainty Quantification), to address the problem of detecting uncertain anomalies in univariate time series data.

LSTMAE-UQ, shown in Fig. 1[2], integrates LSTM cells in the autoencoder architecture and quantifies uncertainties in both data and model. This way, we capture temporal structure of time series data without requiring labels for learning. The combination of LSTM and autencoder can learn to capture relevant features of time series data and aims to reconstruct them properly.

Moreover this approach decouples uncertainties. To this aim, we adapt the model inspired by [11] to output an estimate of data uncertainty as well as epistemic uncertainty using MC Dropout [8].

Loss Function. The model output consists of two values: 1) the reconstruction of the input data, and 2) the estimate of aleatoric uncertainty. We use these two values in a modified version of the mean squared error (MSE) loss function [11]. Reconstruction is used as usual in MSE loss. Special is the usage of aleatoric uncertainty estimate which attributes the MSE a loss resulting in direct influence on algorithm training. For an input vector X, we define the loss in Eq. 1.

$$\mathcal{L}(\theta) = \frac{1}{T} \sum_{i} \frac{1}{2} \hat{\sigma}_i^{-2} ||x_i - \hat{x}_i||^2 + \frac{1}{2} \log \hat{\sigma}_i^2 \tag{1}$$

where: T is the number of Monte Carlo samples, $\hat{x}_i$ is the reconstruction of input value x_i, and σ_i^2 is the variance representing aleatoric uncertainty (the second output of the model).

The minimization objective consists of two parts: 1) the regression part is obtained through stochastic sampling of model parameters (MC Dropout) which represents the difference of reconstructed output and input (weighted by aleatoric uncertainty), and 2) the uncertainty regularization term.

[1] Code is available at https://github.com/p199671/LSTM-AE-UQ.git.

[2] To ease readability, we don't show the averaging over the MC samples for reconstructions and uncertainty for the reconstruction error.

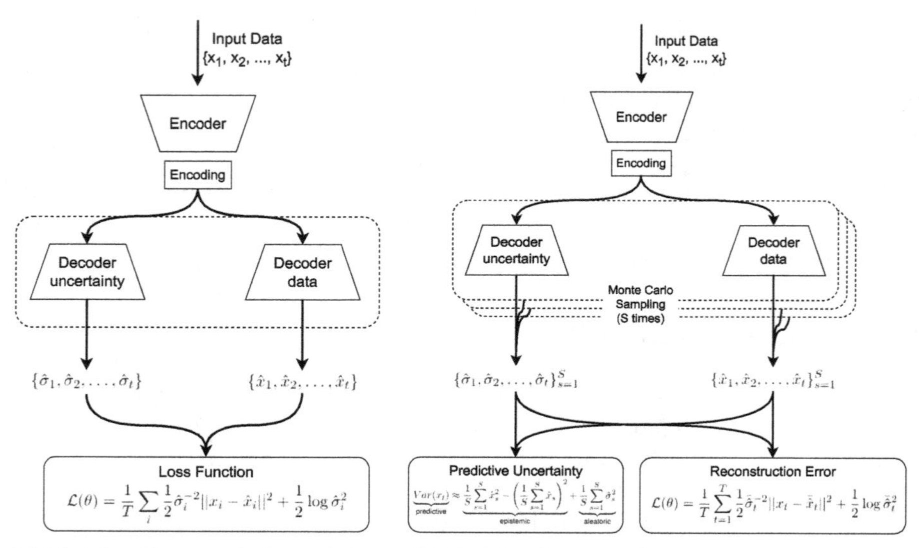

(a) Model architecture during training. (b) Model architecture during prediction.

Fig. 1. LSTMAE-UQ Model Architecture

Additional prediction of uncertainty helps to improve robustness of the network. Note the tempering effect that $\hat{\sigma}^2$ has on the residual discouraging to predict very high or very low uncertainties.

Uncertainty Calculation. Aleatoric uncertainty is learned by the network and served as additional network output. To additionally quantify the epistemic uncertainty, we use Monte Carlo (MC) Dropout. Sampling multiple times from the posterior distribution results in a set $\{\hat{x}_t, \hat{\sigma}_t^2\}_{t=1}^T$ of T sampled outputs.

From that point on, we can derive epistemic uncertainty by calculating the variance over predictions (Eq. 2):

$$\hat{\mu} = \frac{1}{T}\sum_{t=1}^{T}\hat{x}_t^2 - \left(\frac{1}{T}\sum_{t=1}^{T}\hat{x}_t\right)^2 \tag{2}$$

Calculating the mean of uncertainties over all MC samples gives the final aleatoric uncertainty (Eq. 3):

$$\hat{\sigma}^2 = \frac{1}{T}\sum_{t=1}^{T}\hat{\sigma}_t^2 \tag{3}$$

4 Experiments

This section describes the experiments we have conducted in order to evaluate the correlation between uncertainty and anomaly in time series.

4.1 Uncertainty Quantification Evaluation

To investigate uncertainty behaviour during anomalies, we generate different synthetic datasets based on the sine function using the Time Series Generator (GutenTAG) [20].

We create three different pairs of time series - one for each anomaly type. The datasets contain data points according to the mathematical function $sin(x)$. Each includes 100 points per period with overall 10 periods. The training set consists of clean data with little background noise. To each test dataset we inject one anomaly which refers to a larger/smaller amplitude ($200 points$) or an extremum ($1 point$).

Starting with the anomaly of a larger amplitude than normal, observations presented in Fig. 2 show a general trend where larger amplitudes correspond to heightened uncertainties than usual. Notably, the ability of the model to recognise anomaly as uncertainty appears to be dependent on the input size.

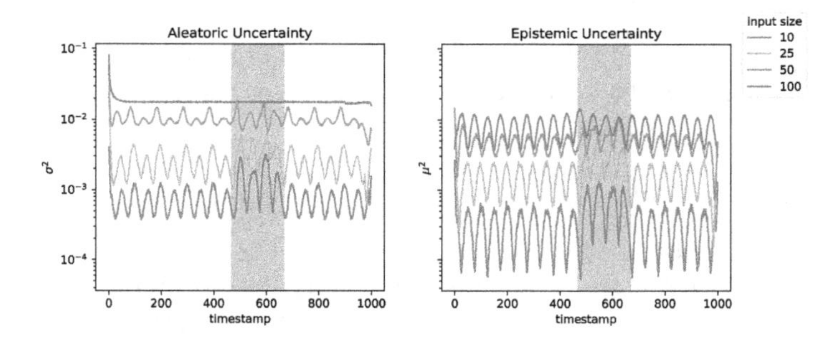

Fig. 2. Results for experiments with anomaly of larger amplitude. The highlighted area in red is the range where the anomaly takes place. (Color figure online)

In the contrary situation (see results in Fig. 3), a smaller amplitude is associated with reduced uncertainties having less intense amplitudes of uncertainty. Also, certain input window sizes don't seem to influence the impact of smaller amplitude anomalies.

During the extremum anomaly (see results in Fig. 4), uncertainty profile shows a distinct behaviour providing an indication of the anomaly's influence. Observations appear mostly for smallest input windows while impact on larger window sizes is minimal.

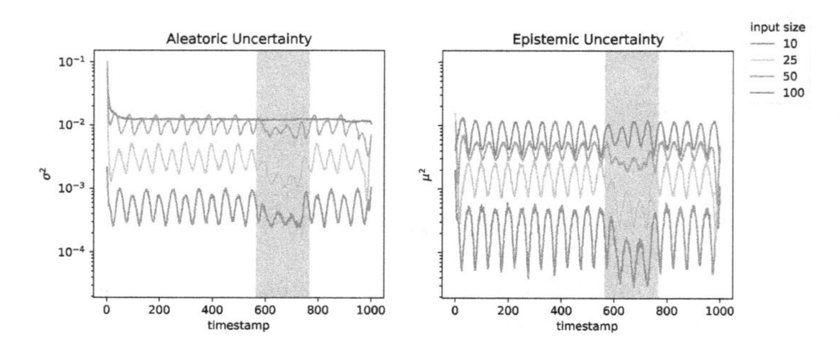

Fig. 3. Results for experiments with anomaly of smaller amplitude. The highlighted area in red is the range where the anomaly takes place. (Color figure online)

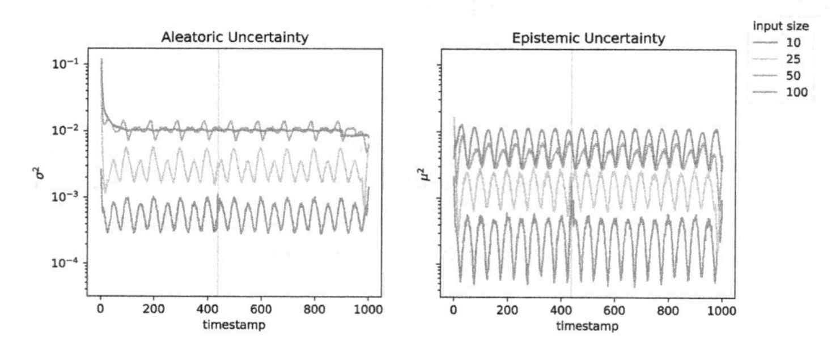

Fig. 4. Results for experiments with anomaly of global extremum. The highlighted area in red is the range where the anomaly takes place. (Color figure online)

4.2 Model Performance

We evaluate the model performance on a variety of real-world univariate time series datasets based on the TSB-UAD[3]. From this, we test on the 18 public real datasets containing 1980 time series, variable in size and domain, with various anomaly types, densities and length.

Figure 5 shows the results of LSTMAE-UQ and its baseline LSTMAE in terms of AUC (Area Under the ROC Curve) and F-score. We note that both algorithms have similar performance across most datasets. On the ECG dataset, the proposed approach is clearly able to outperform its baseline. However, LSTMAE-UQ shows particularly low performance for the YAHOO dataset.

Analysis of the datasets leads to the length of anomaly as main cause. In comparison to other datasets, YAHOO shows the shortest anomaly length [16].

The reason for this could be the consideration of uncertainty in the model. Accounting for uncertainty can cause the model to treat minor anomalies or point anomalies as noise, leading it to ignore them.

[3] https://github.com/TheDatumOrg/TSB-UAD.

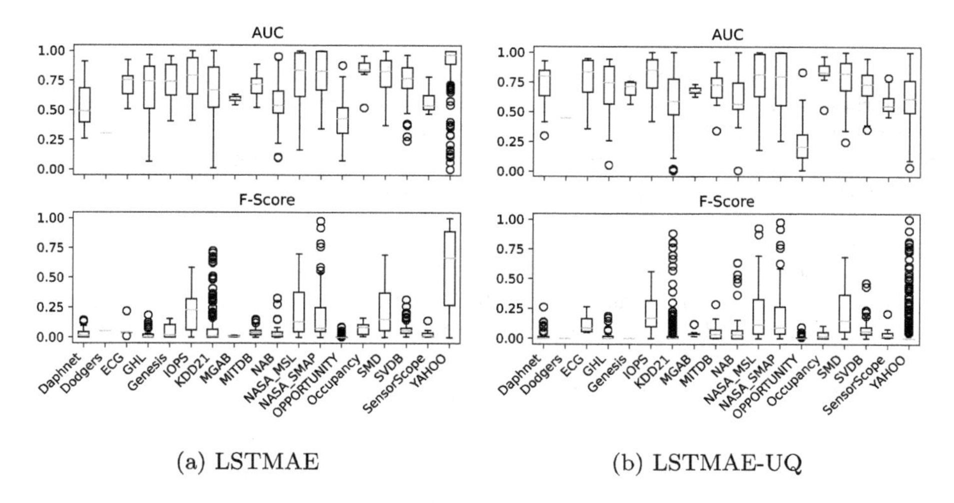

(a) LSTMAE

(b) LSTMAE-UQ

Fig. 5. Performance evaluation of LSTM Autoencoder without and with uncertainty quantification for the real-world datasets of the TSB-UAD.

5 Conclusion

In this work, we developed the LSTMAE-UQ approach to time series anomaly detection under uncertainty. Different to previous approaches, this work takes into consideration both types of uncertainty, aleatoric and epistemic based on the Bayesian posterior approximation method Monte Carlo (MC) Dropout. Experiments conducted on synthetic and real datasets emphasize the need for nuanced approaches in anomaly detection, taking into account the complex interplay between noise, anomalies, and model uncertainty. In future work, we would like to compare our approach with other uncertainty quantification methods and to quantify uncertainty in the IoT environment.

References

1. Abdar, M., et al.: A review of uncertainty quantification in deep learning: techniques, applications and challenges. Inf. Fusion **76**, 243–297 (2021)
2. Chatterjee, A., Ahmed, B.S.: IoT anomaly detection methods and applications: a survey. Internet Things **19**, 100568 (2022)
3. Chauhan, S., Vig, L.: Anomaly detection in ECG time signals via deep long short-term memory networks. In: 2015 IEEE International Conference on Data Science and Advanced Analytics (DSAA), pp. 1–7. IEEE (2015)
4. Chen, J., Pi, D., Wu, Z., Zhao, X., Pan, Y., Zhang, Q.: Imbalanced satellite telemetry data anomaly detection model based on Bayesian LSTM. Acta Astronaut. **180**, 232–242 (2021)
5. Chen, Z., Yeo, C.K., Lee, B.S., Lau, C.T.: Autoencoder-based network anomaly detection. In: 2018 Wireless Telecommunications Symposium (WTS), pp. 1–5. IEEE (2018)

6. Cofta, P., Karatzas, K., Orłowski, C.: A conceptual model of measurement uncertainty in IoT sensor networks. Sensors **21**(5), 1827 (2021)
7. Darban, Z.Z., Webb, G.I., Pan, S., Aggarwal, C.C., Salehi, M.: Deep learning for time series anomaly detection: a survey. CoRR abs/2211.05244 (2022)
8. Gal, Y., Ghahramani, Z.: Dropout as a Bayesian approximation: representing model uncertainty in deep learning. In: International Conference on Machine Learning, pp. 1050–1059. PMLR (2016)
9. Hochreiter, S., Schmidhuber, J.: Long short-term memory. Neural Comput. **9**(8), 1735–1780 (1997)
10. Hüllermeier, E., Waegeman, W.: Aleatoric and epistemic uncertainty in machine learning: an introduction to concepts and methods. Mach. Learn. **110**(3), 457–506 (2021)
11. Kendall, A., Gal, Y.: What uncertainties do we need in Bayesian deep learning for computer vision? In: Advances in Neural Information Processing Systems, vol. 30 (2017)
12. Lakshminarayanan, B., Pritzel, A., Blundell, C.: Simple and scalable predictive uncertainty estimation using deep ensembles. In: Advances in Neural Information Processing Systems, vol. 30 (2017)
13. Legrand, A., Trannois, H., Cournier, A.: Use of uncertainty with autoencoder neural networks for anomaly detection. In: 2019 IEEE Second International Conference on Artificial Intelligence and Knowledge Engineering (AIKE), pp. 32–35. IEEE (2019)
14. Malhotra, P., Ramakrishnan, A., Anand, G., Vig, L., Agarwal, P., Shroff, G.: LSTM-based encoder-decoder for multi-sensor anomaly detection. arXiv preprint arXiv:1607.00148 (2016)
15. Malhotra, P., Vig, L., Shroff, G., Agarwal, P., et al.: Long short term memory networks for anomaly detection in time series. In: ESANN, vol. 2015, p. 89 (2015)
16. Paparrizos, J., Kang, Y., Boniol, P., Tsay, R.S., Palpanas, T., Franklin, M.J.: TSB-UAD: an end-to-end benchmark suite for univariate time-series anomaly detection. Proc. VLDB Endow. **15**(8), 1697–1711 (2022)
17. Sakurada, M., Yairi, T.: Anomaly detection using autoencoders with nonlinear dimensionality reduction. In: Proceedings of the MLSDA 2014 2nd Workshop on Machine Learning for Sensory Data Analysis, pp. 4–11 (2014)
18. Schmidl, S., Wenig, P., Papenbrock, T.: Anomaly detection in time series: a comprehensive evaluation. Proc. VLDB Endow. **15**(9), 1779–1797 (2022)
19. Vidmark, A.: Anomaly or not anomaly, that is the question of uncertainty: investigating the relation between model uncertainty and anomalies using a recurrent autoencoder approach to market time series (2022)
20. Wenig, P., Schmidl, S., Papenbrock, T.: TimeEval: a benchmarking toolkit for time series anomaly detection algorithms. Proc. VLDB Endow. **15**(12), 3678–3681 (2022)
21. Yan, P., et al.: A comprehensive survey of deep transfer learning for anomaly detection in industrial time series: methods, applications, and directions. IEEE Access **12**, 3768–3789 (2024)
22. Yong, B.X., Brintrup, A.: Bayesian autoencoders with uncertainty quantification: towards trustworthy anomaly detection. Expert Syst. Appl. **209**, 118196 (2022)

Comparison of Measures for Characterizing the Difficulty of Time Series Classification

Adam Charane[1](✉), Matteo Ceccarello[2], and Johann Gamper[1]

[1] Free University of Bozen-Bolzano, Bolzano, Italy
adamcharane@gmail.com
[2] University of Padova, Padua, Italy

Abstract. The performance of machine learning algorithms is influenced both by their characteristics and parameterization as well as by the properties of the data they are trained and evaluated on. The latter aspect is often neglected. In this paper, we focus our attention on properties of the data that affect the accuracy of time series classification. We experimentally study how the difficulty of classifying time series is related to well-known model-agnostic data complexity measures. Our experiments show that (a) many of these measures are highly correlated with classification scores such as accuracy and F1 and (b) different families of complexity measures capture different properties of the data.

Keywords: Complexity measures · Time series classification

1 Introduction

In the field of machine learning, particularly in classification, it is well known that the accuracy of classification depends on both the models used and the characteristics of the data. Metrics like error rate and accuracy indicate dataset difficulty but are tied to specific classifiers and don't provide assurance that another model will perform well if one model performs poorly. To address this, researchers have proposed *complexity measures*. These measures evaluate the overall complexity of accurately classifying the dataset by aggregating information from all items, independent of any classification model.

Ho and Basu [5] introduced complexity measures based on geometric dataset properties like linear separation and boundary width between classes. Numerous additional complexity measures have since been proposed [8,9]. Lorena et al. [6] offer an extensive review of these measures, categorizing them and providing open-source implementations for all reviewed measures.

These measures, which we use in this work, encompass a range of categories: *Feature-based measures* (F1, F1v, F2, F3, F4) assess the discriminative power of features in distinguishing classes, particularly effective in binary classification. *Linearity measures* (L1, L2, L3) quantify the linear separability of classes using

R. Wrembel et al. (Eds.): DaWaK 2024, LNCS 14912, pp. 239–244, 2024.
https://doi.org/10.1007/978-3-031-68323-7_19

hyper planes, indicating ease of classification. *Neighborhood measures* (N1-N6) utilize pairwise distances to characterize class overlap and internal structure, favoring datasets with maximized inter-class distances and minimized intra-class distances. *Network measures* (G1, G2, G3) capture structural information from graph representations of datasets, where nodes represent data points and edges represent similarities/dissimilarities. *Dimensionality measures* (D1, D2, D3) quantify data sparsity based on dataset dimensionality, positing sparse datasets as more challenging for classification. *Class imbalance measures* (B1, B2) assess the difference in example numbers per class, highlighting challenges posed by imbalanced datasets favoring majority classes.

In this study, we evaluate the predictive power of the mentioned complexity measures in assessing the difficulty of classifying time series data, by comparing with results of state-of-the-art classification models from the literature. We also show that many complexity measures are redundant, despite being formulated and computed differently.

2 Methodology

Our goal is to verify that high complexity measures, extracted from time series data, result in low classification scores, as high complexity indicates the dataset is difficult to classify. To validate this, we use published results of state-of-the-art models, specifically their F1 scores, which represent the harmonic mean of precision and recall. We ran the experiments with accuracy as well, and results were consistent, thus we only report the evaluation using F1. By examining these F1 scores, we aim to determine if they indeed correlate with the complexity measures, thereby establishing a link between data complexity and model performance.

Because complexity measures are independent of classifiers, relying on a single model would only test if the complexities reflect the difficulty specific to that model. Instead, we need to consider a variety of models. Since the number of possible models is infinite, we focus on state-of-the-art models from the literature. Therefore, we define empirical hardness as:

Definition 1 (Empirical hardness). *Given a set $\mathcal{M}$ of state-of-the-art classification models, the* empirical hardness *of a dataset is the highest F1 score achieved by any model in $\mathcal{M}$.*

In the following, whenever we mention *hardness* we refer to our definition of empirical hardness.

2.1 Data and Models

Data. Our data is collected from the UCR archive [3], a well known benchmark for time series classification with the following characteristics: 42 datasets have binary classes, 17 have three classes, 9 have four classes, and the remaining have 5 or more classes each. We consider 108 datasets out of the total 128. The remaining datasets are excluded from the analysis because of missing data or other issues that prevented the computation of the complexity measures. Among the selected datasets, there are five datasets where the most prevalent class has over 10 times the number of instances compared to the least represented class, while nine datasets exhibit an imbalance exceeding a factor of 5.

Models. The UCR archive is supported by a website that reports the performance of 14 state-of-the-art models across 10 different classification metrics, including F1, accuracy, and recall, as detailed by Bagnall et al. [1]. These results were achieved through exhaustive hyperparameter tuning. For multi-class datasets, the benchmark provides the macro-F1 score, which is the average of the class-wise F1 scores.

The classification models considered in our analysis are categorized into different families: *Distance*, which use distance measures; *Feature*, which extract global features from the raw time series; *Interval*, which extract features from selected intervals of the time series; *Dictionary*, which create histograms of words from subsequences; *Hybrid*, which combine two or more approaches; and *Neural networks*, specifically deep convolutional neural networks adapted for time series.

Additionally, because the benchmark's neural network category only included convolution-based models, we included two more models based on Recurrent Neural Networks: Vanilla RNN and RNN with Attention, bringing the total to 16 models in our analysis. These additional models were included for completeness, as they account for the order of data points. However, they did not outperform the other classifiers in the benchmark.

2.2 Complexity Measures

We utilize 22 well-known complexity measures from the literature, extracted using the open-source R package ECoL provided by Lorena et al. [6]. These measures do not account for the time dependencies inherent in time series data. We first transform our datasets to a feature space using two popular approaches: Catch22 [7], which represents each time series as a 22-dimensional feature vector, and TSFresh [2], which provides a 787-dimensional feature vector. For completeness, we also computed the complexities on the raw time series data using ECoL.

Note that the outputs of these transformations differ. When transforming time series into the feature domain, each time series is transformed individually, retaining its identity as a distinct item. However, when computing complexity measures, the output is a single vector that describes the entire dataset and the interactions between time series, without retaining information about individual time series.

3 Analysis

In this section, we first analyze how the performance of classification models correlates with complexity measures, then we verify if there are redundancies between the measures.

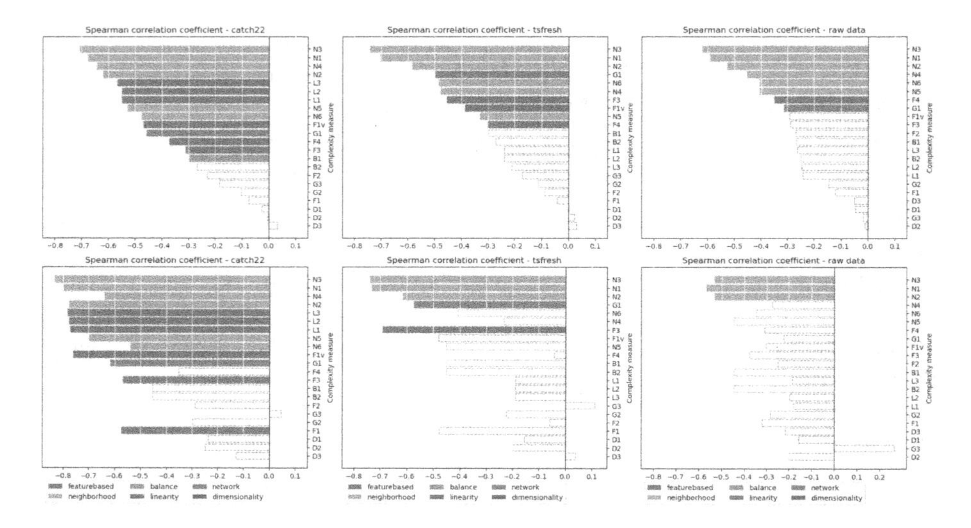

Fig. 1. Correlation of complexity measures with the best F1 score for each set of features and the raw data. Colors encode the complexity measures families. Dashed bars correspond to complexity measures that do not have a statistically significant correlation with the F1 score.

3.1 Correlation Analysis

First, we identify which complexity measures best correlate with the empirical hardness of each dataset using Spearman correlation, comparing the ranking of datasets by empirical hardness and complexity measures. Correlation values close to -1 indicate a strong relationship. We then conduct a statistical test for each complexity measure against empirical hardness, using the Holm correction for multiple tests. Since many complexity measures were originally designed for binary classification, we perform the analysis on both all datasets and binary-only datasets, reporting results for both scenarios.

Figure 1 presents the results of our correlation tests. The top row of plots includes all datasets, while the bottom row focuses on binary datasets. Each column corresponds to different feature sets: Catch22, TSFresh, and raw data. The length of each bar indicates the correlation between the complexity measure and empirical hardness. Bars are solid-colored if the correlation is statistically significant (Holm-corrected p-value ¡ 0.05) and outlined with a dashed line otherwise. Non-significant correlations are included for completeness but should not be considered for further analysis. To facilitate comparison, the bars in the bottom row are ordered to match the top row, sorted by decreasing correlation.

From the top row in Fig. 1 we observe that neighborhood measures, particularly **N3**, show the strongest correlation with empirical hardness across all feature sets. Interestingly, linearity measures are significantly correlated only when using Catch22, likely due to the higher proportion of linear features in Catch22 compared to TSFresh.

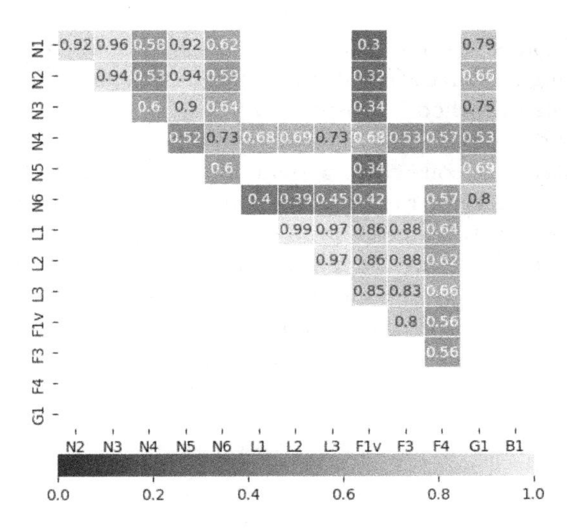

Fig. 2. Heatmap of the correlations between complexity measures that have a statistically significant correlation with the F1 score. Pairs of complexity measures which do not have a statistically significant correlation are not reported.

The last column of Fig. 1 analyzes complexities derived from raw data. Although the complexity measures are less suited for raw time series data, neighborhood measures and **G1** still show significant correlations, albeit weaker. This is because these measures are based on distances between data points, a meaningful property even in raw data. Hence, for the remainder of the analysis, we will focus on the Catch22 and TSFresh feature sets.

For binary datasets, Catch22 shows stronger correlations with empirical hardness, and TSFresh exhibits a similar pattern. However, some correlations in TSFresh are not statistically significant for binary datasets due to the smaller sample size, as indicated by the unfilled bars.

Overall, the most consistent complexity measures are from the neighborhood family, followed by **G1** and certain feature-based complexities, such as **F1v**.

3.2 Relationships Between the Complexity Measures

We now examine whether the complexity measures are correlated with each other, indicating potential redundancy as they might encode the same properties. To investigate this, we selected the 13 complexity measures that showed the strongest correlation with accuracy in the previous section. We then computed the pairwise Spearman correlations between these measures, testing against the Null hypothesis of no correlation and adjusting for multiple comparisons using the Holm method.

Figure 2 displays the results as a heatmap, where intense colors represent strong correlations, and blank cells indicate pairs of measures with non-significant correlations (p-value ¡ 0.05). We observe that measures within the same family are highly correlated, suggesting they capture similar information.

For instance, neighborhood measures, which all start by computing pairwise distances (Gower distance [4]), show high correlations. Notably, N1 and N3 are the most correlated: N1 constructs a spanning tree from the distance matrix, while N3 measures the error rate of a 1-nearest-neighbor classifier. Despite different methods, both assess similar properties, leading to high correlation.

Linearity measures are highly correlated with each other and the feature-based measure F1v. This arises because linearity measures rely on SVM classifier error rates, while F1v identifies a separating vector.

Overall, measures from different families generally show low correlations, indicating they capture distinct data characteristics.

4 Conclusion

In this work, we studied the intrinsic properties of classification datasets and their relationship with the empirical performance of classifiers. We verified the connections between the complexity measures that quantify the properties of classification datasets, and we found that most complexities are measuring different quantities, but some of them are measuring the same property despite the fact that they are defined and computed differently. We also confirmed that the classification is affected by some inherent properties of the data, especially the neighborhood measures, the linearity measures, and G1.

References

1. Bagnall, A.J., Lines, J., Bostrom, A., Large, J., Keogh, E.J.: The great time series classification bake off: a review and experimental evaluation of recent algorithmic advances. Data Min. Knowl. Discov. **31**(3), 606–660 (2017)
2. Christ, M., Braun, N., Neuffer, J., Kempa-Liehr, A.W.: Time series feature extraction on basis of scalable hypothesis tests (tsfresh - a python package). Neurocomputing **307**, 72–77 (2018)
3. Dau, H.A., et al., Hexagon-ML: the UCR time series classification archive (2018)
4. Gower, J.C.: A general coefficient of similarity and some of its properties. Biometrics **27**(4), 857 (1971)
5. Ho, T.K., Basu, M.: Complexity measures of supervised classification problems. IEEE Trans. Pattern Anal. Mach. Intell. **24**(3), 289–300 (2002)
6. Lorena, A.C., Garcia, L.P.F., Lehmann, J., de Souto, M.C.P., Ho, T.K.: How complex is your classification problem?: a survey on measuring classification complexity. ACM Comput. Surv. **52**(5), 107:1–107:34 (2019)
7. Lubba, C.H., Sethi, S.S., Knaute, P., Schultz, S.R., Fulcher, B.D., Jones, N.S.: catch22: canonical time-series characteristics. Data Min. Knowl. Discov. **33**(6), 1821–1852 (2019)
8. Singh, S.: PRISM-A novel framework for pattern recognition. Pattern Anal. Appl. **6**(2), 134–149 (2003)
9. Sohn, S.Y.: Meta analysis of classification algorithms for pattern recognition. IEEE Trans. Pattern Anal. Mach. Intell. **21**(11), 1137–1144 (1999)

Dynamic Time Warping for Phase Recognition in Tribological Sensor Data

Anna-Christina Glock[1(✉)] and Johannes Fürnkranz[2]

[1] Software Competence Center Hagenberg, 4232 Hagenberg im Mühlkreis, Austria
anna-christina.glock@scch.at

[2] Institute for Application Oriented Knowledge Processing (FAW), Johannes Kepler University Linz, Linz, Austria

Abstract. This paper analyzes the potential of dynamic time warping (DTW) for recognizing phases of tribological sensor data. The three classes in these time series—run-in, constant wear, and divergent wear—are distinguished by their long-term trend and curvature. A set of reference data for each class is needed for the classification. Each time series in the reference set represents a typical shape of this class. The classification is done by computing the DTW between a given time series and each reference time series, and assigning it to the class with the minimum distance. In experiments on simulated and real-world time series, we show that DTW is capable of correctly classifying whole time series representing one class. Additional experiments are done to analyze the capability of DTW to classify a time series that is only a part of the entire time series representing one class. During these experiments, limitations arose that demonstrated the importance of the choice of good reference data.

Keywords: Time series classification · Dynamic Time Warping · Sensor Data

1 Introduction

Recently, the availability of large temporal data has increased demand for time series mining and analysis. Time series data, found in various domains like finance and industrial processes, are used in applications such as change point detection and classification. This paper focuses on time series classification, assigning labels based on data characteristics. Numerous algorithms for time series classification have been proposed over the years.

In this paper, we study an application where the task is to recognize different parts of a noise time series that consists of different classes. These classes can be distinguished by long-term trend and curvature. The segmentation of the time series is done by an online change point detection method [2]. We want to explore two questions in this paper:

- **Q1:** What is the classification accuracy of DTW applied to noisy time series data where the classes are distinguished by long-term trend and curvature?

R. Wrembel et al. (Eds.): DaWaK 2024, LNCS 14912, pp. 245–250, 2024.
https://doi.org/10.1007/978-3-031-68323-7_20

- **Q2:** How many data points are needed to reliably classify a segment correctly?

Section 2 reviews time series classification techniques and explains why DTW was chosen. Section 3 details DTW, its application and our dataset. Section 4 presents and analyzes the results. Section 5 provides conclusions and future research directions.

2 Related Work

Time series classification algorithms differ from other classification algorithms because they can use time-dependent features for classification. In our case, this includes long-term trends or data curvature.

Due to the nature of these features and noisy data, large parts of a time series are needed to recognize the features correctly. When using such features, it is important to have a classification algorithm that is based on large parts or even the whole time series is needed. Therefore, algorithms that use short characteristic patterns (like shapelets [6]) or algorithms that rely on a pattern appearing multiple times (like bag-of-words [5]) are not suitable. Deep learning classification approaches (like [3]) are problematic as the required large amount of training data is unavailable. On the other hand, distance-based classification approaches utilize the whole time series. In this paper, we use the Dynamic Time Warping algorithm. Additionally, DTW can calculate the distance between two time series, even if those time series are of different lengths. This is important as the time series of our use case differ quite significantly in length.

3 Method

3.1 Dynamic Time Warping (DTW)

Dynamic time warping (DTW) was first introduced by [7] for a speech recognition task. It is a method for estimating the distance between two sequences which do not need to be of the same length, and significant features might occur at different points in time. This is achieved by calculating a $n \times m$-dimensional cost matrix C between two time series $X = (x_1, x_2, ..., x_n) \in \mathbb{R}^n$ and $Y = (y_1, y_2, ..., y_m) \in \mathbb{R}^m$. After calculating C, the next step is to find all possible warping paths $P = (P_1, ..., P_k)$. One warping path is defined as $P_k = (p_1, ..., p_l)$ where an element $p_l = (i_l, j_l)$ corresponds to an element in C. Once the search is complete, the cost of each warping path is calculated by summing up the individual costs for each element in the warping path. The DTW score is the cost of the warping path with the minimum cost among all warping paths.

3.2 Tribological Use Case

Our data sets stem from tribological experiments, where critical machine parts are examined through a bench test. During these experiments, the wear volume

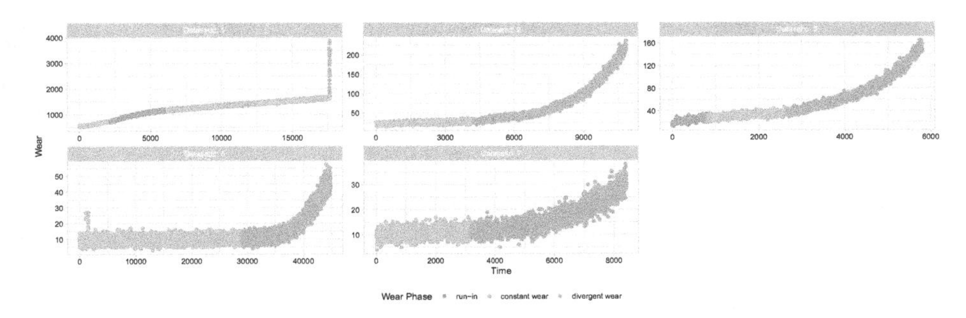

Fig. 1. The five different data sets used in our case study. The run-in phase is shown in pink, the constant wear in green, and the divergent wear in blue. (Color figure online)

is monitored by a technique based on radioactive isotopes [4] The resulting time series has 3 distinctive phases: *run-in* (**E**), *constant wear* (**K**) and *divergent wear* (**A**). All wear phases differ in how much the wear volume increases over time Fig. 1.

As we only have a limited number of real-world data sets with all three classes available, synthetic data were generated by an idealistic model validated by experts for the wear process [2].

3.3 Experiments

In order to validate DTW's performance for our use case, two sets of experiments were conducted on real-world (5 data sets) and synthetic data (1100 data sets) Sect. 3.2. For the first set of experiments, the whole wear phase is classified with different preprocessing (standardizing and filtering) and DTW step patterns (symmetric and slope-constrained [8]). This allows us to test DTW on its ability to classify the time series correctly and shows us its limitations.

This second set of experiments aims to evaluate the number of data points the approach needs to classify each phase robustly and correctly. For this more detailed analysis, only the best preprocessing, the best DTW step pattern from the first set of experiments, and the real-world data set is used. For the partial classification during the experiments a growing data window is used. The growth rate is 180 data points or 5 data points for the divergent wear phase of dataset 1.

For the experiments, reference time series that clearly represent one of the phases are important for a good classification. For each phase, multiple standardized and smooth reference time series were generated, 10 (E), 8 (K), and 14 (A).

The minimal normalized DTW distance is used to classify a segment.

An R implementation was used for the DTW [1].

4 Results

The evaluation of the classification outcomes employs the F1 Score as a key metric and calculates it for each class separately.

4.1 Classification of the Whole Wear Phases

Table 1. Results on artificial data. The best results per preprocessing option are highlighted in bold numbers.

Step Condition	no Preprocessing			Filtered			Standardized			Filtered and Standardized		
	E	K	A	E	K	A	E	K	A	E	K	A
symmetric2	0.53	0.57	0.89	0.00	0.00	0.95	0.18	1.00	0.82	1.00	0.18	0.57
symmetricP0	0.53	0.57	0.89	0.00	0.00	0.95	0.18	1.00	0.82	1.00	0.18	0.57
symmetricP05	0.00	0.46	1.00	**0.00**	**0.33**	**1.00**	0.80	0.82	0.75	1.00	0.57	0.89
asymmetricP0	**1.00**	**1.00**	**0.00**	0.00	0.87	0.00	**1.00**	**0.89**	**0.82**	1.00	0.67	0.57
asymmetricP05	0.00	0.43	1.00	0.00	0.00	1.00	0.78	0.89	0.75	**1.00**	**0.67**	**0.82**

Synthetic Data. The classification results of the synthetic data (Table 1) show that it is feasible to use our DTW-based approach to classify our data. The best results are achieved with the *asymmetric* step pattern for three out of four preprocessing steps. Analyzing the incorrectly classified parts leads to the observation that misclassifications can either be attributed to the distinguishable characteristics of the correct class not being visible enough or the set of reference time series not representing the features of the correct class well enough. These lessons learned are used for the experiments with the real-world data sets.

Real-World Data. In contrast to the synthetic data, the symmetric step pattern performed better for the real-world data sets (Table 2). This is likely due to the distance between the two time series being bigger, leading to a diminished influence of noise and filtering artifacts on the alignment. As with the synthetic data the standardization improves the classification performance significantly more than the filtering alone. For the best result only one divergent wear phase of dataset 1 is not classified correctly possibly due to its shortness.

Answer to Q1: The F1-Score of the DTW classification results (Table 1 and Table 2) show that a successful classification of classes distinguished by long-term trend and curvature is possible with DTW. As the best results are achieved on preprocessed data, this highlights the importance of preprocessing.

Table 2. Results on real-world data. The best results per preprocessing option are highlighted in bold numbers.

Step Condition	no Preprocessing			Filtered			Standardized			Filtered and Standardized		
	E	K	A	E	K	A	E	K	A	E	K	A
symmetric2	0.00	0.77	0.33	0.00	0.67	0.57	0.00	1.00	0.75	**1.00**	**1.00**	**0.92**
symmetricP0	0.00	0.67	0.40	0.00	0.50	0.67	**0.00**	**1.00**	**0.86**	**1.00**	**0.92**	**1.00**
symmetricP05	0.00	0.36	1.00	**0.00**	**0.36**	**1.00**	0.00	0.80	0.86	0.00	0.91	0.86
asymmetricP0	**0.89**	**0.71**	**0.00**	0.80	0.38	0.00	0.00	0.55	0.86	0.00	0.83	1.00
asymmetricP05	0.00	0.33	0.92	0.00	0.29	0.67	0.00	0.91	0.67	0.00	0.91	0.40

4.2 Partial Classification of the Wear Phases

The results of the second set of experiments show that it takes some time until the characteristic features of a class are observable enough for a correct classification. Furthermore, there are single windows that are anomalously classified compared to the surrounding windows. To combat this problem, the classification of 5 consecutive parts is considered, and the most frequent class is used as the classification result (Fig. 2). The results show that the "anomalies" are removed at the cost of a longer classification time.

Answer to Q2: The number of data points needed to reliably classify a wear phase depends on the wear phase. The pattern of the run-in and divergent wear phases needs more time to be distinguishable from the constant wear and each other. In most cases, half of the wear phase was needed until the phase was classified correctly.

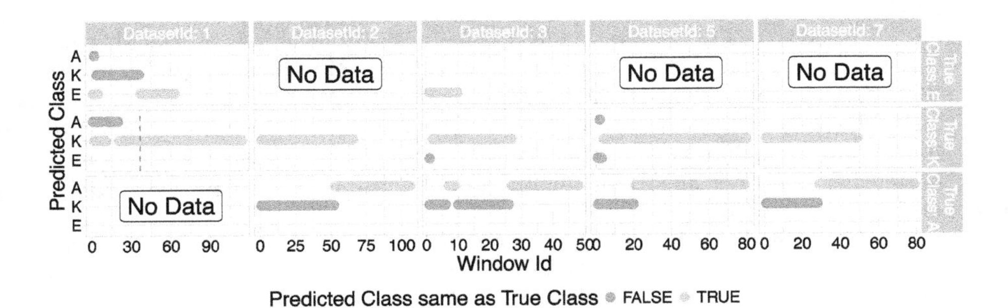

Fig. 2. Each square shows the result of the classification per dataset, true class, and growing window. As the run-in is only visible for datasets 1 and 3, only those have classification results plotted. Another special case is the constant wear in dataset 1. As there are two constant wear phases in this dataset, the square is split into two by a vertical dashed line. The predicted class is determined by choosing the most frequent class of the last 5 windows. As the divergent wear of dataset 1 is only 4 windows long, there are no results for this phase.

5 Conclusion

Our investigation found that DTW successfully classifies noisy time series based on long-term features. However, DTW has limitations: it may misclassify class variants not present in the reference series, requiring careful monitoring and updating. Successful classification also depends on having sufficiently long sequences, which varies by class. Further research is needed to explore DTW distance as a confidence measure. As mentioned in the beginning, our goal is to develop an online classification method using long-term trends and curvature, and we will evaluate combining DTW with an online change point detection method.

Acknowledgements. This work was funded by the Austrian COMET-Program (Project K2 InTribology1, no. 872176), and carried out at the Software Competence Center Hagenberg.

References

1. Giorgino, T.: Computing and visualizing dynamic time warping alignments in R: the DTW package. J. Stat. Softw. **31**(7), 1–24 (2009). https://doi.org/10.18637/jss.v031.i07
2. Glock, A.C., Sobieczky, F., Fürnkranz, J., Filzmoser, P., Jech, M.: Predictive change point detection for heterogeneous data. arXiv preprint arXiv:2305.06630 (2023). https://doi.org/10.48550/ARXIV.2305.06630
3. Ismail Fawaz, H., Forestier, G., Weber, J., Idoumghar, L., Muller, P.A.: Deep learning for time series classification: a review. Data Min. Knowl. Disc. **33**(4), 917–963 (2019)
4. Jech, M., Lenauer, C.: Radionuclide methods. In: Friction. Lubrication, and Wear Technology, ASM Handbook, vol. 18, pp. 1045–1055. ASM International, Ohio (2017)
5. Lin, J., Khade, R., Li, Y.: Rotation-invariant similarity in time series using bag-of-patterns representation. J. Intell. Inf. Syst. **39**(2), 287–315 (2012). https://doi.org/10.1007/s10844-012-0196-5. ISSN 1573-7675
6. Lines, J., Davis, L., Hills, J., Bagnall, A.: A shapelet transform for time series classification. In: Proceedings of the ACM SIGKDD International Conference on Knowledge Discovery and Data Mining, pp. 289–297 (2012). https://doi.org/10.1145/2339530.2339579
7. Sakoe, H., Chiba, S.: A dynamic programming approach to continuous speech recognition. In: Proceedings of the Seventh International Congress on Acoustics, Budapest, vol. 3, pp. 65–69, Akadémiai Kiadó, Budapest (1971)
8. Sakoe, H., Chiba, S.: Dynamic programming algorithm optimization for spoken word recognition. IEEE Trans. Acoust. Speech Signal Process. **26**(1), 43–49 (1978). https://doi.org/10.1109/TASSP.1978.1163055

Data Repositories

Putting Co-Design-Supporting Data Lakes to the Test: An Evaluation on AEC Case Studies

Melanie Herschel[1(✉)], Andreas Gienger[2], Anja P. R. Lauer[2], Charlotte Stein[2], Lior Skoury[3], Nico Lässig[1], Carsten Ellwein[4], Alexander Verl[4], Thomas Wortmann[3], and Cristina Tarin Sauer[2]

[1] IPVS, University of Stuttgart, Stuttgart, Germany
{melanie.herschel,nico.laessig}@ipvs.uni-stuttgart.de
[2] ISYS, University of Stuttgart, Stuttgart, Germany
{andreas.gienger,anja.lauer,charlotte.stein,
cristina.tarin-sauer}@isys.uni-stuttgart.de
[3] ICD, University of Stuttgart, Stuttgart, Germany
{lior.skoury,thomas.wortmann}@icd.uni-stuttgart.de
[4] ISW, University of Stuttgart, Stuttgart, Germany
{carsten.ellwein,alexander.verl}@isw.uni-stuttgart.de

Abstract. Leveraging data from various stakeholders in the architecture, engineering, and construction (AEC) industry is an essential prerequisite to harness the potential of digitization and Artificial Intelligence (AI) in addressing major challenges such as productivity or sustainability. ArchIBALD is a recently proposed information system architecture that caters to AEC applications by its integrative, modular, extensible, and feedback-centered design. Its goal is to support co-design applications, characterized by processes optimized through a holistic view across traditionally isolated AEC steps and data silos.

This paper evaluates the suitability of the ArchIBALD system architecture through three scenario-based co-design case studies. They relate to the design and construction of a real building demonstrator. The evaluation focuses on the functionality and several quality criteria of the information system's architecture and validates that the architecture is amenable to a variety of real-world co-design applications.

Keywords: data integration · co-design · artificial intelligence

1 Motivation: Data Management in AEC

In the Architecture, Engineering, and Construction (AEC) Industry, the various stakeholders typically involved in projects traditionally generate, maintain, and manage their own data. For instance, architects may focus on 3D CAD models of their building design in proprietary software, whereas the foreperson on a construction site may rely mainly on data representing 2D floor plans. The sector

R. Wrembel et al. (Eds.): DaWaK 2024, LNCS 14912, pp. 253–268, 2024.
https://doi.org/10.1007/978-3-031-68323-7_21

Fig. 1. livMatS Biomimetic Shell (©ICD/ITKE/IntCDC University of Stuttgart (Photo: Conné van d'Grachten))

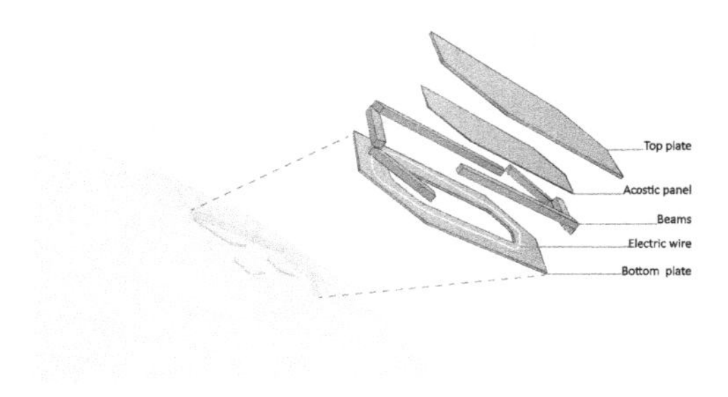

Fig. 2. One cassette component breakdown

Fig. 3. Fabrication setup at müllerblaustein (© ICD/ITKE/ IntCDC University of Stuttgart)

Fig. 4. Cooperative on-site assembly using two large-scale manipulators with a vacuum gripper and screwing effector as end effectors

suffers from a high segmentation with little to no permeability between data silos of different stakeholders. This poses a significant obstacle to leveraging novel technologies, including Artificial Intelligence (AI). These bear the potential to improve, e.g., productivity, diminishing human workforce, or sustainability, three major issues in AEC. As an example, combining data about what should be built (i.e., the architectural design) with data describing what resources are available for the physical fabrication of parts of the design (i.e., prefabrication platforms) potentially allows to automatically infer how to realize a design in the most efficient way or suggest alterations to the design to make it realizable.

The above example showcases how integrating and leveraging data *across sources* that span the different stages of building processes (e.g., design and prefabrication) can be beneficial by enabling *co-design applications*, i.e., applications that interconnect different stages (e.g., informing the design-to-fabrication process) and empower feedback loops (e.g., adapting design based on prefabrication). Recently, information system solutions adapted to the AEC domain are actively being explored [9,13,16,18]. To the best of our knowledge, ArchIBALD [9] is the only information system architecture that both explicitly incorporates the feedback capabilities of data-informed co-design applications and interconnects different stages. As outlined in Sect. 2, we defined this architecture based on a study for data management requirements and desiderata within AEC among 31 expert participants. ArchIBALD extends the "one-way-street" of the common data lake architecture [5, 14] via feedback-channels. These lead to bidirectional data exchange between services, data producers, and data consumers. In addition, ArchIBALD is designed to support data integration and analysis in a modular and extensible way through a service-based layered architecture. It thereby bridges the data silos of individual AEC stages. We refer readers to [9] for details on the ArchIBALD architecture and a brief demonstration of its feasibility.

Contributions. We thoroughly evaluate ArchIBALD's suitability to real-world settings through three case studies that we conduct in the context of the actual construction of a building demonstrator (see Sect. 3). The FIT livMatS Biomimetic Shell[1] is a wood pavilion (Fig. 1) composed of multiple hollow, individually designed cassettes (detailed in Fig. 2). Robotic off-site prefabrication was used to produce the cassettes, using the prefabrication platform shown in Fig. 3. These cassettes were assembled on-site with the help of cyber-physical systems (Fig. 4). Our evaluation of ArchIBALD, covered in Sect. 4, focuses on both the functionality of supporting co-design applications and the software architecture quality criteria of integration, extensibility, reusability, and modularity. Our evaluation methodology aligns with the well-established Architecture Trade-Off Analysis Method (ATAM) for the evaluation of software architectures [3] and an adaptation to information systems [15] we refer to as Information System Architecture Evaluation (ISAE). Besides validating the suitability of the architecture to different AEC co-design applications, the evaluation allows to identify further domain specificities of such applications, which may be addressed through dedicated techniques, only conceivable through gained insights into real-world AEC processes (see Sect. 5). Thereby, observations from the evaluation provide a practical reference for concepts to be explored in future research.

2 ArchIBALD Architecture Development and Definition

We first discuss the requirement analysis (Sect. 2.1) that led to the design choices and ArchIBALD's architecture definition (Sect. 2.2). This conforms to the main steps of the first two phases of the ATAM process, i.e., requirements gathering and architectural views, and yields the first prerequisite for a scenario-based evaluation defined in ISAE. Section 3 covers the second one involving the scenarios.

2.1 Requirement Analysis

With the goal of enabling data-informed co-design in AEC by bridging data silos in a fragmented industry, a key issue is to bridge the gap between heterogeneous data from autonomous sources. From an information system architecture viewpoint, the space of possible solutions encompasses architectures for

[1] https://www.icd.uni-stuttgart.de/projects/livmats-biomimetic-shell/.

data integration [2], including mediator-wrapper architectures [6], data warehouses [8], data lakes [14], polystores [4], data meshes [12], data lakehouses [19], etc.

To identify which paradigm we may base our system architecture on, we first conducted a questionnaire-based survey for data management requirements and desiderata within the building sector among 31 expert participants from the diverse and fragmented domains involved in the process. The study revealed: (R1) the system requires to handle a large variety of heterogeneous data (e.g., semi-structured, unstructured data, sensor data); (R2) providing, using, and maintaining the integrated data should not disrupt the existing workflows used at various stages; (R3) extracting metadata, structure, knowledge, etc. should be handled by the system, as the experts at the different stages (architects, civil engineers, etc.) do not have the expertise in these areas; (R4) some applications relying on the integrated data require real-time feedback, others can rely on offline processing; (R5) data updates are frequent so versioning and propagation of differences are important, and (R6) access control should be possible.

2.2 Design of the ArchIBALD Architecture

Above requirements guided the development of the ArchIBALD architecture, summarized next, and the definition of the evaluation criteria (Sect. 4).

Design Choices. Based on the results of our requirements analysis, we compared different architecture options (including those listed in Sect. 2.1) and identified the data lake architecture as the most suitable basis for our setting, where we target AEC co-design applications rather than Business Intelligence or reporting applications more common in building maintenance. The data lake architecture satisfies (R1) and (R2), as data providers at various stages of the building process can provide their data with little effort. (R6) can also be commonly handled by the system used to implement the data lake. To satisfy (R3), data lakes allow the implementation of services to handle the data wrangling. To address (R4) and (R5), the services need to be engineered and coordinated accordingly. To support this, ArchIBALD extends the usual "one-way-street" from data ingestion into the data lake to data provisioning to the applications by allowing "shortcuts" and "bridges" via feedback channels, leading to a bidirectional exchange of information between services, data producers, and data consumers. Further performance gains for (R4) can be obtained by avoiding the full integration of all data by implementing services with a specific scope, e.g., some bridge a specific data model heterogeneity while others offer added-value services such as matching geometries of components in the design to machines capable of transporting these components to the construction site. This approach is designed to not only reduce the complexity of individual services, it also aims at simplifying the implementation and maintenance of the individual components and increasing reuseability by being able to compose services in different configurations.

Architecture Overview. Figure 5 provides a general overview of the ArchIBALD architecture [9]. Each layer (rounded rectangles) includes multi-

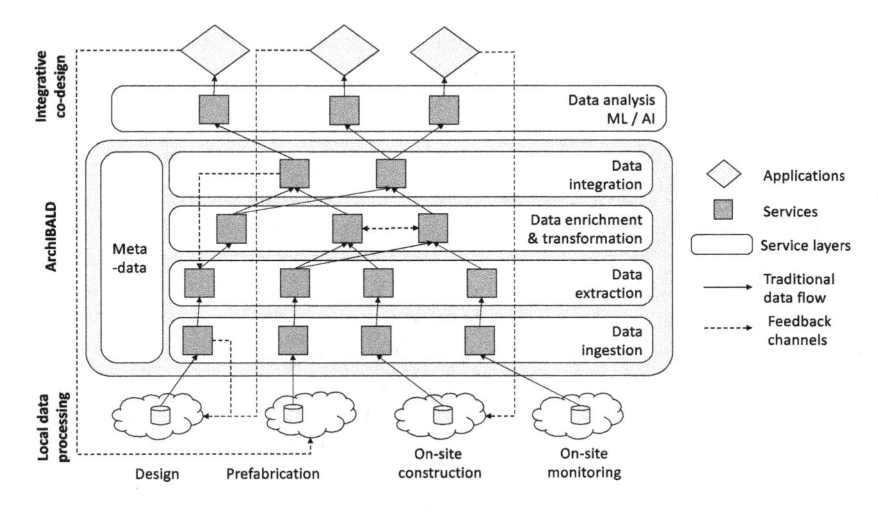

Fig. 5. ArchIBALD architecture overview

ple services (rectangles) offering different functionalities within the scope of the respective layer. For instance, the data integration layer may include services to automatically recognize multiple representations of the same real-world entity (i.e., entity resolution approaches such as [7]) or domain-specific task-to-resource matchers [9].

Data are produced in the different phases of the building process (represented at the bottom). These data need to be ingested by the data lake. *Data ingestion* obtains data from heterogeneous and typically autonomous sources, putting raw datasets of various data formats into the control sphere of the data lake. *Data extraction* transforms these raw data to pre-determined data models (part of the available *metadata*), alleviating the heterogeneity that further data lake services have to handle. The data models underlying the extraction services are designed in an application-dependent way. During *data enrichment and transformation*, the data are "enhanced", e.g., to improve data quality, infer additional features, transform data values to a format required by upstream services etc. Once the data from the different phases of the building process are adequately transformed, *data integration* interconnects and unifies data from different sources. At this stage, the data offer a global perspective of the data otherwise confined to data silos in different phases of the building process. This allows joint *analysis* of the data that, depending on the targeted *co-design applications*, may include, e.g., data mining, machine learning (ML), or artificial intelligence (AI) techniques. Note that while Fig. 5 depicts data producers at the bottom and data consumers (the applications) at the top, data producers and consumers are not necessarily distinct. This represents one type of feedback channel, as the newly gained insights can cause, e.g., design to change, the changed file being ingested again, etc. However, in this case, the data processing through the data lake is still a one-way street (solid arrows).

ArchIBALD integrates feedback channels. Figure 5 illustrates these as dashed arrows. Service-to-service feedback channels can connect services at the same level as well as services from a higher level to a lower level. For instance, at the data enrichment and transformation level, assume one service automatically infers missing data (aka data imputation) about geometries to achieve better completeness. Another service may check the consistency of data, e.g., if material properties and specification of building component dimensions (their geometry) match the weight associated to the component. Then, sending over inferred geometry data to the consistency checking service is helpful, as is sending potentially inconsistent data to the data imputation service (to possibly "overwrite" the errors). As an example of a service-to-service feedback across levels, consider the data "improved" by data transformation that can be fed back to the data ingestion level such that the raw files, which possibly serve as input to other services, reflect the cleaned data as well. Moving on to the service-to-producer feedback channels, these implement human-in-the-loop interactions. For instance, once the file handled at the data ingestion level is modified (e.g., based on enrichment and transformation services), we can notify architects that a previously missing geometry has been added, which they could refine.

3 Scenario-Based Case Studies: Context and Overview

After the overview of the ArchIBALD architecture, we now describe the scenario-based case studies underlying our evaluation. These scenarios were defined and realized in the context of a building demonstrator (i.e., covering the scenario-related steps of Phase 1 and Phase 2 of the ATAM and resulting in the second "input" necessary for evaluation following the ISAE). We first introduce the building demonstrator before describing the co-design applications defining the three scenarios we use for evaluation.

3.1 The livMatS Biomimetic Shell

The livMatS Biomimetic Shell, introduced by the FIT Freiburg Center, serves as an architectural incubator for innovative research concepts. A collaborative project between Excellence Clusters at the University of Stuttgart and the University of Freiburg, it explores an integrative approach to sustainable architecture. This synthesis of research methodologies achieves 50% reduction in the environmental life cycle impact compared to conventional timber constructions. The segmented timber shell, characterized by high resource efficiency, is fully deconstructible and reusable. Enabled by integrative development, including computational design, robotic prefabrication, automated construction processes, and human-machine interaction, it provides an expressive, flexible space showcasing alternative paths for sustainable construction. As illustrated in Sect. 1, the construction of this building demonstrator, which spans 200 m^2, included the prefabrication of 127 individually designed hollow cassettes, which then needed to be properly handled and assembled on-site, (partially) using two large-scale hydraulic manipulators with end effectors specifically designed to that effect.

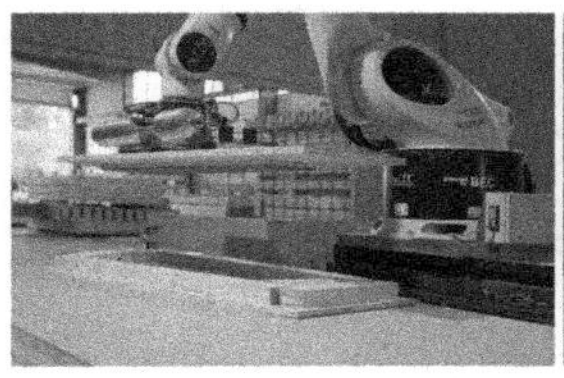

Fig. 6. Different end effectors of the prefabrication platform (© ICD/ITKE/IntCDC University of Stuttgart)

3.2 Co-Design of Robotic Prefabrication

Hollow cassettes, each consisting of a lower deck plate, edge beams, and upper deck plate, represent a material-efficient approach that was first used in the BUGA Wood Pavilion Heilbronn 2019 as a temporary, open structure [17]. It got an upgrade for a permanent, closed building designed for year-round use. This research improved the timber construction method by using more sustainable timber materials and adjusting component sizes to produce as little waste as possible during robotic manufacturing. The idea of resource-efficient, custom components was extended into various aspects, including the integration of acoustic elements, lighting, insulation, façade connections, and grip holes for automated assembly at the design stage (see cassette breakdown in Fig. 2).

High material efficiency demands flexible and precise fabrication processes, achieved through the use of robotic manufacturing techniques with sub-millimetre precision for elements up to 3.5 m. The fabrication environment includes a 40-foot-long, transportable 7-axis robot platform, stationed temporarily at "müllerbaustein HolzBauWerke". Three end effectors, including a large vacuum gripper for plate elements, a smaller vacuum gripper with a nail gun for beam elements, and a glue gun for assembly, are employed to complete the diverse tasks of the cassette prefabrication process (see Fig. 6).

Due to the large scale of the demonstrator, the fabrication process was divided into fabrication jobs. Conceiving a fabrication job involves meticulous consideration of component specifications, the fabrication environment, and tool design: Component design determines assembly sequencing, environment design sets size limits and travel times, and tool design impacts task discretization. These three factors collectively shape task conception, sequencing, and overall fabrication time. In this project, robots maximized their reachability, concurrently assembling four hollow cassette elements in a single job. Each job, averaging 245 tasks, drew from a pool of 11 task types.

The co-design challenge in this case study involves the fabrication optimization of timber hollow cassettes for more efficient and environmentally friendly fabrication. This optimization relies on the integration of the design and fabrication domains, taking into account specifications of design elements, the fabrication environment, and end-effectors design.

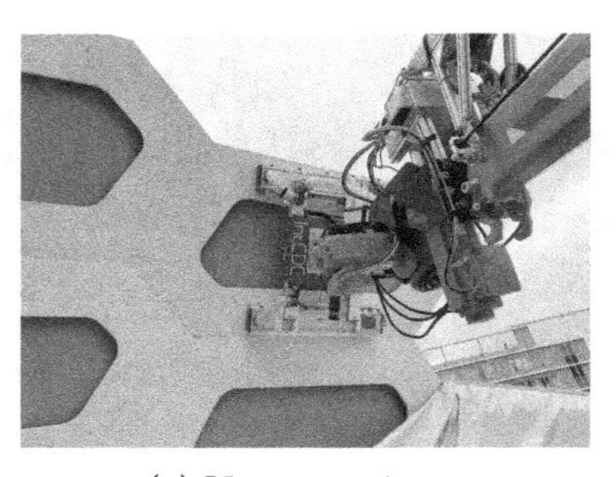

(a) Vacuum gripper

(b) Screwing effector

Fig. 7. End effectors of the hydraulic manipulators for on-site assembly.

3.3 Co-Design of End-Effectors for On-Site Assembly

On-site positioning and assembly of the individualized cassettes requires a specific design of the vacuum gripper and screwing effector, illustrated in Fig. 7. The vacuum gripper, attached to the first manipulator, picks up a hollow cassette and positions it at a target pose. When reaching the target pose, the second manipulator positions the screwing effector along the edges and pin holes. The pin holes are milled into the cassette and serve as attachment points for the pins on the gripper to brace the cassettes. Upon bracing completion, the screwing process starts, repeated for all edges and screwing positions until the cassette is completely mounted to the existing building structure. Afterwards, the vacuum gripper is released and the assembly process is repeated for the next cassette.

Given that most hollow cassettes have a unique design, an optimal design of the end effectors needs to be determined, as they need to handle all the various cassettes. Identifying such an optimal end effector design relies on design data of all the building elements and data descriptive of the end effector configuration possibilities.

More precisely, cassettes vary in their size, the size of the opening on the lower deck plate, their weight, the type of screws used, the screw positions, and angles. These design parameters of the cassettes are relevant to properly design the two end effectors. For instance, the weight of the cassette is relevant for designing the rectangular surface area of the vacuum gripper. Furthermore, the vacuum gripper requires full contact to the wooden surface of the cassette, which is given by the size of the cassette and its opening. For the screwing effector, the type, number, and positions of the screws as well as the screwing angle are determined by a static analysis of the structure and influence the design of the screwing effector. This includes the design of the two screwing units, which is determined by the type and size of the screws. The edge length of the cassettes and the position of the pin holes impacts the length of the screwing effector as well as the position of the pins.

In summary, the problem of co-designing end effectors for on-site assembly encompasses the determination of an optimized end effector design (both for the vacuum gripper and the screwing effector) used on-site, given the design with its constraints of the hollow cassettes.

3.4 Co-Design of On-Site Planning and Execution

In addition to the end effector design, we need to plan the positioning of the large-scale hydraulic manipulators on the construction site, as shown in Fig. 4 and their motion to realize automatic on-site assembly. This avoids unnecessary (time-consuming and costly) repositioning and collisions.

Optimal positioning guarantees that the target pose of a manipulator that is necessary for assembly is reachable for all cassettes, while not exceeding the payload. After determining the base position of the manipulators, a collision-free path for each manipulator is planned. The necessary information about the target poses of the manipulators tool center point (TCP) is derived from the design data of the cassettes and the building. These data are further used to generate meshes of the cassettes and the building geometry for each construction step in order to check for collisions during path planning. This requires the kinematic model and the geometry of the manipulators as additional input [10].

Hence, this case study considers the problem of co-designing on-site planning and execution by planning a collision-free path, represented by a sequence of poses, for the TCP of each manipulator. This necessitates the joint consideration of the kinematic model of manipulators including the payload characteristics, the geometry of both the building and elements for collision avoidance, and the starting and final pose derived for each cassette from the building design data.

4 Evaluation

Our case-study based evaluation of the suitability of the ArchIBALD architecture considers both functional and non-functional quality criteria.

Functionality. We assess the functionality with respect to the intended use in the individual scenarios, defined in Sect. 3, i.e., we evaluate how well the expected application functionalities are covered by the architecture.

Non-functional Criteria. Our requirement analysis (see Sect. 2.1) led us to the definition of an integrative information system architecture designed to be amenable to a wide range of heterogeneous data sources, co-design applications, and implementation variants, as discussed in Sect. 2.2. We, therefore, focus our evaluation on the related quality criteria from the literature [1,15]:

- **Integration (coordination)** is the ability to link and coordinate data and processes over systems.
- **Extensibility (modifiability)** is the ability to add, update, or extend a system with new functionalities with ease.
- **Reusability (variability)** is the ability of the architecture to be reusable and adaptable to produce new, derived architectures.
- **Modularity** is the ability of (parts of) the architecture to be decomposable, recomposable, and reconfigurable.

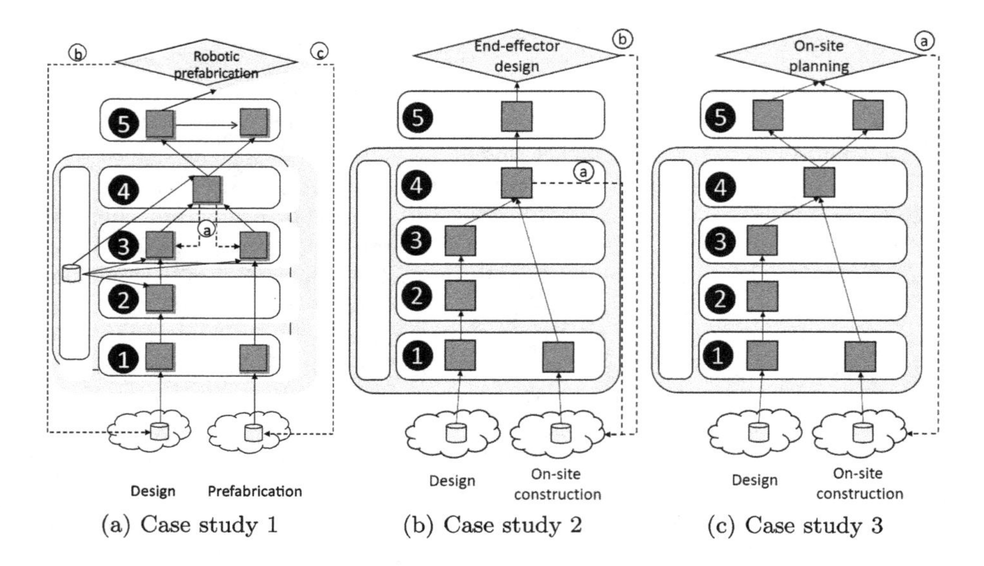

(a) Case study 1 (b) Case study 2 (c) Case study 3

Fig. 8. Overview of the mapping of scenarios to the ArchIBALD architecture

To conduct our evaluation based on the aforementioned case studies, we discuss how the ArchIBALD architecture accommodates through a mapping and reflect on how the mapping relates to the quality criteria. The mappings to the general architecture of the different case studies is shown in Fig. 8.

4.1 Case Study 1: Co-Design of Robotic Prefabrication

Mapping to Architecture. This case study introduces a condensed metadata layer, translating to a data schema fitting into the ArchIBALD metadata component. The pre-defined schema models the intricate interactions among building systems, design elements, materials, actors, tools, and the environment, as per Lee's ontology [11], present in design-to-fabrication processes. The schema covers four categories: (a) task identification, elucidating interactions involving building systems, design elements, and materials; (b) task detailing, focusing on interactions between design elements and actors/tools in the fabrication process; (c) task sequencing, delineating interactions among building components, actors, and the environment; and (d) task execution, representing interactions involving actors, tools, materials, and design elements.

The design-to-fabrication process relies on input data from both the design and fabrication domains ❶. The design domain contributes a detailed 3D model encompassing all cassettes and sub-elements—top and bottom plates, beams, acoustic panels, and electric wires. The fabrication domain provides digital shadows for each actor, outlining their corresponding end-effectors.

Based on the ingested design data, a task identification process extracts tasks from design elements to generate the appropriate fabrication tasks for the fabrication process of the design elements ❷.

To enrich the "partial" data from both design and prefabrication to eventually have data for all attributes of the four categories defined by the data schema, data enrichment services ❸ first forward the data they receive such that they can be properly linked as part of the integration layer. Linked data are then fed back ⓐ to allow enrichment services to properly assign tasks to the appropriate actor and to enrich the tasks with the corresponding actor information for the physical fabrication (i.e., inferring category (b) attribute values of the data schema). The inference relies on extracting 3D planes from the design elements and translating them to robotic axis values or executable robotic frames. This feedback loop enables iterative adjustments to the enriched data, as the design-to-fabrication process is optimized as discussed further below.

As mentioned before, data from the design and fabrication domain are linked, which maps to the integration layer ❹. This integration is guided by the global schema provided as metadata.

The linked data are then analyzed to perform task sequencing as a final preparation for the execution of fabrication tasks by the prefabrication platform ❺. This involves checking possible sequences to ensure they achieve a valid and executable fabrication process. When converged, the sequenced tasks are transferred to the physical fabrication actor. Mapped to the ArchIBALD architecture, this corresponds to a feedback channel ⓒ.

This triggers a second "loop" of data processing in this case study, as each executed task is monitored to quantify the performance of an actual task execution. Once available, these data are also injected and processed involving, again, services mapped to ❶, ❸, and ❹ in Fig. 8 (a). At the analysis/ML/AI level ❺, these additional data enable an additional service that predicts fabrication times for generated task sequences, such that task assignment and scheduling are refined. This is fed back to the fabrication domain ⓒ for improved execution, while this fabrication prediction model and monitored tasks are also relevant information for the design domain ⓑ, which can take this information into account for future design-to-fabrication processes.

Discussion. This first case study demonstrates that the ArchIBALD architecture can accommodate the functionality necessary to handle data as implemented in this case study. This case study heavily uses feedback channels and loops, showcasing the importance of this feature. We further observe that ArchIBALD offers the ability to link and coordinate data and processes over systems pertinent to this case study, relying on the data schema embedding into the metadata component. In terms of extensibility, this first use case shows that the general system architecture definition is easily amenable to specializations necessary to faithfully reflect the processing implemented as part of the case study. The architecture represented in Fig. 8(a) differs from the general architecture, thereby showcasing the variability of the ArchIBALD architecture, producing a derived architecture. This is linked to modularity that decomposes system functionalities that can be configured as needed by a use case. For instance, we observe that the configuration for the first case study does not require extraction services for the prefabrication domain.

4.2 Case Study 2: Co-Design of End-Effectors

Mapping to Architecture. At the level of data ingestion ❶, the design data of the building, e.g., a Rhino 3D building model, and additional information about the types and position of screws is provided by the designer. The vacuum gripper and the screwing effector used during on-site construction inject data about their configurable parameters, current values, and constraints. For the gripper, these parameters are given by the size of the vacuum plates and the distance in between. For the screwing effector, the parameters are the size of the screwing units, the screwing angle, and the dimensions of the screwing effector including the pin positions.

Data extraction involves two types of services applicable to the ingested design data ❷. One service extracts relevant cassettes for on-site assembly from the building model. Further services can transform the building design model from the Rhino .3dm file to other file formats, e.g., COLLADA or Wavefront .obj. This service simplifies the import of data by other programs in level ❺, e.g., CAD or Python, in order to visualize or analyze the geometry data.

The extracted design data can subsequently enter the data enrichment and transformation stage ❸. This involves enriching the design data with information that is relevant to the gripper and screwing effector design. This includes the determination of the mass for each cassette derived by the material density and volumes of the geometries. For the design of the screwing effector, the edge lengths and pinhole positions of each cassette are determined by analyzing the vertices positions and normal directions of the triangle mesh.

The relevant design data are linked with data ingested from the vacuum gripper and screwing effector during data integration ❹. This includes the connection of geometric constraints, e.g., the distance between the pinhole positions of the cassette and the distance between the pins of the screwing effector or the surface geometry of the bottom plate and the vacuum plates.

The integrated data form the basis for the analysis situated at the ArchIBALD architecture level of data analysis/ML/AI. They are the input to the optimization problem that determines an optimal end effector design based on both architectural design and end effector design constraints and requirements ❺. For instance, the vacuum gripper must be able to pick up and hold the weight of the cassette, while its surface area is in full contact with the bottom plate. The design of the screwing effector is optimized such that all pin holes and screwing positions are reachable for the dimensions of the screwing effector.

This case study includes two feedback channels. Feedback ⓐ allows the on-site construction engineer to visualize the designs and to examine/adapt the end effector design with CAD programs. This eventually results in feasible and optimized end effectors for on-site construction (feedback channel ⓑ).

Discussion. We observe that the ArchIBALD architecture accommodates all the functional requirements of this case study, as the mapping of relevant functionalities necessary to solve the optimization problem underlying the co-design application to the architecture is possible and complete. Concerning the individual quality criteria we consider, we see that the architecture effectively supports the integration of data from design and on-site construction. Additionally, it is adjustable to incorporate new functions, such as different enrichment, analysis, and optimization algorithms introduced above. By supporting this second case study as well, we have evidence that the general ArchIBALD definition is reusable and adaptable, customizing it to the application needs. We observe that this case study relies on a different configuration of services than case study 1 (and case study 3, see below), while the services defined here are partially reused in case study 3. This demonstrates modularity of the architecture design.

4.3 Case Study 3: Co-Design of On-Site Planning and Execution

Mapping to Architecture. For data ingestion, we reuse services already discussed for case study 2 ❶ to ingest design data as a Rhino 3D model with additional information about the assembly sequence and the kinematic model and load characteristics of the on-site manipulators. We further ingest the "result" of case study 2, i.e., data describing the optimized end-effector design.

Data extraction reuses the cassette extraction service of use case 2, which needs to be extended to also include the cassettes of the building structure at the different assembly steps ❷. The service for transforming the building design model from the Rhino 3D model to other file formats introduced in use case 2 is also reused here. Indeed, the resulting COLLADA or Wavefront .obj file formats are input formats that collision-checking algorithms in Matlab or Python use.

At the level of data enrichment and transformation ❸, the target poses for the TCP of the manipulators are inferred from the geometry data of the cassettes in order to grip and position the vacuum gripper and to position the screwing effector along the edges. This takes into account the geometries and the positioning of the optimized end effectors on the cassette. The target poses represent the position and orientation of the TCP with respect to a global building coordinate system. Furthermore, we need to determine the mass of each cassette, reusing another service from case study 2.

The relevant data are linked during data integration ❹. For each assembly step, the target pose for positioning the cassette or screwing effector is connected with the cassette geometry, the geometry of the building structure and the responsible manipulator.

As part of data analysis/ML/AI ❺, the integrated data are used to perform a reachability analysis. By solving the inverse kinematics problem and analyzing the load characteristics with the mass of the cassette, the reachability of the target poses for the two manipulators is analyzed, and their base position is optimized such that all cassettes can be positioned and screwed together. After determining the base position, the kinematic model and a polygon mesh of the manipulator are utilized to find a collision-free path toward the target pose using a bidirectional rapidly-exploring random trees algorithm. This requires joint consideration of the geometry data of the cassettes and the building structure. The path planning algorithm returns a sequence of joint positions.

This case study involves one feedback channel ⓐ to "implement" the result of the optimization on the construction site. The determined base positions are fed back to on-site stakeholders for manipulator setup. To execute the on-site assembly process, the sequence of joint positions for placing the cassettes and the screwing effector is passed on to the manipulator control unit that executes an online trajectory generation method.

Discussion. The ArchIBALD architecture fulfills all the functional requirements of this case study. It is feasible to map the relevant functionalities of the co-design application to the architecture. As in case study 2, the architecture enables a seamless integration of design and on-site construction data. The architecture is suited to encompass new functions, e.g., path planning algorithms, showcasing extensibility. The successful mapping of the third case study emphasizes the reusability of the ArchIBALD architecture. The reuse of several services from case study 2 validates modularity.

5 Conclusion

We presented three case studies used to evaluate the integrative ArchIBALD architecture, designed to support a variety of co-design applications in AEC by bridging common data silos through data integration and offering extensibility, reusability, and modularity. The evaluation validates that the system is suitable to its intended use, i.e., all case studies could be faithfully mapped to the ArchIBALD architecture. This is ensured by its integration capabilities, its inherent extensibility to new functionalities, its reusability by deriving different architectures for the different case studies, and its modularity allowing reuse of functions for different case studies.

While the evaluation validates that ArchIBALD is suitable for co-design applications in AEC, it also reveals several points that may need further investigation to design even more robust information system architectures for AEC. First, in case study 1, we observed that after an initial execution of a task sequence on the prefabrication platform, the services needed to process data are different from those that were used in the first iteration of the "forward" data processing along layers. More efficient processing requires a stronger "awareness"' and tracking of the processing state to better orchestrate services. A second observation is that the output of case study 2 serves as input to case study 3, indicating that we need to model interactions and feedback channels across different co-design applications as well, which is currently not covered by the ArchIBALD architecture, which mainly focuses on service configurations confined to a single case study. Both these aspects should be considered when further extending ArchIBALD to better fit AEC applications. We believe that the information system community also benefits from the details of the AEC case studies reported in this paper as practical reference for devising novel concepts.

Acknowledgement. Supported by the Deutsche Forschungsgemeinschaft (DFG, German Research Foundation) under Germany's Excellence Strategy - EXC 2120/1 - 390831618.

References

1. Clements, P., Kazman, R., Klein, M., et al.: Evaluating Software Architectures. Tsinghua University Press, Beijing (2003)
2. Doan, A., Halevy, A.Y., Ives, Z.G.: Principles of Data Integration. Morgan Kaufmann, Burlington (2012)
3. Dobrica, L., Niemela, E.: A survey on software architecture analysis methods. IEEE Trans. Softw. Eng. **28**(7), 638–653 (2002)
4. Duggan, J., et al.: The bigdawg polystore system. ACM Sigmod Rec. **44**(2), 11–16 (2015)
5. Fang, H.: Managing data lakes in big data era: what's a data lake and why has it became popular in data management ecosystem. IEEE International Conference on Cyber Technology in Automation, Control, and Intelligent Systems (2015)
6. Garcia-Molina, H., et al.: The TSIMMIS approach to mediation: data models and languages. J. Intell. Inf. Syst. **8**(2), 117–132 (1997)
7. Gazzarri, L., Herschel, M.: End-to-end task based parallelization for entity resolution on dynamic data. In: IEEE International Conference on Data Engineering (2021)
8. Inmon, W.H.: Building the Data Warehouse. Wiley, Hoboken (2005)
9. Lässig, N., Herschel, M., Reichle, A., Ellwein, C., Verl, A.: The ArchIBALD data integration platform: bridging fragmented processes in the building industry. In: Intelligent Information Systems - CAiSE Forum (2022)
10. Lauer, A.P.R., et al.: Automated on-site assembly of timber buildings on the example of a biomimetic shell. Autom. Constr. **156**, 105118 (2023)
11. Lee, S., Isaac, S., Bock, T.: An ontology for process information modeling. In: International Symposium on Automation and Robotics in Construction and Mining (2015)
12. Machado, I.A., Costa, C., Santos, M.Y.: Advancing data architectures with data mesh implementations. In: Intelligent Information Systems - CAiSE Forum (2022)
13. Mele, T.V., Liew, A., Echenagucia, T.M., Rippmann, M., et al.: COMPAS: a framework for computational research in architecture and structures (2017)
14. Nargesian, F., Zhu, E., Miller, R.J., Pu, K.Q., Arocena, P.C.: Data lake management: challenges and opportunities. Proc. VLDB Endowment **12**(12), 1986–1989 (2019)
15. Niu, N., Da Xu, L., Bi, Z.: Enterprise information systems architecture-analysis and evaluation. IEEE Trans. Ind. Inform. **9**(4), 2147–2154 (2013)
16. Poinet, P., Stefanescu, D., Papadonikolaki, E.: Collaborative workflows and version control through open-source and distributed common data environment. In: International Conference on Computing in Civil and Building Engineering (2020)
17. Wagner, H.J., Alvarez, M., Groenewolt, A., Menges, A.: Towards digital automation flexibility in large-scale timber construction: integrative robotic prefabrication and co-design of the BUGA Wood Pavilion. Constr. Robot. **4**(3–4), 187–204 (2020)

18. Wu, L., Li, Z., AbouRizk, S.: Automating common data integration for improved data-driven decision-support system in industrial construction. .J. Comput. Civ. Eng. **36**(2), 04021037 (2022)
19. Zaharia, M., Ghodsi, A., Xin, R., Armbrust, M.: Lakehouse: a new generation of open platforms that unify data warehousing and advanced analytics. In: Conference on Innovative Data Systems Research (CIDR) (2021)

Creating and Querying Data Cubes in Python Using PyCube

Sigmundur Vang(✉), Christian Thomsen, and Torben Bach Pedersen

Department of Computer Science, Aalborg University,
Selma Lagerløfs Vej 300, 9220 Aalborg, Denmark
{siva,chr,tbp}@cs.aau.dk

Abstract. Data cubes are used for analyzing large data sets usually contained in data warehouses. The most popular data cube tools use graphical user interfaces (GUI) to do the data analysis. Traditionally this was necessary since data analysts were not expected to be technical people. However, in the subsequent decades the data landscape changed dramatically requiring companies to employ large teams of highly technical data scientists in order to manage and use the ever increasing amount of data. These data scientists generally use tools like Python, interactive notebooks, pandas, etc. while modern data cube tools are still GUI based. To bridge this gap, this paper proposes a Python-based data cube tool called pyCube. pyCube is able to semi-automatically create data cubes for data stored in an RDBMS and manages the data cube metadata. pyCube's programmatic interface enables data scientists to query data cubes by specifying the metadata of the desired result. pyCube is experimentally evaluated on Star Schema Benchmark (SSB). The results show that pyCube vastly outperforms different implementations of SSB queries in pandas in both runtime and memory while being easier to read and write.

Keywords: Data cubes · Python · OLAP

1 Introduction

Data scientists are important to organizations by enabling data driven decision making aided by different data models and analysis methods. A data model that has historically been heavily used for data analysis is the data cube. End users were technically relatively unskilled at the time of data cubes' rise in popularity, resulting in a need for GUI clients to make it fast and easy to construct complex analytical queries using data cubes.

Today, organizations largely employ data scientists who in addition to analytical skills also have good technical skills and are comfortable with using tools like Python, pandas, Spark, etc. in interactive notebooks. To meet new requirements caused by the volume, velocity and variety of big data, there has been a switch from the rigid data cube model to the flexible data lake concept where it

R. Wrembel et al. (Eds.): DaWaK 2024, LNCS 14912, pp. 269–283, 2024.
https://doi.org/10.1007/978-3-031-68323-7_22

is easy to add new datasets. The data lake makes it possible for data scientists to use many different types of data analysis and machine learning methods.

Even though data analysis now is done in numerous and flexible ways, data cubes remain useful in many scenarios. Yet almost all data cube tools are still GUI based, which conflicts with the digital workspace of modern data scientists who need a tool that enables them to create and query data cubes without requiring them to leave their interactive Python notebook. This will increase the productivity of data scientists in two ways: (1) it eliminates the attention cost of switching between a GUI application and an interactive notebook and (2) it eliminates any compatibility issues between the two. For this we propose pyCube: a Python-based data cube tool for data scientists that can be used with Python code. pyCube can semi-automatically create data cubes, using data from RDBMSs, and automatically manages the data cube metadata. The data cubes can be queried using pyCube's programmatic interface and the results are returned as pandas [14] dataframes, which are well-known by data scientists.

The paper is organized as follows. Related work is described in Sect. 2, and formal definitions are given in Sect. 3. pyCube's declarative API is presented in Sect. 4. Section 5 explains how pyCube computes results. An experimental evaluation is given in Sect. 6. Section 7 concludes and points to future work.

2 Related Work

There already exist tools that enable programmatic access to data cubes. One is the Multidimensional eXpressions (MDX) [18] query language, whose syntax resembles SQL while its semantics does not. An MDX query `SELECT`s multiple axes, each consisting of a set of tuples, `FROM` a cube `WHERE` certain data is included. The tuples are made from dimension members that uniquely identify a cell in the cube. While powerful, MDX quickly becomes complex and difficult to read and write. As a consequence it is most often used in machine to machine communication. The syntax and semantics of pyCube are inspired by MDX, but unlike MDX there is not a heavy emphasis on sets and tuples in pyCube; instead, ordered lists are used. This way, pyCube constructs cube views by specifying the dimension values instead of selecting cells from the intersection of dimensions as is done in MDX. This increases the readability and writability of pyCube compared to MDX. MDX is fully declarative, while pyCube constructs cube views declaratively in a procedural environment, which is an approach superficially similar to Spark SQL [3]. Thus pyCube can be seen as MDX with variables in the same way as Spark SQL can be seen as SQL with variables.

TinyOlap [15] and *atoti* [1] already offer data cubes in Python. Both the data cube model and its processing are in-memory for TinyOlap and atoti, whereas pyCube pushes queries down to a DBMS. Multilevel hierarchies and joins are specified manually in atoti, while pyCube suggests multilevel hierarchies and automatically handles joins. TinyOlap uses a custom-made SQL-like query language and Python's array notation for querying. In atoti, the results must present dimension values by rows and measures by columns. Using pyCube's declarative

API, users can write queries that are easier to read and write, while offering users flexibility in structuring results similarly to how axes are used in MDX.

Enabling data cube analysis can also be done by reducing the development time of building dashboards and reporting applications for multidimensional data. One way to achieve that is by using OlaPy [13], which is an OLAP engine written in Python. With CSV and SQLAlchemy [4] compliant databases as data sources and support for MDX and XMLA [10], OlaPy makes it easy to develop reporting and analytical applications. Cube Dev's *Cube* [5] and Databrewery's *cubes* [6] provide methods for modeling the data in a semantic layer agnostic to the underlying physical implementation. This enables organizations to centralize the data modeling, creating an intermediate representation of the data model between all data sources and all reporting applications. Custom-made or off-the-shelf front-end applications can then perform data cube analysis by sending API requests to the semantic layer, which translates the query to the appropriate form. While the above tools help organizations build custom applications, many data scientists do not want to analyze data through a graphical application due to potential interoperability issues. The focus of pyCube is on the many use cases where data scientists work directly with data through Python.

Finally, multidimensional data analysis can also be facilitated by raising the abstraction level of the end-user's query interaction. A popular method is text-to-SQL [7,9], where the query interface leverages large language models (LLMs) to convert natural language queries to SQL queries. Another approach is by defining queries to be stated in terms of the data scientist's analytical goals rather than their data requirements [17]. Conversely, query answers consist of data narratives, models and highlights, rather than just plain cuboids. Even in the light of these valuable methods, data scientists typically need to control the exact data representation when creating analytics reports. pyCube helps them do that with concise Python code when working with data cubes.

3 Preliminaries

This section gives formal definitions for all relevant concepts. Some definitions are inspired by [2].

A LEVEL SCHEMA $ls = (k, A)$ is a tuple, where $k \subseteq A$ is the LEVEL KEY consisting of one or more attributes, and A is a set of attributes. A LEVEL INSTANCE of ls is a function that maps values of k to tuples in $dom(a_1) \times \cdots \times dom(a_n)$ for $a_1, \ldots, a_n \in A$ where $dom(a_i)$ denotes the domain of a_i, i.e., the set of possible values for a_i. A DIMENSION SCHEMA $ds = (LS, \preceq)$ is a tuple, where LS is a set of level schemas including the level schema $ALL_{ds} = (\{all\}, \{all\})$ and $\preceq$ is a partial order on the level schemas $ls \in LS$. The ALL_{ds} level schema includes the *all* attribute, which is a special attribute that only takes the value $\top$, and the level instance of ALL_{ds} is given by $ALL_{ds}(all) = (\top)$. The ALL_{ds} level schema of ds is the top level schema, such that $ls \preceq ALL_{ds}$ for all level schemas $ls \in LS$. Let LI be an ordered set of level instances with exactly one level instance for each level schema in LS. A DIMENSION INSTANCE $d = (LM, R)$

of ds is a tuple, where $LM = \{LM_1, LM_2, \ldots, LM_{|LI|}\}$ is an ordered set of disjoint multisets of attribute values for each level instance $li \in LI$ called LEVEL MEMBERS, and $R = \{R_1, R_2, \ldots R_{|LS|}\}$ is an ordered set of sets of ROLL-UP functions, one set for each set of level members, such that if $r(lm_i) = lm_j$ for some $r \in R_i$ then $ls_i \preceq ls_j$, for every $lm_i \in LM_i$ and $lm_j \in LM_j$, together with their associated level schemas ls_i and ls_j, where $1 \leq i, j \leq |LS|$. Let LM_ℓ denote the set of level members associated with level instance $\ell \in LI$.

A MEASURE SCHEMA $ms = (a, \text{AGG})$ is a tuple, where a is an attribute with values in $\mathbb{R}$ and AGG is an aggregate function which takes in a multiset of real numbers and returns a value $r \in \mathbb{R}$. Measure schemas can be combined with numeric operators to create calculated measure schemas. Given two measure schemas $ms_1 = (a_1, \text{AGG})$ and $ms_2 = (a_2, \text{AGG})$, the expression $ms_1 * ms_2$ results in a new measure schema $(a_1 * a_2, \text{AGG})$. Note that the numeric operator is bound to the attribute values and not the attributes themselves and that the aggregation function in ms_1 and ms_2 must be the same.

A DATA CUBE SCHEMA $cs = (DS, MS)$ is a tuple, where DS is a set of dimension schemas, and MS is a set of measure schemas. Let D_I be a set of dimension instances with exactly one dimension instance for each dimension schema in DS and LM_i be the set of level members in the ith dimension instance. Furthermore let $B_{D_I} = \times_{j=1}^{|D_I|} LM_j$ be the cartesian product of the level members across all dimension instances, and let $B_{MS} = (\mathbb{R} \cup \{\perp\})^{|MS|}$ be the numerical values of the measures where $\perp$ denotes no value. A DATA CUBE INSTANCE $c : B_{D_I} \to B_{MS}$ is a function that maps combinations of level members from B_{D_I} to the numerical values of the measures from B_{MS}. An AXIS $ax = (d, \ell, a, lm)$ is a 4-tuple, where d is a dimension instance, $\ell = (k, A) \in LS$ is a level schema, $a \in A$ is an attribute on ℓ and lm is a list of level members from LM_ℓ. An expression of the form TRUE, FALSE, or $a\ \omega\ lit$ is a predicate where a is an attribute, $\omega \in \{<, \leq, =, \neq, \geq, >\}$ is a numeric operator and lit a literal. If p_1, p_2 are predicates, then (p_1), $p_1 \wedge p_2$, and $p_1 \vee p_2$ are predicates. The evaluation order is defined by the standard operator bindings. A DATA CUBE VIEW $v = (AX, MS_v, p, c)$ is a 4-tuple where $AX = \{ax_0, ax_1, \ldots, ax_n\}$ and $MS_v = \{ms_1, ms_2, \ldots, ms_m\}$ are ordered sets of axes and measure schemas respectively, p is a predicate and c is the corresponding data cube instance. The axes in a data cube view contain information on how the data in the data cube instance c should be presented while the measure schemas contain information on what measures should be shown. The first measure schema in MS_v is the default measure. The predicates are applied over axes. Data cube views are the main construct in pyCube. In a pyCube session there are usually many data cube views which are constructed from querying data cube instances. In contrast data cube instances are defined once at the beginning of a pyCube session and never change afterwards. Each data cube instance c has a default data cube view $v = (AX, MS_v, p, c)$ where AX contains a 4-tuple (d, ℓ, a, lm) for each dimension instance d in c where ℓ is the lowest level schema in the dimension instance, a is the attribute that contains the set LM_ℓ and lm is a list containing all level members from LM_ℓ, MS_v contains one measure schema from c and $p =$ TRUE.

4 Use Case

This section shows how to interact with pyCube using a scenario for a data scientist called Helle. Helle works for a Danish company that sells children clothes. The company collects sales data into a PostgreSQL database structured using a snowflake schema [8]. The sales are modelled as facts, with every fact having 4 dimensions: supplier, store, product, and sale date and 2 measures: total sales price and the unit sales. Helle gets the task of analysing the 2022 sales. She wants to solve the task using data cubes in an interactive Python notebook. However it has been difficult for Helle to incorporate data cubes in data analysis documents using only Python due to the previously limited data cube capabilities in Python. Therefore she uses pyCube in an interactive notebook to solve the task.

4.1 Initializing PyCube

```
1  >>> import pyCube
2  >>> postgres_eng = pyCube.engines.postgres(dbname="salesdb", user="helle",
3                          password="secret", host="127.0.0.1", port="5432")
4  >>> session = pyCube.create_session(engine=postgres_eng)
5  >>> session.views
6  ['Sales', 'HR']
7  >>> sales_view = session.Sales
8  >>> sales_view.measures
9  [total_sales_price, unit_sales]
10 >>> sales_view.dimensions()
11 [Supplier, Store, Product, Date]
12 >>> sales_view.Date.hierarchies()
13 [day, month, year, ALL]
14 >>> sales_view.Date.year.attributes()
15 ['year']
16 >>> sales_view.Date.year.year.members()
17 [2021, 2022, 2023, 2024]
18 >>> sales_view.Date.year.year[2022].children()
19 [January, February, March, April, May, June, July, August, September, ...]
```

Listing 1. Exploring the data cube using pyCube

Helle writes the expressions in Listing 1. Helle imports pyCube and initializes it to use a PostgreSQL database by creating a postgres engine with the relevant connection details in Lines 1–3 and creates a pyCube session in Line 4. The cubes are automatically inferred by pyCube by first identifying the fact table and then traversing the foreign keys in a snowflake manner. For full details see [16]. The cubes inferred by pyCube are data cube instances. Cubes are accessed exclusively through a view in pyCube, as data cubes do not hold information about the representation of its containing data. Default cube views are created for each cube. Helle lists the names of the available views in Line 5, which returns `'Sales'` and `'HR'`. Helle saves the `Sales` view in the `sales_view` variable in Line 7. First of all Helle wants to examine the metadata of the `Sales` view. Line 8 lists the measure schemas defined in the view, which returns `[total_sales_price, unit_sales]`. Line 10 lists the dimension schemas contained in the view which, returns `[Supplier, Store ↪ , Product, Date]`. Helle notices a `Date` dimension schema and inspects the levels of the schema with Line 12, which returns `[day, month, year, ALL]`. The returned result is a list of level schemas, which should be interpreted as `day` $\longrightarrow$ `month` $\longrightarrow$ `year` $\longrightarrow$ `ALL`. Helle inspects the attributes on the `year` level schema with Line 14, which returns `[year]`. Helle accesses attributes and attribute values on the `year` level and `year` attribute using the expressions `sales_view.Date.year.year` and `sales_view.Date.year.` $\hookrightarrow$ `year[2022]`, respectively. Accessing attribute values whose names either contain spaces or begin with numbers uses a dictionary-like syntax, i.e., using square brackets around the value, since Python only allows variable names beginning with letters. Helle can access attribute values that begin with letters with the expression `sales_view.Product.category.category.Blouse`, which uses dot notation.

The `members` method, when invoked on an attribute a, returns a list of all attribute values in a. Helle uses the `members` method in Line 16 to access all years in the database. The `children` method, when invoked on an attribute value v, returns a list of all attribute values that have v as parent. Helle uses the `children` method in Line 18 to access all months in 2022.

4.2 Analyzing the Data in the View

Helle is able to group the data in the `sales_view` $= (AX, MS_v, p, c)$ view using the `axis`, `measures` and `where` methods in pyCube as shown in Listing 2.

```
1 sales_view\
2   .axis(0, sales_view.Date.year.year[2022].children())\
3   .axis(1, sales_view.Product.category.category.members())\
4   .where((sales_view.Date.day.day <= 7)
5          & (sales_view.Supplier.nation.nation == "Denmark"))\
6   .measures(sales_view.UnitSales)\
7   .output()
```

Listing 2. Analyzing the data using pyCube methods

The `axis(`i, lm`)` method specifies the values on the axes $AX = \{(d_0, \ell_0, a_0, lm_0), (d_1, \ell_1, a_1, lm_1), \ldots, (d_n, \ell_n, a_n, lm_n)\}$ in a view. Recall that AX is an ordered set. The parameters are a natural number i, specifying what axis to modify, and a list of level members lm. The level members must be absolute paths, which consist of a cube, dimension, the highest (non-$\top$) level and one or more level members. Level members in lower levels can be reached through level members in higher levels. In general, level members in a level l can only be accessed from level members in a level directly above l. The `members` and `children` methods return absolute paths. The dimension, level and attribute can be inferred from the level members. The result of the `axis` method is a view. Helle specifies the first axis to be the months in 2022 with Line 2 in Listing 2. Line 3 specifies the second axis to be the product categories. Note that if axis n is specified then axis $n-1$ should also be specified for $n > 0$ when `output()` is invoked. pyCube includes the `columns`, `rows`, `pages`, `sections` and `chapters` methods as aliases for the `axis` method where the i parameter have been set to 0, 1, 2, 3 and 4, respectively. Therefore the `columns(sales_view.Date.year.year[2022].children())` and `rows(` $\hookrightarrow$ `sales_view.Product.category.category.members())` expressions are equivalent to Lines 2 and 3, respectively. The `where(`p`)` method, where p is a predicate of the form given in the preliminaries, specifies the predicate in the data cube view. Helle limits

Table 1. Some of the result from invoking `view.output()`

	January	February	March	April	May	June	July	...
	UnitSales	UnitSales	UnitSales	UnitSales	UnitSales	UnitSales	UnitSales	...
Blouse	754	659	844	827	793	856	883	...
Pants	378	129	928	453	392	427	471	...
...	...	...	...	...	...	...	...	...

the view to only include the first week in every month in Line 4. Predicates can be combined using logical and (`&`) or logical or (`|`). Helle furthermore limits the result to only include Danish suppliers with the `where` method in Line 5. Note that since `&` and `|` have higher precedence than the (in)equality in Python, the predicates on either side of the logical and or the logical or need to be surrounded with parentheses. The `measures(`$ms_1, ms_2, \ldots$`)` method specifies the values in the measure schemas $MS_v = \{ms_1, ms_2, \ldots, ms_m\}$ according to the order given in the parameters. The parameters of `measures` is any number of measure schemas ms_i. Helle includes the unit sales measures in Line 6. The `output()` method on Line 7 populates the view with values by sending an SQL query to the PostgreSQL database and reformatting the SQL result set into a pandas dataframe which is partially shown in Table 1. The SQL generation is explained in Sect. 5.1.

The retrieved values are obtained using the unit sales measures.

5 Populating the View

When populating a view $v = (AX, MS_v, p, c)$ with values (i.e., when using the `output` method on v) an SQL query is generated from v. The query retrieves the values of the cells in c. Subsequently the values are formatted into a pandas dataframe df. Finally df is returned to the user. The SQL generation is described in Sect. 5.1 and the formatting of the dataframe is described in Sect. 5.2.

5.1 Generating the SQL Query

Given a data cube view $v = (AX, MS_v, p, c)$, where $AX = [(d_0, \ell_0, a_0, lm_0), (d_1, \ell_1, a_1, lm_1), \ldots, (d_n, \ell_n, a_n, lm_n)]$, $MS_v = \{(a^0, \text{AGG}_0), (a^1, \text{AGG}_1), \ldots, (a^m, \text{AGG}_m)\}$, p is a predicate and c is the data cube instance, then the SQL generated from `v.output()` is given in Listing 3. The SQL generation is divided into three functions which are explained in the following.

The `get_from_clause_subset(`d`)` function generates an SQL query subset to join the dimension tables in d with the fact table. An example is shown in [16]. The `get_from_clause_subset` function also accepts a list of dimensions, which is equivalent to sequentially invoking the method on each individual element in the list. The `inclusion_where_clause_subset(`AX`)` is a function that produces a series of $\ell_i.a_i$ `IN` ↪ `(`lm_i`)` expressions for every ax_i in AX. The expressions are separated with ANDs. The `pred_where_clause_subset(p)` is a function that produces a series of $\ell_{p_i}.a_{p_i}$ ω_i lit_i expressions for every p_i in p. The expressions are separated either with an

AND or an OR depending on the user-provided predicates. Parentheses are placed according to the evaluation order. The dimension schemas included in p but not in AX are indicated by $d_{p_1}, d_{p_2}, \ldots, d_{p_i}$. The key on level ℓ is denoted by $\ell.k$.

```
SELECT ℓ0.a0, ℓ1.a1, ..., ℓn.an,
       AGG0(ft.a^0) AS a^0, AGG1(ft.a^1) AS a^1, ..., AGGm(ft.a^m) AS a^m
FROM ft get_from_clause_subset([d0, d1, ..., dn, dp0, dp1, ..., dpi])
WHERE inclusion_where_clause_subset(AX) AND (pred_where_clause_subset(p))
GROUP BY ℓ0.a0, ℓ0.k, ℓ1.a1, ℓ1.k, ... ℓn.an, ℓn.k
```

Listing 3. Generated SQL query

The generated SQL selects the attributes in all axes first and all measures second, both in the order they appear in the view v. The measures are given the attribute name as aliases. The fact table is joined with all dimensions used in the axes AX and all tables used in the predicate p. The WHERE clause consists of the inclusion_where_clause_subset and pred_where_clause_subset methods with an AND between them. The pred_where_clause_subset method is surrounded by parentheses to ensure the correct precedence between conjunctions and disjunctions. Finally the attributes in the axes AX are used in the group by clause. The keys are also included in the group by clause to ensure that the correct attributes are grouped together. An example of a pyCube expression is shown in Listing 4.

```
sales_view\
  .columns(sales_view.Date.year.year[2022].children())\
  .rows([sales_view.Product.category.category.Blouse,
         sales_view.Product.category.category.Pants])\
  .pages([sales_view.Store.city.city.Aalborg])\
  .where((sales_view.Date.month.month == "January")
         | (sales_view.Date.month.month == "February"))\
  .measures(sales_view.TotalSalesPrice, sales_view.UnitSales)\
  .output()
```

Listing 4. Example pyCube expression

The pyCube expression shown in Listing 4 generates a dataframe where months of the year 2022 are on the columns, the "Blouse" and "Pants" categories are on rows and the "Aalborg" city is on pages. For brevity, only a single level member is specified in the pages method. The where method further specifies that the columns should only be the months of January and February and the measures method specifies that two measures should be used: TotalSalesprice and UnitSales. The pyCube expression generates a view (AX, MS, p, c) where $AX = [(Date, month, month, lm_0), (Product, category, category, [Blouse, Pants]), (Store, city, city, [Aalborg])]$, $MS = \{(TotalSalesPrice, SUM), (UnitSales, SUM)\}$, $p = (month = January \vee month = February)$ and c is the data cube instance. The list lm_0 consists of all months in 2022. The result set from the generated SQL is shown in Table 2(a). TSP and US are abbreviations for TotalSalesPrice and UnitSales, respectively. See [16] for an elaborated example.

5.2 Converting Result Sets to Dataframes

Listing 5 shows the conversion of an SQL result set into a pandas dataframe conforming to the metadata specified by a pyCube expression.

```
1  df = pd.read_sql(query, conn)
2  final_df = df.pivot(columns=columns, index=rows, values=measures)
3  final_df = final_df.reorder_levels(list(range(1, len(columns)+1)) + [0],
4                                     axis=1)
```

Listing 5. Converting SQL result set

Three pandas methods are used: `read_sql`, `pivot` and `reorder_levels` [14]. Given an SQL query and a connection to a database, the `read_sql` method returns the result set formatted as a dataframe. The connection is created using a Python database adapter. The `pivot` method returns, when given the `columns`, `index` and `values` parameters, a dataframe pivoted according to the parameters. The `columns` and `index` must be one or more valid column names from the result set generated from the `output` method. If the `columns` or `index` parameters are given multiple names, then `pivot` creates a hierarchical structure on the columns or rows, respectively. Finally the `reorder_levels` method in Lines 3–4 ensures that the measures are furthest down in the column hierarchy. If `df` on Line 1 contains Table 2(a), then the `columns`, `index` and `measures` parameters on Line 2 contain $[Month, City]$, $[Category]$ and $[TSP, US]$, respectively, and the entire Listing 5 produces the dataframe shown in Table 2(b).

6 Experiments

To measure the benefit data scientists attain by choosing pyCube over pandas [14] for their data cube use cases, a comparison between pyCube and pandas is made. atoti and tinyOlap are not used in our experiments. atoti is available under a commercial license and an evaluation license with restrictions, while tinyOlap used 18.8GB of memory when storing only a small subset of the smallest dataset used in the experiments. The source code for pyCube and the experiments is available from https://github.com/FittiSigmund/cube/.

6.1 Experimental Setup

Pandas is separated into three baselines: (1) JoinFactsFirst (JFF), (2) JoinDimensionsFirst (JDF) and (3) SQL Join (SQL J). JoinFactsFirst and JoinDimensionsFirst load all relevant tables into memory one by one as pandas dataframes.

Table 2. (a) Result set for the SQL and (b) DataFrame generated by Listing 4

Month	Category	City	TSP	US
January	Blouse	Aalborg	94651	754
January	Pants	Aalborg	84659	378
February	Blouse	Aalborg	46895	659
February	Pants	Aalborg	12054	129

(a)

	Aalborg			
	January	January	February	February
	TSP	US	TSP	US
Blouse	946513	754	468954	659
Pants	846598	378	120546	129

(b)

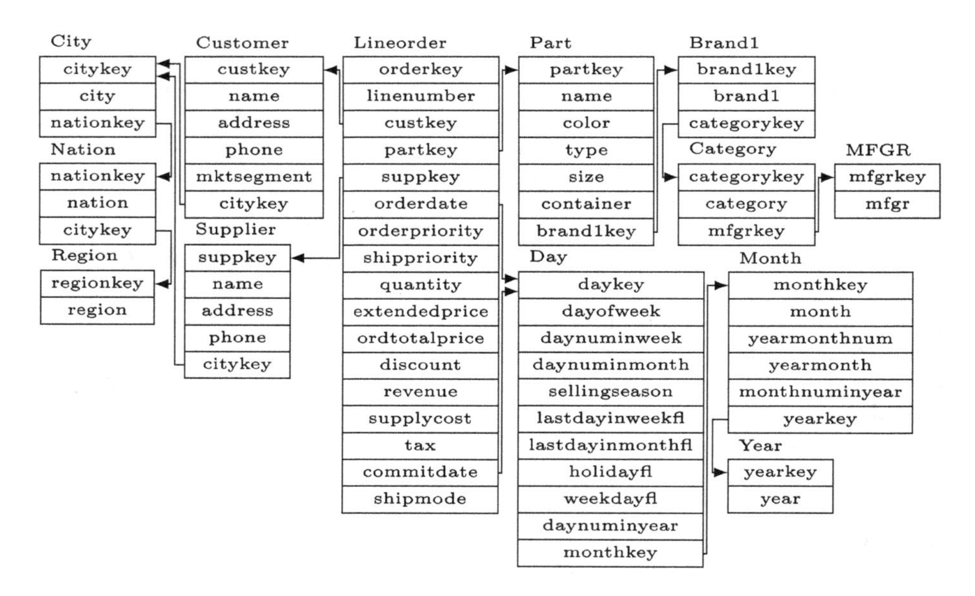

Fig. 1. The snowflake schema for SSB

JoinDimensionsFirst furthermore only loads the relevant columns of each table. JoinDimensionsFirst denormalizes all hierarchies by joining the relevant dimension tables before joining the hierarchies with the fact table. JoinFactsFirst joins the fact table and the lowest level first. Then the second lowest level is joined with the intermediate join result. This process is repeated for all relevant levels in a hierarchy and for all relevant hierarchies. Note that the joins are computed one at a time using the `merge` method on dataframes. Every intermediate result is stored as a new dataframe. `SQL` Join formulates all joins as a single `SQL` query and stores the result as a dataframe. After joining the tables, all baselines filter the resulting dataframes in the same manner. Some experiments include calculated measures which are computed using the `apply` method on dataframes. Finally the dataframes are reshaped into the final result using the `pivot_table` method on dataframes. The baselines represent different levels of skill required by the data scientist with JoinFactsFirst being the easiest and `SQL` Join the hardest. The joins, calculated measures, filtering and reshaping of the dataframes are done in the way the pandas user guide recommends [12].

The used database schema (shown in Fig. 1) is a snowflaked Star Schema Benchmark (SSB) [11]. Data is generated from SSB's data generator and converted into the correct schema. The fact table has 6,000,000 rows multiplied by a scale factor. The total data size used in the experiments is 581MB for a scale factor of 1. The SSB is based on the popular TPC-H benchmark. The queries in SSB are grouped into four groups named query flights. All queries from SSB are used and are implemented four times: once using pyCube and once for each of the baselines using pandas. All query and implementation combinations

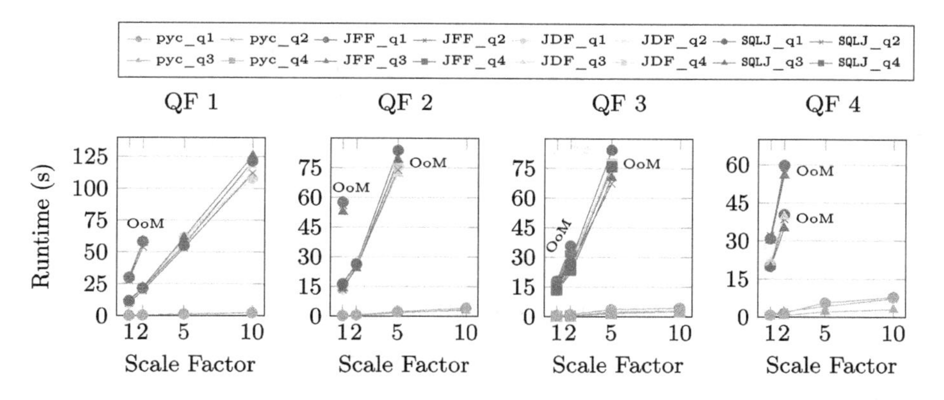

Fig. 2. Laptop runtimes growth over scale factors

are executed five times in randomized order and the highest and lowest values are discarded. The final result is the average of the remaining values. Unless otherwise noted, all experiments are run on a Arch linux machine with kernel version 6.4.12-arch1-1, four cores of 11th Gen Intel Core i5-1135G7 running at a 2.40 GHz clock frequency and 32 GB RAM using PostgreSQL 15.4, Python 3.11 and pandas 2.0.1. The system is similar to what a data scientist would use.

6.2 Data Retrieval Speeds

Figure 2 shows the runtime of pyCube and the baselines for scale factors 1, 2, 5 and 10. The runtime is in seconds of wall clock time and is measured using Python's `time` module. Queries that ran out of memory are marked with `OoM`.

pyCube vastly outperforms all baselines for all SSB queries and is in some instances two orders of magnitude faster. The baselines scale worse than pyCube with baseline runtimes for QF 1 ranging from 10 to 120 s for scale factors 1 and 10, respectively, while pyCube runtimes range from 0.2 to 2.5 s. Only pyCube is able to perform all SSB queries for all scale factors without running out of memory. JoinDimensionsFirst and `SQL` Join are able to perform QF 1 for scale factor 10. Otherwise all baselines failed to perform any QF for scale factor 10. Furthermore all baselines were unable to perform QF 4 for scale factor 5 while scale factor 2 was the highest JoinFactsFirst was able to achieve before running out of memory. Figure 3 shows the runtime of the individual queries for scale factors 1 and 10. The runtimes for scale factors 2 and 5 can be seen in the extended version of the paper [16]. The runtime is split into time spent in Python and time spent in the database. The time spent in the database constitutes the time when control has been passed on to the Python database adapter until the adapter returns with a result. The remaining time is Python time. The runtimes for pyCube are too small to produce a noticeable bar so the combined total runtime is given where the bar should have been. The vast majority of time in pyCube is spent in the database. This is because pyCube only spends its Python

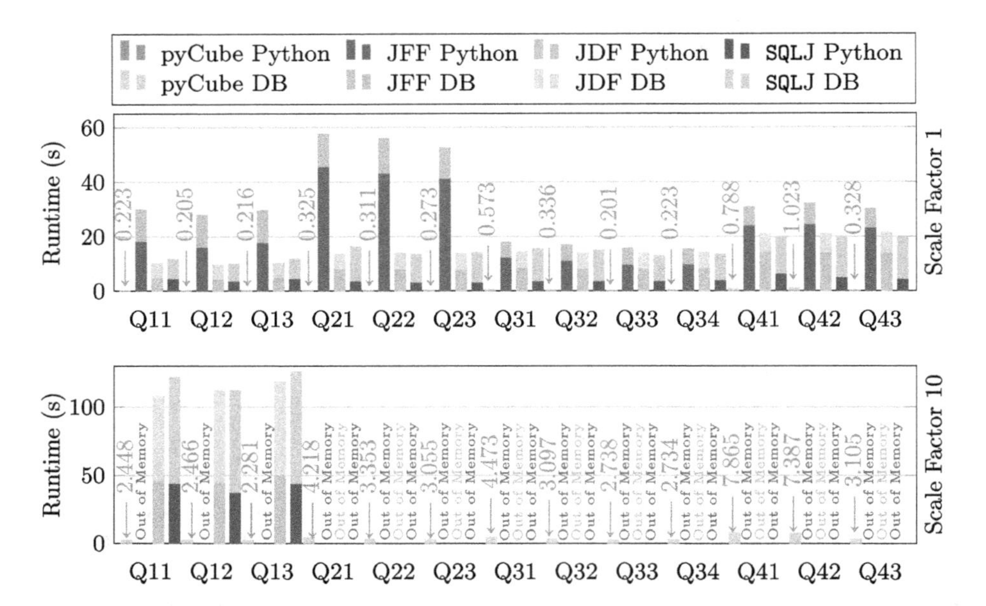

Fig. 3. Comparing the runtime performance of pyCube and the baselines

time generating SQL and converting the result set to a dataframe. The SQL query generated by pyCube includes all tables to be joined, all predicates in the WHERE clause and the aggregation function in the SELECT clause. This lets the DBMS to do the usual optimizations such as predicate pushdown, on-the-fly aggregation and join-order selection.

The baselines are slowed down by loading large amounts of data into memory and by filtering and reshaping the large dataframes in Python. This can especially be seen for JoinFactsFirst where most of the time is spent in Python as seen in Fig. 3. This is because the hierarchies are joined with the fact table first which creates new copies of the fact data for every join. Furthermore joins are eagerly evaluated in pandas. JoinFactsFirst and JoinDimensionsFirst have more or less the same Python and database time split. However, JoinDimensionsFirst is generally faster than JoinFactsFirst since the smaller dimension tables are joined before joining them with the fact table. SQL Join is significantly slower than pyCube. This is because SQL Join does not include predicates in the WHERE clause or any aggregation function in the SELECT clause. The DBMS joins all tables for SQL Join and more time is spent waiting for the database. The Python and database time split is generally the same for scale factors 2, 5 and 10.

6.3 Memory Usage

Figure 4 shows the memory usage of pyCube and the baselines measured using the time tool. The queries that ran out of memory are noted with OoM. The memory shown is the resident set size of a process in GB which includes all stack and

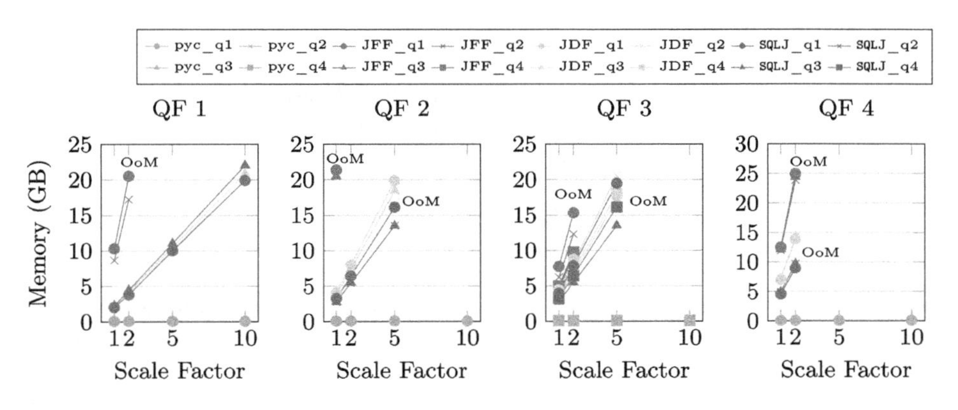

Fig. 4. Laptop memory growth over scale factors

heap memory of a process in addition to the shared libraries in memory. It does not include swapped out memory. The charts show the same pattern found in Fig. 3 with pyCube outperforming the baselines across all queries for all scale factors. The baselines' memory scales linearly with the scale factor while pyCube's memory remains constant. As a result JoinFactsFirst runs out of memory on scale factors 5, 2 and 5 for QFs 1, 2 and 3, respectively. JoinDimensionsFirst and `SQL` Join run out of memory on scale factor 10 for QFs 2 and 3 while all baselines run out of memory on scale factor 5 for QF 4. The experiments were repeated on a high-end server running Ubuntu, with 16 cores of AMD Epyc 7302P processor running at a 3.2GHz clock frequency and 264GB RAM to see how much memory the baselines needed in order to perform the SSB queries on higher scale factors. JoinFactsFirst uses the most memory with the highest being about 200GB on QF 2 and in general approaches 100GB on scale factor 10 for all QFs. JoinDimensionsFirst and `SQL` Join's memory ranges between 30GB and 70GB for scale factor 10 on QFs 2 to 4. These queries cannot fit on laptops that data scientists use. For full details, see [16].

6.4 Code Comparison

Listing 6 shows SSB query 4.1 implemented with a single statement in pyCube.

```
view.columns(view.date.year.y_year.members())\
    .rows(view.customer.nation.n_nation.members())\
    .where((view.customer.region.r_region == "AMERICA")
           & (view.supplier.region.r_region == "AMERICA")
           & ((view.part.mfgr.m_mfgr == "MFGR#1")
              | (view.part.mfgr.m_mfgr == "MFGR#2")))\
    .measures(profit=view.lo_revenue - view.lo_supplycost)\
    .output()
```

Listing 6. Query 4.1 of SSB in pyCube

Table 3 compares the code sizes of the implementations. The definition of a statement in the grammar specification for Python is used. The pyCube query

Table 3. Code size differences between pyCube and the JDF baseline

	pyCube	JDF	JDF reduced
Characters	412	2725	2571
Statements	1	32	22
Tables	13	13	13

has 6.6 times fewer characters than the JDF query and 31 fewer statements. The statement comparison may seem skewed since expressions in pyCube are chained into a single statement. However pyCube allows expressions to be chained while maintaining readability (as shown in Listing 6), which pandas only can do to a limited extent. JDF reduced in Table 3 is the smallest form of the JDF query without making the implementation overly convoluted. The reduction is achieved by chaining `merge` methods such that the dimensions are joined using one statement each and the fact table is joined with the dimensions using an additional statement. Even with the reduction, the pyCube query still has 6.2 times fewer characters and 21 fewer statements. Both the JDF and the reduced JDF query can be seen in the extended version of the paper [16]. A major reason for the increased complexity of reading and writing the JDF query is due to the higher user responsibilities in pandas than in pyCube. In pandas, the user needs to manually fetch the data from the database and then merge, filter and pivot the dataframes correctly. In fact, 13 of the 32 statements in the JDF query come from the many tables that need to be fetched from the database. This is because the data is structured in a snowflake schema. However, even if the data was structured in a star schema, there still would need to be five `read_sql` statements (four dimensions and one fact table) and four invocations of the `merge` method just for loading the data correctly into memory. The `read_sql` and `merge` methods are well designed and easy to write but writing nine invocations of two methods back-to-back is repetitive which will increase the chances of error from the user. Additionally, when merging dataframes, all relevant keys must be provided and column names need to be given suffixes in case of column name clashes. In contrast, the user responsibility in pyCube is specifying the metadata and values of the desired dataframe using methods with meaningful names and then calling `output` as is shown in Listing 6. This results in significantly more compact code that is easier to read, write, debug and maintain.

7 Conclusion and Future Work

This paper has presented pyCube: a Python-based data cube tool. pyCube has been designed to match the suite of tools usually employed by data scientists. The user interface of pyCube has been shown alongside how pyCube processes user queries. pyCube has been experimentally evaluated. The results show that pyCube outperforms pandas both in runtime and in memory for data cube analysis. Future work includes expanding pyCube: (1) to be able to distributively manage and query data cubes using data from flat files, (2) by providing more ways to handle metadata and (3) by handling schema updates gracefully.

References

1. ActiveViam: atoti's. https://atoti.io. Accessed 15 Sept 2023
2. de Aguiar Ciferri, C.D., et al.: Cube algebra: a generic user-centric model and query language for OLAP cubes. Int. J. Data Warehous. Min. **9**(2), 39–65 (2013). https://doi.org/10.4018/jdwm.2013040103
3. Armbrust, M., et al.: Spark SQL: relational data processing in spark. In: Sellis, T.K., Davidson, S.B., Ives, Z.G. (eds.) Proceedings of the 2015 ACM SIGMOD International Conference on Management of Data, pp. 1383–1394. ACM (2015). https://doi.org/10.1145/2723372.2742797
4. Bayer, M.: Sqlalchemy. In: Brown, A., Wilson, G. (eds.) The Architecture of Open Source Applications Volume II: Structure, Scale, and a Few More Fearless Hacks. aosabook.org (2012). http://aosabook.org/en/sqlalchemy.html
5. Cube: Cube's. https://cube.dev/. Accessed 15 Sept 2023
6. Databrewery: Databrewery's cubes. http://cubes.databrewery.org/. Accessed 15 Sept 2023
7. Kim, H., So, B., Han, W., Lee, H.: Natural language to SQL: where are we today? Proc. VLDB Endow. **13**(10), 1737–1750 (2020). https://doi.org/10.14778/3401960.3401970
8. Kimball, R., Ross, M.: The Data Warehouse Toolkit: The Complete Guide to Dimensional Modeling, 2nd. edn. Wiley, Hoboken (2002). https://www.worldcat.org/oclc/49284159
9. Ma, P., Wang, S.: MT-Teql: evaluating and augmenting neural NLIDB on real-world linguistic and schema variations. Proc. VLDB Endow. **15**(3), 569–582 (2021). https://doi.org/10.14778/3494124.3494139. http://www.vldb.org/pvldb/vol15/p569-ma.pdf
10. Microsoft: XMLA reference. https://learn.microsoft.com/en-us/analysis-services/xmla/xml-for-analysis-xmla-reference?view=asallproducts-allversions. Accessed 17 Mar 2024
11. O'Neil, P.E., O'Neil, E.J., Chen, X.: The star schema benchmark (SSB). Pat **200**, 50 (2007)
12. Pandas: Pandas user guide. https://pandas.pydata.org/docs/user_guide/index.html. Accessed 01 Oct 2023
13. SAS, A.: Olapy's documentation. https://olapy.readthedocs.io/en/latest/. Accessed 17 Mar 2024
14. Team pandas development: pandas-dev/pandas: Pandas (2020). https://doi.org/10.5281/zenodo.3509134
15. TinyOlap: Tinyolap. https://tinyolap.com/. Accessed 18 Mar 2024
16. Vang, S., Thomsen, C., Pedersen, T.B.: Creating and querying data cubes in python using pyCube. Technical report (2023). https://doi.org/10.48550/arXiv.2312.08557
17. Vassiliadis, P., Marcel, P., Rizzi, S.: Beyond roll-up's and drill-down's: an intentional analytics model to reinvent OLAP. Inf. Syst. **85**, 68–91 (2019). https://doi.org/10.1016/J.IS.2019.03.011
18. Whitehorn, M., Zare, R., Pasumansky, M.: Fast track to MDX, 2nd edn. Springer, Cham (2006). https://doi.org/10.1007/1-84628-182-2

An E-Commerce Benchmark for Evaluating Performance Trade-Offs in Document Stores

Dimitri Van Landuyt(✉), Marie Levrau, Vincent Reniers, and Wouter Joosen

KU Leuven, B-3001 Leuven, Belgium
{dimitri.vanlanduyt,marie.levrau,vincent.reniers, wouter.joosen}@kuleuven.be

Abstract. In document stores, application developers are faced with complicated schema design trade-off decisions between embedding or referencing of document relationships. However, existing performance benchmark systems for document stores either work with simplistic and unrepresentative documents, or involve non-document store-specific workloads, and none of the existing benchmarks allows evaluating the performance impact of the fine-grained trade-offs between embedding and referencing in realistic application cases.

In this paper, we present the design of a novel benchmark which consists of a data model and corresponding workload consisting of ten representative queries. The evaluated trade-offs are based on a study of two specific open source e-commerce systems.

Keywords: Document store · Performance trade-off · e-commerce · benchmark

1 Introduction

Performance benchmark systems allow to quantitatively evaluate the performance characteristics of a database. Application developers can use benchmark systems to evaluate database technology, to optimize deployment configurations, and to identify performance regressions in different configurations of the same database, or over time. The importance of proper benchmarking can not be understated: even minor performance differences may lead to significant cost savings in large scale and elastic databases.

Data schema flexibility is a distinctive characteristic of NoSQL [14]. While relational databases use fixed data schemas encoded in normalized, tabular structures, NoSQL databases are more open in how the data records are structured and processed [2]. Such flexibility provides a number of advantages as it allows for more (i) specialization of database technology, (ii) dynamism and evolution of schemas [4], (iii) application-specific optimization [13].

R. Wrembel et al. (Eds.): DaWaK 2024, LNCS 14912, pp. 284–290, 2024.
https://doi.org/10.1007/978-3-031-68323-7_23

Document stores, also called *document-(oriented) databases*, persist data records in document structures, and are the most popular NoSQL database type. Each document, identified with a unique key, consists of field-value pairs, where the values can be of any type such as strings, integers, nested documents and arrays of nested documents.

When it comes to implementing document relationships, the application designer can either entirely insert the referred document –*embedding* or *nesting*– in a document, or include the identifier of the referred document –a tactic called *referencing*. Multiple factors are at play when deciding between embedding and referencing: (i) the ratio between read and write queries, (ii) query and document complexity and size, (iii) the deployment and database distribution which affect data locality, (iv) data availability, (v) frequency of change, and (vi) whether documents must be updated atomically and consistently. Trade-offs between these factors are generally well-known and discussed in developer-oriented publications [11].

NoSQL benchmarks commonly adopt a generalistic approach [12] to ensure applicability to a wide range of database types. While useful for technology selection and comparison purposes, such an approach fails to properly assess features that are specific to document stores. Existing benchmark systems for document databases [7] do not allow investigating in depth the performance trade-offs between embedding and referencing, which regardless may lead to significantly different performance outcomes.

The main objective is to design and implement a benchmark for document databases that supports evaluating realistic performance trade-offs between document embedding and referencing. We specifically present a benchmark that is rooted upon a realistic data model in the e-commerce domain. The overall design approach involves three steps: (1) we select a realistic application case which is sensitive to the highlighted design trade-offs, (2) in the context of this application case, realistic data models are defined that address possible trade-offs between embedding and referencing, (3) a realistic workload consisting of queries is defined, that is optimized for showing the performance impact of such design trade-offs. The benchmark is based on the study of two existing open source e-commerce platforms: GrandNode2 [6] and Veniqa [10] which each make different trade-offs in their document schema. The workload defines 10 distinct queries that are each tailored to evaluate the trade-offs of interest.

The remainder of this article is structured as follows: Sect. 2 presents the overall design of the benchmark and Sect. 3 concludes the article.

2 Benchmark Design

We first motivate the selected application domain (Sect. 2.1), then illustrate the different data models and query implementations (Sect. 2.2) and finally, discuss the prototype implementation (Sect. 2.3)

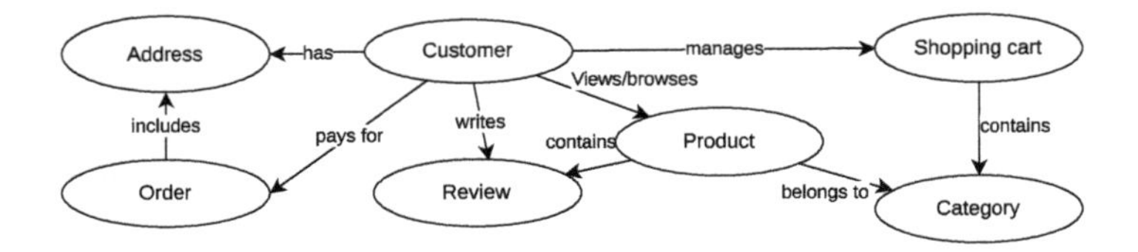

Fig. 1. E-commerce domain model representing key concepts and their relationships.

2.1 E-Commerce Application

The document database benchmark presented in this article is based on a real-world application case, an e-commerce platform. Figure 1 illustrates the domain model of an e-commerce platform[1]. In its most canonical form, this is a shopping website that allows customers to purchase a variety of products online. Products are categorized and entail information, such as name, brand, description, etc. Customers in turn have personal information such as their name, contact information, shipping and billing addresses. While browsing these products, customers can add them to their personal shopping cart. After that, the shopping cart be used to place an order. In addition, customers can upload reviews about specific products, which in turn may help other customers in their decision to purchase these items.

We select the e-commerce application domain for two main reasons. Firstly, it is frequently used as an example case in the domain on document database research [1,3,5,9,15,16]. Secondly, a number of open-source implementations are available upon which the selected workload and data model can be based, thus strengthening the argument of realism and relevance. Furthermore, we observed that different implementations already make different trade-offs in terms of embedding or referencing.

We have investigated two available e-commerce implementations– Veniqa [10] and GrandNode2 [6]. Both are open-source e-commerce platforms. Although both platforms essentially handle the same type of data, we found a number of interesting differences in their respective data models. In the Veniqa data model, the shopping cart document embeds the product ID and item description, item brand and additional meta-data (embedding), whereas in GrandNode2, only the product ID is included (referencing). The Veniqa site always displays the product information when the customer opens its shopping cart, whereas GrandNode2 provides a link to the product page which means that additional product information is only fetched when the user actively selects the product. This shows that the choice between document embedding or referencing is highly application-specific, and co-dependent on other design decisions.

[1] For brevity, we have excluded additional relevant aspects related to payment, management of stock and inventory, dynamic pricing, and personalized recommendations.

Table 1. Illustration of design trade-offs between embedding and referencing in the e-commerce use case.

Document-relationship	Embedded data model	Hybrid data model	Referencing data model
customer-product	Embeds product information in shopping cart	Embeds product information together with the product IDs in shopping cart	References product IDs in shopping cart
customer-address	Embeds address document	Embeds regularly-accessed address fields together with address ID	References address ID
customer-order	Embeds order documents	Embeds regularly-accessed order fields together with order IDs	References order IDs
product-category	Embeds category documents	Embeds category fields together with category ID	References category ID
product-review	Embeds review documents	Embeds review documents together with review IDs	References review IDs
product-product	Embeds related product documents	Embeds attributes of related products together with related product IDs	References related product IDs
order-product	N/A	Embeds regularly-accessed product fields together with product IDs	References ordered product IDs
order-customer	N/A	Embeds regularly accessed customer attributes together with customer ID	References customer ID

2.2 Data Models and Benchmark Queries

In order to analyze the performance trade-offs between embedding and referencing, alternative variant data models can be considered that each represent different trade-offs. The benchmark implements two document schemas or data models, each representing a different extreme: the '*embedded data model*' (2nd column in table 1) which embeds all document relationships, and the '*referenced data model*' (4th column) which references all document relationships. Table 1 contrasts these two data models (second column and fourth column) to a hybrid data model (third column), which is more akin the choices made in Veniqa and GrandNode2.

The benchmark implements ten e-commerce queries (labeled from **Q1** to **Q10**). Table 2 lists these queries and outlines their implementation in both the embedded and the referenced data model. Each query is representative of a different type, and together, the resulting query set represents a well-balanced combination of insert, read, update, delete and search operations (CRUDS).

2.3 Benchmark Implementation

The benchmark is built in Python and currently supports MongoDB. The code is available via [8]. The Python `faker` package is used for generating synthetic data records such as names, addresses, etc. The benchmark user can specify how many `customer`, `product` and `order` documents are to be generated and loaded. The numbers of reviews in a `product` document, and of `items` in a

Table 2. Overview of the supported benchmark queries.

ID: Query	Implementation in the embedded data model	Implementation in the referenced data model
Q1: InsertProduct()	Insert a new product document in the product collection	
Q2: InsertOrder()	Insert a new order in the orders array of the customer document	Insert a new order and insert the corresponding order ID in the orders array of the customer
Q3: AddItemToCart()	Add an item, incl. product name and brand, to the shopping cart of a customer	Add an item, incl. the product ID, to the shopping cart of a customer
Q4: ReadProduct()	Read a product document from the product collection	Read a product document from the product collection. Involves looking up the category, reviews and related products
Q5: ReadOrder()	Read an order document from a customer document	Read an order document from an order collection. This involves also looking up the ordered products and the customer
Q6: ListAllCategories()	View all categories by looking up the category information in each product document and removing all duplicates	View all categories in the category collection
Q7: ListProductsInCategory()	Fetch all products within a specific category by looking up the category information in each product document	Search all products within a specific category by looking up the category ID and then selecting the product documents with that category ID
Q8: UpdateStockQuantity()	Update the stock quantity of a product document	
Q9: DeleteProduct()	Delete a product from the product collection	Delete a product from the product collection. This involves deleting referenced review documents
Q10: RemoveItemFromCart()	Remove an item from a shopping cart in a customer document	

`shopping cart` follow a normal distribution[2]. The amount of related `products` in a `product` document and the number of `ordered products` is randomly chosen using Python's random number generator (`random`), where again the range is left as a configuration option. The benchmark implements the ten queries (**Q1** to **Q10**) in MongoDB with the help of different MongoDB operators. The benchmark executes a workload, which consists of a set of queries from the `Queries.py` file, on the embedded and referenced database. The user can further configure the ratio of the queries used in the workload and the number of operations that workload should execute sequentially. When executing a workload, the benchmark keep track of the execution time at the basis of individual operations (query latency) and writes these to a CSV output file.

3 Conclusion

Few dedicated benchmark systems have been created that are capable of evaluating document-store specific features. In this article, we presented a benchmark

[2] Using `numpy`, where median and standard deviation are configuration options.

that is based upon a realistic e-commerce reference case, and which defines 10 queries to evaluate the trade-offs between embedding and referencing.

The benchmark currently integrates two distinct data models that each represent one extreme form of embedding and referencing. In future work, the benchmark will be extended in with support for more hybrid variants and serve as a basis for further experimentation. Extending the benchmark with the capability to evaluate the impact of specific document store features and configuration options –different indexing strategies (e.g., secondary indexing), deployment distribution choices, and features such as data partitioning, sharding or replica sets– is also considered future work.

Acknowledgments. This research is partially funded by the Research Fund KU Leuven and Cybersecurity Research Program Flanders.

References

1. Aboutorabi, S.H., Rezapour, M., Moradi, M., Ghadiri, N.: Performance evaluation of SQL and MongoDB databases for big e-commerce data. In: 2015 International Symposium on Computer Science and Software Engineering (CSSE), pp. 1–7 (2015). https://doi.org/10.1109/CSICSSE.2015.7369245
2. Brewer, E.A.: Towards robust distributed systems. In: PODC, vol. 7, pp. 343477–343502. Portland, OR (2000)
3. Cabral, J.V.L., Noguera, V.E.R., Ciferri, R.R., Lucrédio, D.: Enabling schema-independent data retrieval queries in MongoDB. Inf. Syst. **114**, 102165 (2023). https://doi.org/10.1016/j.is.2023.102165
4. Chillón, A.H., Ruiz, D.S., Molina, J.G.: Towards a taxonomy of schema changes for NoSQL databases: the Orion language. In: Ghose, A., Horkoff, J., Silva Souza, V.E., Parsons, J., Evermann, J. (eds.) ER 2021. LNCS, vol. 13011, pp. 176–185. Springer, Cham (2021). https://doi.org/10.1007/978-3-030-89022-3_15
5. Garcia, D., Garcia, J.: TPC-W e-commerce benchmark evaluation. Computer **36**(2), 42–48 (2003). https://doi.org/10.1109/MC.2003.1178045
6. GrandNode2: Free, Fast, Flexible, Feature-Rich Open-Source E-Commerce Platform. https://github.com/grandnode/grandnode2
7. Gyryk, A.: YCSB-JSON: implementation for couchbase and MongoDB (2018). https://dzone.com/articles/ycsb-json-implementation-for-couchbase-and-mongodb. dZone Performance
8. Levrau, M., Van Landuyt, D.: Document schema benchmark: prototype implementation (2024). https://people.cs.kuleuven.be/dimitri.vanlanduyt/performance_benchmark_docstores.zip
9. Maulidin, A.J., Renaldi, F., Umbara, F.R.: Online integration of SQL and NoSQL databases using RestAPIs: a case on 2 furniture e-commerce sites. In: 3rd International Conference on Computer and Informatics Engineering (IC2IE), pp. 261–266 (2020)https://doi.org/10.1109/IC2IE50715.2020.9274613
10. Pandey, V.: Veniqa (2019). https://github.com/Viveckh/Veniqa
11. Peacock, A.: Embedded vs. referenced documents in MongoDB: how to choose correctly for increased performance (2022). https://betterprogramming.pub/embedded-vs-referenced-documents-in-mongodb-how-to-choose-correctly-for-increased-performance-d267769b8671

12. Reniers, V., Van Landuyt, D., Rafique, A., Joosen, W.: On the state of NoSQL benchmarks. In: Proceedings of the 8th ACM/SPEC on International Conference on Performance Engineering Companion, pp. 107–112. ACM, L'Aquila (2017). https://doi.org/10.1145/3053600.3053622
13. Schram, A., Anderson, K.M.: MySQL to NoSQL: data modeling challenges in supporting scalability. In: Proceedings of the 3rd Annual Conference on Systems, Programming, and Applications: Software for humanity, pp. 191–202 (2012)
14. Stonebraker, M.: SQL databases v. NoSQL databases. Commun. ACM **53**(4), 10–11 (2010)
15. Van Landuyt, D., Benaouda, J., Reniers, V., Rafique, A., Joosen, W.: A comparative performance evaluation of multi-model NoSQL databases and polyglot persistence. In: Proceedings of the 38th ACM/SIGAPP Symposium on Applied Computing, pp. 286–293 (2023)
16. Zhang, C., Lu, J., Xu, P., Chen, Y.: UniBench: a benchmark for multi-model database management systems. In: Nambiar, R., Poess, M. (eds.) TPCTC 2018. LNCS, vol. 11135, pp. 7–23. Springer, Cham (2019). https://doi.org/10.1007/978-3-030-11404-6_2

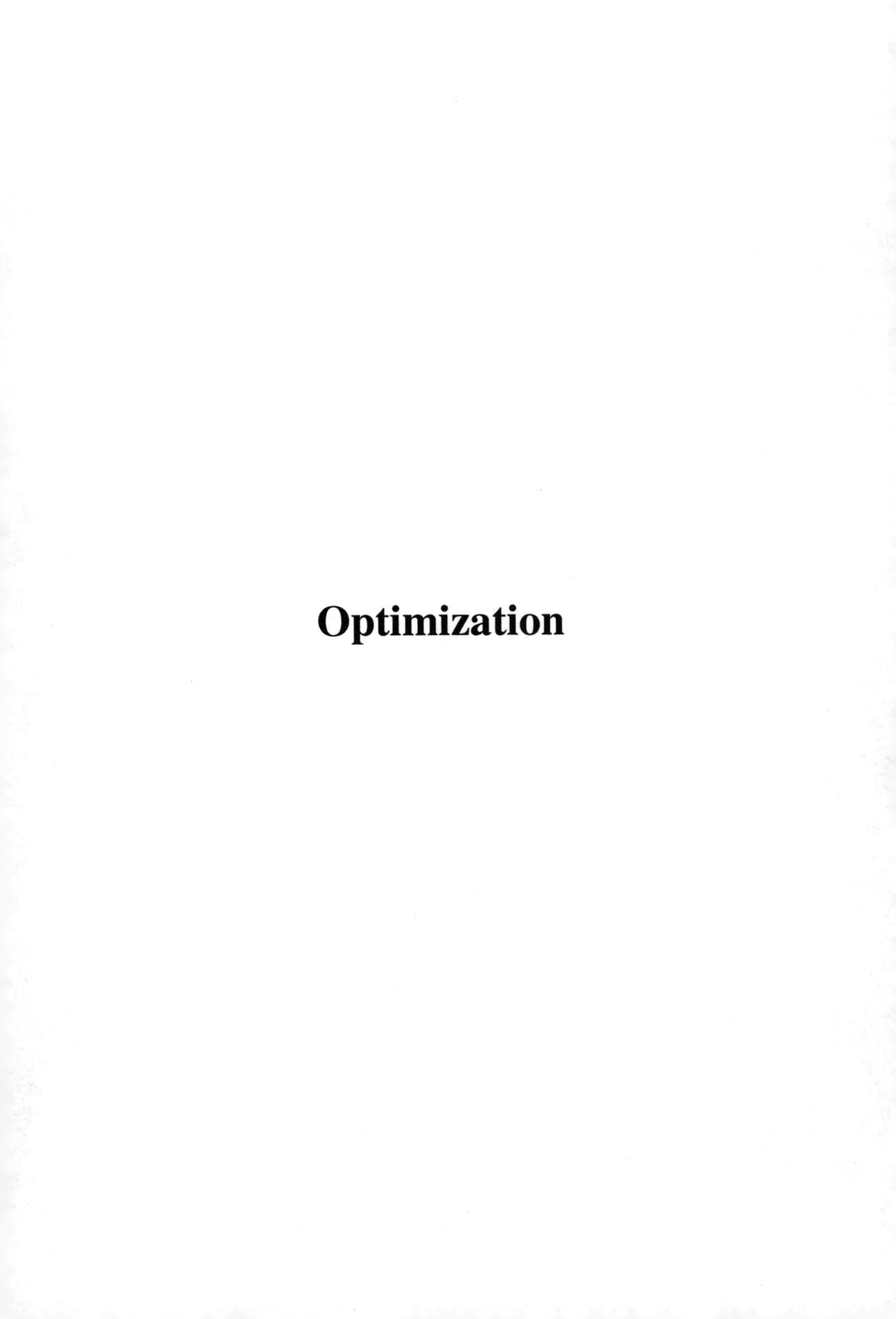

Optimization

Effective Reward Schemes for Tardiness Optimization

Lihi Idan[1,2](✉)

[1] Texas A&M University, College Station, TX, USA
[2] Harvard University, Cambridge, MA, USA
lidan@hbs.edu

Abstract. Over recent years, reinforcement learning has become a prominent method for the optimization of sequential decision-making problems. One group of sequential decision-making problems that has benefited significantly from reinforcement-learning-based optimization techniques is scheduling problems. However, most existing reinforcement learning works on scheduling optimization aim at optimizing a single, makespan-based objective. While the makespan—the overall time from the start of the first task to the end of the last task—is indeed important in some endeavors, other endeavors benefit more from the optimization of other types of objectives. In this work, we focus on Tardiness-based objectives and present a new reward scheme that aims at simultaneously optimizing multiple notions of Tardiness.

1 Introduction

Over recent years, reinforcement learning has become a prominent method for the optimization of sequential decision-making problems. One group of sequential decision-making problems that has benefited significantly from reinforcement-learning-based optimization techniques is scheduling problems. However, most existing reinforcement learning works on scheduling optimization aim at optimizing a single, makespan-based objective. While the makespan—the overall time from the start of the first task to the end of the last task—is indeed important in some endeavors, other endeavors benefit more from the optimization of other types of objectives. In this work, we focus on Tardiness-based objectives, such as minimizing the number of late tasks or the amount of overall late time. We present a new reward scheme that aims at simultaneously optimizing both notions of tardiness, *i.e.* both the number of late tasks and the amount of late time.

2 Related Work

The use of reinforcement learning for scheduling optimization was first examined in [9], which applied a value-based reinforcement-learning method for learning

R. Wrembel et al. (Eds.): DaWaK 2024, LNCS 14912, pp. 293–298, 2024.
https://doi.org/10.1007/978-3-031-68323-7_24

scheduling heuristics in the domain of space shuttle payload processing. Recently, there has been a resurgence of interest in the use of reinforcement learning for scheduling optimization. [10] designed a reinforcement-learning-based cloud task scheduler using the DDPG algorithm. [3] used a DQN deployed on an edge computing framework for job shop scheduling in a smart manufacturing setting. [2] applied a multi-agent, value-based approach for optimizing the charging scheduling of electric vehicles. [4,5,8] used a combination of a reinforcement-learning agent and a graph-neural network structure to optimize the job shop scheduling problem; they represented the problem as a disjunctive graph and used an actor-critic algorithm to learn embeddings of nodes which are then used to select actions. All the works mentioned above use the makespan as the primary objective to be optimized. On the contrary, [7] explored the use of a different reward function based on machine utilization rate instead of the system's makespan.

3 Technical Problem Statement

We consider J activities, and denote by $\mathcal{J}$ the list of activities' indices. Each activity is composed of N_j tasks, $\{t_{j,k}\}_{j\in\mathcal{J},k\in\mathcal{N}_j}$. We denote by $\mathcal{N}_j$ the list of tasks' indices of activity a_j. Each task takes $d_{t_{j,k}}$ minutes. We assume the existence of a global clock and denote the current time by δ. Tasks within an activity must be performed sequentially, *i.e.* $t_{1,1}$ must be completed before $t_{1,2}$ starts; thus, a task, $t_{j,k}$ is "available", denoted by $av_{t_{j,k}} = 1$, only if its predecessor task in the activity a_j has finished running. The next task that is ready to be scheduled of an activity a_j is denoted by $a_j.cur$. Each activity, a_j, has a "SLA time", a_j^{sla}, the deadline until which a_j must complete its last task; a release time, a_j^{start}, the time in which the first task of a_j can start; and a completion time, $a_j^{complete}$, the time in which a_j was completed. $a_{max}^{complete}$ and a_{min}^{start} denote the activity with the maximal completion time and minimal start time, respectively. a_j^{all} denotes the sum of durations of tasks in a_j, and a_j^{done} denotes the portion of a_j^{all} that was completed by δ. We denote by J, T, R the number of activities, tasks, and resources in the system, respectively. We are interested in optimizing the "Tardiness Ratio":

$$T^l = \frac{S^l}{J - C^l} \tag{1}$$

C^l corresponds to the number of activities that failed to meet their deadline:

$$C^l = \sum_{j=1}^{J} X_j, \qquad X_j = \begin{cases} 1 & \text{if} \quad a_j^{sla} - a_j^{end} < 0 \\ 0 & \text{otherwise} \end{cases} \tag{2}$$

S^l corresponds to the overall late time of all the activities in the system:

$$S^l = \sum_{j=1}^{J} min(0, a_j^{sla} - a_j^{end}) \tag{3}$$

The Tardiness Ratio is a multi-objective that captures the fact that we would like to both minimize the number of late activities and at the same time minimize the overall late time even if an activity already missed its deadline.

4 Reward Function

The core idea underlying reward function (Algorithm 1) is providing the agent, at each step, with a reward that aggregates the "tardiness status" with respect to time δ, $c_{j,\delta}$, for each activity. For each activity, a_j, $c_{j,\delta}$ combines two success predictors: the time until a_j's SLA deadline, and the work remaining for a_j until completion. At each time step, δ, if an activity hasn't been completed yet, we compute its "tardiness status" as:

$$c_{j,\delta} = \frac{\frac{a_j^{sla}}{(1+(c_{i,j}^2(\delta-1)))}}{\frac{a_j^{all}}{a_j^{done}}}_{i=1 \text{ if } a_j^{sla}-\delta<0 \ else \ i=2} \tag{4}$$

where $\{c_{i,j}^2\}$ are hyperparameters aimed at *controlling the notion of time as seen by the agent* thus allowing the practitioner a complete control on the reward shaping process by serving three main functions. First, by decaying the current time, δ, $\{c_{i,j}^2\}$ prevent the deterioration of the reward function in long-horizon trajectories: in long-horizon trajectories, the difference between two consecutive steps tends to be very small, hence the agent receives no meaningful signal. Second, by using a different hyperparameter, c_2^2, to compute the tardiness status of a running activity that has not passed its SLA deadline yet, and a different hyperparameter, c_1^2, to compute the tardiness status of a running activity that has already passed its SLA deadline, where $c_2^2 < c_1^2$, we signal the agent that an activity that has passed its SLA deadline is an unwanted event. In a way, from the moment we pass the activity's deadline, our time is "more critical", and is thus promoted faster, than before we passed the deadline and thus the activity's tardiness status will decrease at a higher rate than an activity that its SLA deadline has not passed. Finally, the fact that both hyperparameters are activity-specific facilitates the assignment of different decaying factors for different activities based on the client-specific, activity-specific churn probability.

If an activity was completed, we compute its "tardiness status" as:

$$c_{j,\delta} = \frac{a_j^{sla}}{(1+(c_{i,j}^4(a_j^{complete}-1)))}_{i=1 \text{ if } a_j^{sla}-\delta<0 \ else \ i=2} \tag{5}$$

We assume $c_1^4, c_2^4 < c_1^2, c_2^2$, with the goal of enabling the agent to grasp the idea that the completion of an activity is a positive event. By using c_1^4 for computing the status of an activity that was completed after its SLA deadline, and using c_2^4 for computing the status of an activity that was completed before its SLA deadline, where $c_2^4 < c_1^4$, we teach the agent that the completion of an activity that its SLA deadline hasn't passed yet is valued more than completion of an

Algorithm 1. ComputeReward

Input: $1 < c^1, c^3 < 2, 0 < \{c^2_{1,j}, c^2_{2,j}, c^4_{1,j}, c^4_{2,j}\}_{j \in J} <= 1$: Hyperparameters
A: Array of activities, R_{t-1}: Previous step's reward
Output: R: Reward

1: $R^* = 0$
2: **for** $j \in range(0, len(A))$ **do**
3: **if** $\neg(A[j].completed)$ **then**
4: **if** $\delta == 0$ **then**
5: $p1 = a^{sla}_j * c^1$
6: **else**
7: **if** $a^{sla}_j - \delta <= 0$ **then**
8: $p1 = \frac{a^{sla}_j}{(1+(c^2_{1,j}*(\delta-1)))}$
9: **else**
10: $p1 = \frac{a^{sla}_j}{(1+(c^2_{2,j}*(\delta-1)))}$
11: **end if**
12: **end if**
13: **if** $a^{done}_j > 0$ **then**
14: $p2 = \frac{a^{all}_j}{a^{done}_j}$
15: **else**
16: $p2 = a^{all}_j * c^3$
17: **end if**
18: **else**
19: $p2 = 1$
20: **if** $a^{sla}_j - a^{complete}_j < 0$ **then**
21: $p1 = \frac{a^{sla}_j}{(1+(c^4_{1,j}*(a^{complete}_j-1)))}$
22: **else**
23: $p1 = \frac{a^{sla}_j}{(1+(c^4_{2,j}*(a^{complete}_j-1)))}$
24: **end if**
25: **end if**
26: $c_{j,\delta} = \frac{p1}{p2}$
27: $R^* = R^* + c_{j,\delta}$
28: **end for**
29: $R = R^* - R_{t-1}$
30: $R_{t-1} = R^*$
31: **return** R

activity that its SLA deadline has already passed. Note, however, that both events are valued more than an activity that hasn't been completed yet, since for each activity, a_j, $c^4_{1,j}, c^4_{2,j} < c^2_{1,j}, c^2_{2,j}$. Importantly, the ratio between $\{c^2_i\}$ and $\{c^4_i\}$, as well as between c^4_1 and c^4_2 determines the weight given to C^l compared to S^l in our multi-objective reward function. For instance, the smaller c^4_2 compared to c^4_1, the more weight we give to C^l—Count Tardiness. On the other hand, the more similar $\{c^2_i\}$ and $\{c^4_i\}$ are, the more weight is given to S^l—Sum Tardiness.

The primary goal of c^1 is enabling the agent to differentiate between the reward at time 0 and at time 1. We did not find its value to be significant, apart from instances with short trajectories. c^3's goal is to ensure a positive reward at steps where the first task of an activity starts running; if $c^3 = 1$, the reward in such steps will be 0. To ensure a positive reward in such steps, c^3 should be chosen according to the following equation:

$$c^3 a_j^{all} + c^2_{i,j} \delta c^3 a_j^{all} - c^2_{i,j} c^3 a_j^{all} > a_j^{all} + c^2_{i,j} a_j^{all} \delta \tag{6}$$

The tardiness statuses of all the activities are summed up, the sum is denoted as R. Finally, we take the difference between the current sum at time t, and the previous sum at time $t-1$. The reward at time t is then $R_t - R_{t-1}$.

5 Experimental Results

Datasets. We follow other prominent scheduling works such as [8] and evaluate our approach using simulated instances of various sizes, $J \times N_j \times R$, as shown in Table 1. For each type of instance ($J \times N_j \times R$), the model is validated on 30 instances in which J,N_j, and R are set according to the instance's type (*i.e.* fixed) while activities' SLA time, the resources assigned to each task, and tasks' durations are randomly generated.

Implementation Details. We implement our agent using Stable Baselines [1]. We use the Proximal Policy Optimization (PPO) algorithm [6] as our learning algorithm. We train the policy network using 4 independent trajectories; the number of epochs is set to 10; the clipping parameter is set to 0.2; the discount factor is set to 0.99. Our network has two hidden layers with a dimension of 32.

Baselines and Results. We use two of the most common scheduling heuristics as baselines: Shortest Processing Time (SPT) and Earliest SLA Time (EST). SPT works by choosing the task with the shortest duration. EST works by choosing the task that belongs to the activity with the nearest deadline.

Table 1 compares our agent's results with those obtained by our baselines. As can be seen, our agent significantly outperforms our baselines on most instances. As is evident from Table 1 it seems as if as the number of activities increases, the differences between our model's results and our baselines' results increase. A similar trend is not seen for the number of tasks. An interesting trend is seen for R: from Table 1 it seems as if the best results are obtained on instances where there is a large-enough number of resources but only if the resource-to-activity ratio is low enough. For instance, in instances with 20 activities and

Table 1. A comparison of our model's results to the results obtained by multiple baselines using the Tardiness Ratio

Instance	EST	SPT	Ours
$20 \times 10 \times 6$	24549	7259	6421
$20 \times 10 \times 10$	4833	1915	1782
$30 \times 10 \times 6$	96798	11609	9412
$30 \times 10 \times 10$	19186	10735	8932
$50 \times 10 \times 6$	205269	55101	42936
$60 \times 10 \times 15$	27921	24269	19799

10 resources the resource-to-activity ratio is relatively high, and hence, since multiple resources are assigned to each task, there are very few resource-based constraints; on those instances, the improvement in the Tardiness Ratio yielded by the use of our agent compared to SPT is modest (though the improvement is significant compared to EST). However, in bigger and more constrained instances such as instances with $\geq$50 activities and 6 or 15 resources, the resource-to-activity ratio is relatively low; on these instances, the use of our agent leads to a tremendous improvement in the Tardiness Ratio compared to all of our baselines.

References

1. Hill, A., et al.: Stable baselines (2018). https://github.com/hill-a/stable-baselines
2. Lee, X.-L., Yang, H.-T., Tang, W., Toosi, A.N., Lam, E.: An adaptive charging scheduling for electric vehicles using multiagent reinforcement learning. In: Hacid, H., Kao, O., Mecella, M., Moha, N., Paik, H. (eds.) ICSOC 2021. LNCS, vol. 13121, pp. 273–286. Springer, Cham (2021). https://doi.org/10.1007/978-3-030-91431-8_17
3. Lin, C.C., et al.: Smart manufacturing scheduling with edge computing using deep Q network. IEEE Trans. Industr. Inf. **15**(7), 4276–4284 (2019)
4. Ni, F., et al.: A multi-graph attributed reinforcement learning based optimization algorithm for large-scale flow shop scheduling problem. In: SIGKDD (2021)
5. Park, J., et al.: Learning to schedule job-shop problems. IJPR (2021)
6. Schulman, J., et al.: Proximal policy optimization algorithms. arXiv:1707.06347 (2017)
7. Tassel, P., et al.: A reinforcement learning environment for job-shop scheduling. arXiv preprint arXiv:2104.03760 (2021)
8. Zhang, C., et al.: Learning to dispatch for job shop scheduling via deep reinforcement learning. In: NeurIPS, pp. 1621–1632 (2020)
9. Zhang, W., Dietterich, T.G.: A reinforcement learning approach to job-shop scheduling. In: IJCAI (1995)
10. Zhao, Z., Shi, X., Shang, M.: Performance and cost-aware task scheduling via deep reinforcement learning in cloud environment. In: Troya, J., Medjahed, B., Piattini, M., Yao, L., Fernández, P., Ruiz-Cortés, A. (eds.) ICSOC 2022. LNCS, vol. 13740, pp. 600–615. Springer, Cham (2022). https://doi.org/10.1007/978-3-031-20984-0_43

A Novel Technique for Query Plan Representation Based on Graph Neural Nets

Baoming Chang(✉), Amin Kamali, and Verena Kantere

University of Ottawa, 75 Laurier Ave E, Ottawa, ON K1N 6N5, Canada
{bchan081,skama043,vkantere}@uottawa.ca

Abstract. Learning representations for query plans play a pivotal role in machine learning-based query optimizers of database management systems. To this end, particular model architectures are proposed in the literature to transform the tree-structured query plans into representations with formats learnable by downstream machine learning models However, existing research rarely compares and analyzes the query plan representation capabilities of these tree models and their direct impact on the performance of the overall optimizer. To address this problem, we perform a comparative study to explore the effect of using different state-of-the-art tree models on the optimizer's cost estimation and plan selection performance in relatively complex workloads. Additionally, we explore the possibility of using graph neural networks (GNNs) in the query plan representation task. We propose a novel tree model BiGG employing Bidirectional GNN aggregated by Gated recurrent units (GRUs) and demonstrate experimentally that BiGG provides significant improvements to cost estimation tasks and relatively excellent plan selection performance compared to the state-of-the-art tree models.

Keywords: Query Plan Representation · Tree Model · Graph Neural Network

1 Introduction

Query optimization is a critical component of a database management system due to its difficulty and importance in query execution performance. The process of accurately and efficiently estimating the cost of generated candidate query execution plans and selecting the optimal plan is always a challenge in the field of query optimization. A query execution plan is typically represented as a tree, where the nodes contain information about operators used to access, join, or aggregate data, and the edges contain dependencies between the parent and child nodes. An optimal plan tree enables the database management system to access and manipulate data efficiently. Traditional query optimizers employ cost models to estimate the amount of processing data using plan trees. These cost models rely heavily on statistical methods such as Histograms [5], which

R. Wrembel et al. (Eds.): DaWaK 2024, LNCS 14912, pp. 299–314, 2024.
https://doi.org/10.1007/978-3-031-68323-7_25

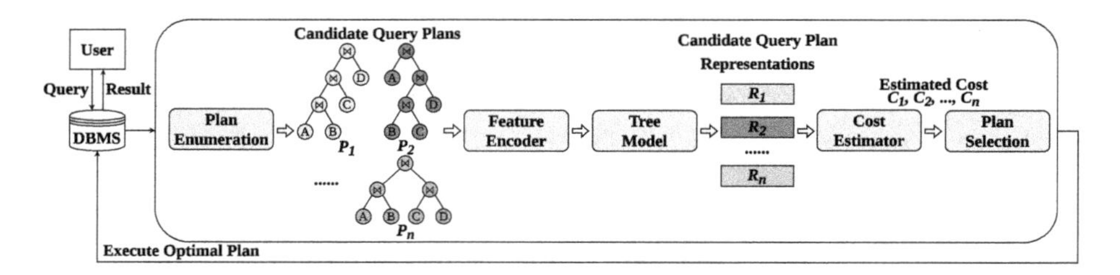

Fig. 1. Machine learning-based query optimizer framework.

are susceptible to large errors because of their inability to effectively capture characteristics such as join-crossing correlations.

The development of machine learning provides a promising solution to improve query optimization and has been proposed in a wide range of applications in this field. Recent machine learning-based query optimizers employ a similar structure, as shown in Fig. 1. In these frameworks, node information from a query plan tree is first encoded into node features. Then, a tree model transforms and aggregates these into graph-level representations for each candidate query plan, and a cost estimator predicts execution costs based on these representations, enabling the optimizer to select the most efficient plan. The challenge in this process is to effectively transform the tree-structured query plan into a graph-level vector while preserving as much of the original node features and structural information as possible. The quality of representation generated by the tree model is crucial to ensure the cost estimator model makes accurate predictions and allows the optimizer to make informed decisions for specific tasks involving plan selection, such as join order optimization and query rewriting.

Current research in query optimization often lacks direct performance comparisons of the representation abilities of these tree models, focusing instead on the general performance of the whole query optimization process [25]. Therefore, in this paper, we evaluate the mainstream tree models query plan representation under more complex workloads, comparing and analyzing their performance in cost estimation tasks. In addition, unlike previous studies [25], instead of pairwise comparisons, we compare and analyze the tree model's plan selection capability in a way that is closer to actual usage scenarios by selecting an optimal plan from multiple candidate query plans to explore how different tree models affect the performance of plan selection.

Our research also explores the possibility of using graph neural networks (GNNs) [26] as tree models. Although GNNs are being developed and have achieved success across various graph-based domains, their application within query plan cost estimation remains insufficiently explored. In this context, our paper explores the potential of employing GNNs for query execution plan representation learning. We also propose a novel tree model architecture for query plan representation based on bidirectional GNN [18] and a GRU-based aggregation method [1], which can capture the intricacies of query plan trees more accurately and robustly, setting a new stage for query plan representation learning.

To summarize, our main contributions are:

- We conduct a comparative analysis of the performance of mainstream tree models in cost estimation under complex conditions.
- We evaluate the impact of tree models in the task of selecting the optimal plan from multiple candidate plans, closely mirroring actual application scenarios, and provide an intuitive comparison and analysis.
- We thoroughly explore the possibility of using GNNs as a query execution plan tree model.
- We propose a novel query plan tree model BiGG based on bidirectional GNN and GRU, demonstrating its superior representational capabilities and performance compared to existing tree models.

In the rest of the paper: Sect. 2 presents the problem of query plan representation and current tree models. Section 3 outlines related works along with their merits and shortcomings. Section 4 introduces the proposed architecture of BiGG. Section 5 presents the comparative study of existing and GNN-based tree models' performance on cost estimation and plan selection tasks, and Sect. 6 concludes the paper and discusses future work.

2 Problem Statement

In a query optimizer based on machine learning, query plan representation is a module that takes the physical query plan as input and uses a feature encoder and a tree model to generate vectors that subsequent machine learning models use for learning a downstream task. The generated query plan graph vector will condense important information about the physical query plan. How to accurately represent both structural and node information in the tree model's output can directly impact the overall performance of the query optimizer. In this study, we focus on the representation capability of query plan tree models in the same application scenario using the same structure of feature encoder and cost estimator, so that the representation quality of a query plan can entirely rely on the tree model, allowing us to compare the performance of these tree models directly.

Due to the tree structure characteristics of the query plan, it is difficult for tree models to use traditional machine learning methods to learn and aggregate node information directly, for they usually cannot accept inputs in the form of trees. Consequently, the database community has started developing machine-learning models designed explicitly for tree structures [14,21]. While the advanced models have marked performance improvements, they still have two main limitations:

- **Information Dilution.** As information traverses from the leaf nodes-where specific relation-based operations are stored-towards the root, it tends to get diluted. This dilution process may negatively impact the accuracy of predictive cost estimations. The depth of the tree exacerbates this issue, leading

to possible loss or dilution of critical information as it is passed upward or downward through the tree. The challenge of information dilution becomes more noticeable for complex queries that result in deeply structured trees.
- **Preserving Structural Information.** The plan tree contains essential structural and logical information. Preserving the integrity of this information while aggregating data from the entire graph is crucial and challenging.

Therefore, the design of the tree model used for processing query plans should consider the above two limitations to further improve the model's representation ability according to the structural characteristics of the query plan tree.

3 Related Work

Query plan representation is mainly studied under one of the following categories: studies proposing novel tree models and their comparative analysis.

Tree Models. RNN-based models like long short-term memory (LSTM) [20] and gated recurrent units (GRUs) [3] are commonly used in query plan representation. Some works apply the self-attention mechanism to tree models, such as Saturn [10], which aggregates LSTM's hidden layers based on attention as the query plan representation, or QueryFormer [24], which uses a Transformer [22] to encode the query plan. However, these methods have to convert tree structure plans into sequential nodes as input, inevitably leading to the loss of tree structure information. Machine learning models designed explicitly for tree structures, for instance, tree-structured LSTM (Tree-LSTM) [21], tree convolutional neural network (Tree-CNN) [14], allow information transfer between child nodes and parent node so that the learned node features can also contain structural information. While these approaches improve upon non-tree models, they still do not demonstrate good results in solving the problem of information dilution and effective aggregation of node features. Recently, some works [2,6] attempt to apply GNNs in this context. However, there is no GNN model specifically designed for query plan tree graphs yet; relevant works use the GNNs only to capture the query's join relationship in order to assist the representation learning by other tree models rather than being directly used for learning query plan trees.

Comparative Study. A significant contribution to this field is a study by Yao Z. et al. [25]. The researchers conducted a detailed analysis comparing the performance of mainstream feature encoders and tree models under various scenarios. However, their study asserts that the tree model has no obvious impact on the overall performance of the optimizer, which is not confirmed by the findings of the current research. Their analysis also has a limitation: it lacks a comprehensive assessment of tree models in the plan selection task, since their evaluation only relies on pairwise index selection [4], which does not adequately reflect the complexities in practical query optimization scenarios. Furthermore, they do not isolate the impact of the tree model architectures on this downstream task's performance.

4 Model Architecture

In this section, we first introduce the feature encoding method we used and then describe the structure of our proposed tree model BiGG.

4.1 Feature Encoding

A query execution plan consists of information on the operators used to access and join data from various sources, including the physical implementation of the operators, their sequence, and the tables and columns accessed at each step. The amount of data flowing through each plan node, determined by the cardinality, is a decisive factor of the execution cost. In addition, the local and join predicates involved in the query plan may exhibit various levels of skewness and pairwise correlations, which in turn impact the accuracy of the cardinality estimates. To transform this complex information into fixed-length node features as input to the tree model, we developed a plan encoder inspired by RTOS [23], while making critical changes to suit our experimental purposes. As shown in Fig. 2, each node feature in the query execution plan tree is composed of three parts: Node Type Embedding, Table Embedding, and Predicate Embedding.

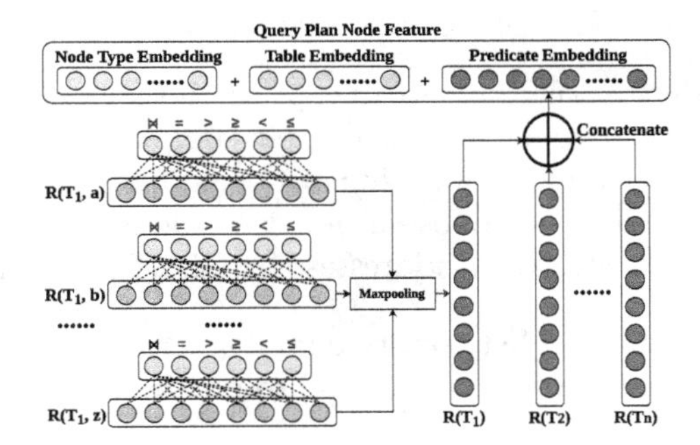

Fig. 2. Learning-based plan node representation.

Node Type Embedding. The node contains a specific operator type, such as Hash Join or Index Scan. We perform one-hot encoding on the node types and pass through a fully connected (FC) layer to obtain the node type embedding.

Table Embedding. We apply one-hot encoding on all tables in the node's operation to obtain the Table Embedding.

Predicate Embedding. We consider the two main types of columns involved in the predicate:

For numerical columns with values, the predicate operations in the node are classified into 6 cases: $\bowtie$, $=$, $>$, $\geq$, $<$, $\leq$. Each column is represented by a feature vector of the same length. If the predicate involves a specific operation on a certain column, the corresponding vector will be encoded according to the specific predicate type. If the column exists in a join predicate, the corresponding position will be 1 otherwise 0. For the other five cases, the predicate value is normalized to [0, 1] using the maximum and minimum values of the column in the database and plus 1 as the value of the corresponding position. If the predicate value exceeds the value range of the column in the database, according to the operator type, if there is no tuple that satisfies the predicate, the corresponding position is set to -1 otherwise 2. Uninvolved cases remain 0.

For other value-type columns, such as string type, it cannot be mapped to a value with interval meaning by a simple method (e.g. hash) [23]. Thus, the join predicate is encoded in the same way as above. For the other five cases, we use Word2vec [13] to translate characters into numerical values for their respective positions in the column feature.

Each column has a dedicated FC layer to process its encoded feature vector and generate learned column embeddings. Then, max pooling is performed on all column embeddings belonging to the same table to obtain the embedding expression of each table with the same length. To avoid information loss during the aggregation of column embeddings in the node feature encoding stage, we directly concatenate embeddings of all tables as the predicate embedding.

4.2 Bidirectional GNN for Query Plan Tree

The basic idea of GNNs is that nodes can capture the characteristics and contextual information of neighbour nodes and their complex relationships via edges in the graph. Acknowledging the benefits of GNNs, due to query plans' characteristics discussed above, we propose a new tree model based on bidirectional GNN and GRU, which effectively addresses the following two main issues:

1. Information Transmission between tree nodes
2. Information Aggregation from node-level to graph-level

Information Transmission. In GNNs, message passing between nodes depends on both the existence of edges in graphs and their directionality. In the tree structure, this issue is mainly divided into two situations: single-directed and undirected edges. If single-directional edges are used, such as from the child node pointing to the parent node, information is always transmitted from the child node to the parent node. A node can only learn the information of its adjacent nodes in a GNN layer, which means any node in the tree will only learn information about its child nodes. When learning a deeper tree, for the information of the leaf nodes to be transferred to the root node, the same number of GNN layers as the depth of the tree is needed, which is generally unacceptable. The single-directed edge also makes it challenging to transfer the leaves' information to the root node, passing the entire graph without diluting or losing it, which makes using root node features as query plan graph-level representation infeasible. It also lets child nodes never learn the node feature and structural

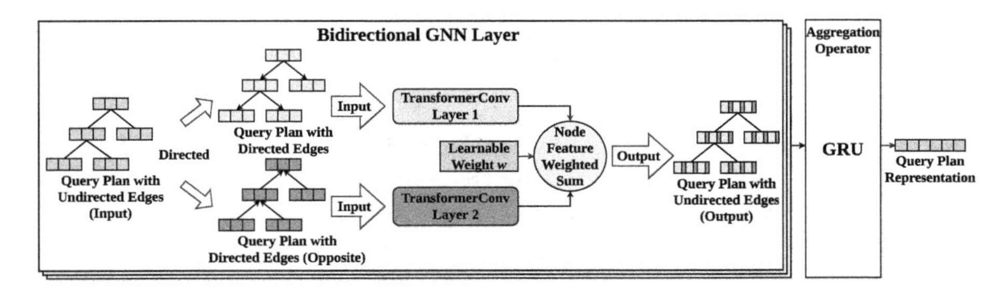

Fig. 3. Framework of tree model BiGG using bidirectional GNN aggregated by GRU.

information about its parent node and above nodes, so the relevant structural information will inevitably be lost in the stage of aggregating node features.

Therefore, using undirected edges has the potential to significantly speed up the information transfer between nodes. However, our experiments have proven that simply using undirected edges cannot perform well in tree graphs. It destroys the structural relationship between parent and child nodes and makes the model unable to learn the hierarchical dependency between them. Inspired by DirGNNConv [18], we innovatively treat the query plan trees as two one-way graphs with opposite edge directions. In each layer of our tree model BiGG, as shown in Fig. 3, we divide the original query plan into two tree graphs with opposite edge directions as input to two independent TransformerConv [19] layers respectively and integrate the corresponding nodes from the two output learned graphs through a learnable parameter, which makes it possible to transmit information in both directions while still utilizing the direction information of the edges and retaining relevant structural information. Given a query plan tree graph G with N nodes, n as the nodes in the node feature set $Nodes$, E as the undirected edge index set, $E_{\text{child to parent}}$ and $E_{\text{parent to child}}$ as the edge index sets with opposite directions, p as a learnable parameter, the bidirectional GNN layer is defined as Algorithm 1:

Algorithm 1. Update node features in bidirectional GNN

1: **Input:** $G = (Nodes, E)$
   $Nodes = \{n_i \mid \forall i \in \{1, \ldots, N\}\}$
   $E = E_{\text{child to parent}} \cup E_{\text{parent to child}}$
2: $Nodes_{\text{child to parent}} = \text{TransformerConv1}(Nodes, E_{\text{child to parent}})$
3: $Nodes_{\text{parent to child}} = \text{TransformerConv2}(Nodes, E_{\text{parent to child}})$
4: **for** $i = 1$ to N **do**
5: $\quad n_{\text{new}_i} = p \times n_{\text{child to parent}_i} + (1 - p) \times n_{\text{parent to child}_i}$
6: **end for**
7: **Output:** $G = (Nodes_{\text{new}}, E)$

Additionally, the graph attention mechanism [22] in TransformerConv can further improve the model's ability to capture the tree's local graph topology and global dependencies. To verify this perspective, our experiments also evaluated the influence of undirected edges, single-directed edges, and weighted directed edges on the model's capability to represent query plans.

Information Aggregation. Conventional graph aggregation methods often produce inferior results when processing tree structure graphs. These methods usually simply aggregate node features by global pooling, which ignores the structural information in the graph. In our model we innovated by applying GRU to aggregate the GNN-learned query plan node features after post traversal of the tree. The rationale for this design decision is based on the observation that the order in which DBMS executes the query plan tree nodes and the order of nodes obtained by post-order traversal of the tree are similar [10]. This design learns the dependencies between nodes while more closely conforming to the actual execution sequence, thereby allowing our tree model to obtain the query plan's graph-level embedding while retaining the node features and structural information in the query plan tree to a certain extent. Our experiments have proven that GRU, as the aggregation operator, performs far better than other aggregation methods in query plan representation.

5 Experimental Study

Cost estimation is a crucial step in the query optimizer, as it directly determines how the model selects the optimal plan from candidate plans. However, these cost estimates are subject to arbitrary errors rooted in various sources, such as inaccurate cardinality estimates or simplifying assumptions. Therefore, in this work, we opt to learn the plan execution latency from runtime.

In addition to evaluating tree model cost estimation error and correlation, model performance in plan selection is also crucial. In the plan selection phase of query optimization, due to data and model uncertainties [6], the model's cost predictions may have inevitable errors, potentially affecting the final optimal plan selection. In practical applications, if a model can select the actual optimal plan is far more important than making more accurate predictions. Therefore, we evaluate the model's ability to correctly select the optimal plan with the shortest actual execution latency among multiple candidate query plans. We simulate real application scenarios to explore and compare the impact of employing different tree models on the overall plan selection accuracy of the optimizer.

To assess the impact of different tree models in query plan representation, we use the same feature encoding model described in Sect. 4.1 and the same multi-layer perceptron (MLP)-based cost estimator for all experiments. In this manner, we can evaluate the representation ability of different tree models through the accuracy of cost estimates and plan selection.

5.1 Experimental Setup

The experiments are performed on a Linux server with 8-core Intel Silver 4216 Cascade Lake 2.1 GHz CPU, 64 GB memory, and a 32 GB NVIDIA V100 Volta

GPU. PostgreSQL 15.1 is used as the RDBMS for compiling and executing the workloads. The prototype code is written in Python 3.10 with the machine learning library PyTorch [15]. All experimental results are the average after 10-fold cross-validation.

Selected Existing Tree Models. We select five state-of-the-art tree models for further evaluation as follows:

1. *LSTM:* LSTM is a variant of recurrent neural network (RNN) architecture designed for sequences of data and capable of remembering long-term dependencies [20]. Since it cannot directly handle the tree, we flatten the query plan according to post-order traversal [10] as input. The hidden layer of the last node is treated as a graph-level representation of the query plan.
2. *GRU:* GRU is a type of RNN that simplifies the LSTM architecture by combining the forget and input gates into an update gate, improving efficiency and performance on tasks involving sequential data [3]. Model input and output are obtained in the same way as LSTM.
3. *LSTM + Self-Attention:* Saturn [10] proposed an improvement of LSTM for query plan representation by applying the self-attention [22] mechanism to weigh each hidden layer and aggregate all hidden layers according to their weight as the plan-level representation.
4. *Tree-LSTM:* Tree-LSTM is an adaptation of the conventional RNN for tree-structured graphs, which can aggregate information from the leaf nodes to the root across different tree branches by generalizing the traditional LSTM cell by accepting inputs from multiple channels [21]. The root node representation is treated as the graph-level representation of the query plan.
5. *Tree-CNN:* Tree-CNN is designed to handle tree-structured data [14], and was first applied to query plan representation in NEO [12]. It slides a triangular kernel from the root to the leaf nodes, which allows learning relationships between each combination of two child nodes and their parent node. Dynamic pooling is used to aggregate all features into a graph-level representation of the query plan [12]. This model can only be applied to binary trees. Non-binary trees require preprocessing, involving adding more nodes and layers, to become compatible with the model.

Dataset. We use workloads based on a 10 GB database generated by TPC-DS 3.2 [17], an industrial standard benchmark commonly used to evaluate the performance of cost estimation with 25 tables and 429 columns. TPC-DS has more relations and allows for more complex query patterns compared with other benchmarks, such as IMDB (JOB) [8] and TPC-H [16]. Therefore, using TPC-DS we can generate more complex join queries to evaluate the representation learning performance of the tree models in more complex scenarios. Due to the limited complexity of the query generation template that comes with TPC-DS, we used a random query generator to generate queries based on TPC-DS relations. The queries are determined by parameters provided to the query generator, including the number of joins, join types (such as inner-join, outer-join, anti-join), the number of join and local predicates specified within *Where* clauses along with the types of operators used. The following are the datasets used in the two tasks:

1. *Dataset for cost estimation.* We generated 22k queries and each query has up to 10 joins, sampled from referential integrity one-to-many and artificial many-to-many joins, with each join having up to 3 join predicates and each table with up to 5 local predicates. PostgreSQL's default query plans for each query and their execution time are seen as samples and labels. 18k of the query plans are used for training, 2k for validation and 2k for testing.
2. *Dataset for plan selection.* We generate 10k queries using the same query generator for evaluating plan selection performance. Each query was compiled in PostgreSQL using the 13 most influential optimization hints [6] inspired by Bao [11]. These hint sets introduced specific constraints on the join and access operators utilized within the query plans. Each sample in the dataset is a collection of candidate query plans generated based on the 13 hints from the same query. The dataset contains 10k queries yielding about 130k query plan trees and is divided into 8k queries for training, 1k for validation, and 1k for testing.

Evaluation Metrics. We employ four evaluation metrics for the experiment, as shown in the following. Metrics 1–3 are applied to the cost estimation task, while metric 4 is dedicated to the plan selection task.

1. *Prediction Error:* We use Q-Error to evaluate the accuracy of estimated latency. Given the estimated latency y_{el} and actual latency y_{al}, the Q-Error is defined in Eq. 1, ranging from 1 (perfect accuracy) to ∞.

$$Q\text{-}Error = \frac{\max(y_{el}, y_{al})}{\min(y_{el}, y_{al})} \tag{1}$$

2. *Correlation:* We use Spearman's rank correlation to measure the relationship between model-estimated latency and actual latency, with values closer to 1 indicating a stronger correlation. Unlike the widely used Pearson's coefficient, Spearman correlation is not such sensitive to outliers and data scales for using a monotonic function, making it more suitable to evaluate indicators such as latency that may have large orders-of-magnitude differences.
3. *Inference Overheads:* We measured the average time the tree model used to predict a latency estimate for each encoded query plan during testing, which can be used to compare the computational overhead of each tree model.
4. *Plan Suboptimality:* Given a set of candidate execution plans p generated based on the same query, ranking these candidate plans according to the actual execution latency $AL(.)$ and identifying the plan with the shortest execution latency as the actual optimal plan, denoted as p_o. Concurrently, the optimal plan selected by the optimizer with the shortest predicted latency is donated as p_m. In this context, we define Plan Suboptimality as Eq. 2 which takes a value in the range of $[1, \infty)$ and the value closer to 1 reflects the model's better ability to identify the optimal plan among multiple options.

$$\text{Plan Suboptimality} = \frac{AL(p_m)}{AL(p_o)} \tag{2}$$

Loss Function. We used the Mean Squared Error algorithm based on the cost estimator model's output and labels after being preprocessed by natural log transformation and min-max scaling. To reduce the significant skewness in execution latency values, we first apply a natural log to all labels and make them more suitable targets for a machine learning model to learn [6]. Then, min-max scaling is performed to normalize the processed label to the range [0, 1], aligning it with the same output range of the Sigmoid activation function used in the final layer of the cost estimator. Given all the labels in the dataset after natural logarithmic transformation $y_{\log_e}$, a cost estimator output y_{out}, the corresponding actual latency y_{al}, the loss function is defined as follows.

$$\text{Loss} = \sum_{i=1}^{n} \left(\frac{\log_e(y_{al_i}) - \min(y_{\log_e})}{\max(y_{\log_e}) - \min(y_{\log_e})} - y_{out_i} \right)^2 \quad (3)$$

Model Training and Parameter Tuning. In the model training phase, the hyper-parameters are individually tuned for each tree model. The learning rate and other hyper-parameters of the remaining models, such as the feature encoder or cost estimator, are jointly tuned. Parameter tuning is done using the Asynchronous Successful Halving Algorithm (ASHA) [9]. Adaptive Moment Estimation (ADAM) [7] is used as the optimizer for model training. Dropout and Early stopping are used to avoid over-fitting.

5.2 Existing Tree Model Cost Estimation Performance

We evaluate the impact of using different tree models under the same framework on the overall cost estimation performance of the query optimizer. We present the results of the current state-of-the-art tree models in the upper part of Table 1.

Even in workloads with larger scale and more complex join relationships, five existing tree models, including GRU, LSTM, LSTM + Self-Attention, Tree-LSTM, and TCNN, do not have a significant difference in query plan representation capabilities as stated in the study by Yao Z. et al. [25]. However, in the special scenario of representing complex query plans, the differences in Q-Error and Spearman's Correlation performance between different models become more apparent. We outline interesting observations regarding the impacts of various characteristics of the models on plan tree representation learning as follows.

Observation: Tree-LSTM has the Best Performance. Compared with other models, Tree-LSTM performs best in all evaluation indicators. The highest Spearman's correlation and the lowest average Q-Error indicate that it can accurately capture and express critical information in most cases, even when processing complex query plans. In terms of Q-Error distribution, it has a significantly smaller tail-end error than other tree models, indicating that it is more robust in dealing with outliers and extreme situations. Experimental results demonstrate the effectiveness of its hierarchical structure-aware processing in accurately predicting complex query plan costs. However, this comes at the cost of having the highest inference overheads, as shown in Fig. 4(a).

Table 1. Cost Estimation Accuracy (Q-Error) in TPC-DS workload. The first block includes the state-of-the-art tree models we evaluated. The second block includes the tree models based on GNN layers with different graph edge directions and aggregation methods. The overall best results across blocks are underlined.

Tree Model (+Aggregation Method)	Graph Edge Direction	Median Q-error	90th Q-error	99th Q-error	Spearman's Correlation	Top 50% Mean Q-error	Top 99% Mean Q-error
GRU	–	1.895	22.582	296.242	0.776	1.378	5.727
LSTM	–	1.954	21.979	311.205	0.765	1.398	5.824
LSTM + Self-Attention	–	1.885	20.710	313.775	0.778	1.370	5.489
Tree-LSTM	–	1.840	19.413	235.727	0.783	1.352	5.063
TCNN	–	1.912	25.758	430.101	0.761	1.380	6.674
GNN + AddPool	Single directed	1.971	25.184	471.242	0.765	1.403	6.697
GNN + GRU	Single directed	1.884	20.960	324.524	0.776	1.369	5.684
GNN + GRU	Undirected	1.880	20.320	298.524	0.775	1.371	5.409
Bidirectional GNN + AddPool	Weighted directed	1.882	23.387	459.604	0.771	1.370	6.502
Bidirectional GNN + GRU	Weighted directed	**1.762**	**16.537**	**199.137**	**0.805**	**1.327**	**4.558**

Observation: GRU has a Better Performance than LSTM. GRU and LSTM have similar principles to capture long-term dependencies, and both demonstrate relatively good and stable performance in the experiment. However, we observe that GRU, which has a simpler architecture and fewer model parameters, has better performance than LSTM across all evaluation metrics. Due to the reduced overhead and enhanced performance, GRU may offer advantages in query-plan representation tasks. Due to its simplified design, it requires less training data than LSTM to learn the dependencies between query plan nodes. Additionally, its mechanism of reset gate and update gate [3] may enable GRU to more effectively forget unimportant information while learning the query plan tree of post-order expansion, leading to more accurate predictions.

Observation: Self-attention Enhances LSTM. Comparing the results of LSTM and LSTM+Self-Attention in the experiment, we observe that integrating LSTM and self-attention as a tree model has better cost estimation performance than the basic LSTM. This means that using the self-attention mechanism to aggregate hidden states of all nodes can more effectively aggregate node features and better represent the query plan than relying solely on the final node's hidden state. Such findings show the potential of self-attention mechanisms in improving the tree model, demonstrating their capability to enrich model understanding and performance in complex query plan representation tasks.

5.3 GNN-Based Tree Model Cost Estimation Performance

We explore and evaluate the impact of using different edge structures and aggregation methods on the representation ability of GNN-based tree models. We present the experiment results of these models in the lower part of Table 1.

Observation: Using GRU as an Aggregation Method Works Better than Conventional Methods. Compared with using global add pooling, the experiment results indicate a significant improvement in the performance of GNN models across all evaluation metrics by employing GRU as the aggregation operator. The improvement verifies our perspective discussed in Sect. 4.2. Expanding the tree structure graph through post-order traversal enables GRU to learn dependencies between nodes in the actual order of how nodes are executed in the DBMS and aggregate the node features utilizing GRU's gating mechanism. By leveraging the GRU for aggregation, the generated query plan representation can preserve more structural information and more valuable node features, thereby enhancing the model's ability to predict cost estimation accurately.

Observation: Utilizing the Direction of the Edges in the Query Plan Tree Improves the Representation Performance of GNN Models. Our experiments demonstrate that simply using single-directed or undirected edges does not have a noticeable impact on the representation performance of the GNN-based tree models. However, models employing bidirectional GNN with weighted directed edges all show significant cost estimation performance improvements, which both promote message passing between nodes in the tree structure graph and allow nodes to learn the parent-child relationship through weighted directed edges, thereby enabling the model to capture more accurately structural and logical information of query plan even in complex workload representation tasks.

Observation: Bidirectional GNN + GRU Aggregation Operator Performs Much Better than Other Tree Models. Our experimental findings confirm that the novel architecture that we introduced, combining bidirectional TransformerConv and GRU as the aggregation operator, significantly outperforms other GNN-based as well as the state-of-the-art non-GNN tree models. Although it requires a higher inference overhead for running two complete GNN layers, it undoubtedly demonstrates the great potential and research value of GNN technology in query plan representation.

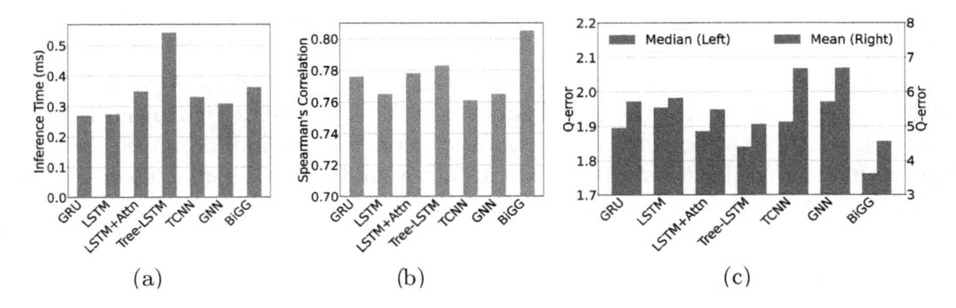

Fig. 4. Performance comparison of different tree models of the cost estimation task in TPC-DS workload. (a) is the inference time of each tree model, (b) is Spearman's rank correlation coefficient and (c) is the mean and median Q-error experimental results.

5.4 Plan Selection Performance and Analysis

We evaluate the impact of different tree models on plan selection performance in scenarios where PostgreSQL generates multiple candidate plans using different hints. The experimental results on Plan Suboptimality are shown in Table 2.

Table 2. Plan Suboptimality performance of tree models in TPC-DS workloads. The best results in each column are underlined.

Tree Model	Median Plan Subopt	90th Plan Subopt	99th Plan Subopt	Top 50% Mean Plan Subopt	Top 90% Mean Plan Subopt	Top 99% Mean Plan Subopt
GRU	1.049	2.478	97.991	1.013	1.139	1.581
LSTM	1.052	2.447	87.656	1.013	1.143	1.637
LSTM + Self-Attention	1.051	2.593	102.315	1.013	1.146	1.881
Tree-LSTM	1.047	2.551	72.518	1.012	1.137	1.824
TCNN	1.054	2.610	93.157	1.014	1.161	1.923
GNN + AddPool	1.052	2.543	100.953	1.013	1.148	1.700
Bidirectional GNN + GRU	**1.045**	**2.094**	**85.441**	**1.011**	**1.122**	**1.516**

Observation: The Impact of Tree Models on the Plan Selection of the Optimizer is Not Significant. Experimental data shows that the impact of different tree models on the plan selection task of the optimizer is not obvious. Although the models employ various strategies to learn query plan representations, the choice between pure tree models does not result in a sharp contrast in the optimizer's ability to select the plan closest to the optimal plan. This subtle impact suggests that factors outside the tree model, such as the design of other components within the optimizer or the specific optimizations based on the workload's characteristics, may also play a crucial role in the overall robustness and effectiveness of the optimizer.

Observation: Improvements in Cost Estimation Performance Do Not Imply Improvements in Plan Selection. In the experimental results shown in Table 1, the tree model we proposed based on bidirectional GNN has a cost estimation performance significantly ahead of other tree models. Although this trend is still reflected in the plan selection evaluation metric, it can no longer widen the gap with the performance of other models. This indicates that merely enhancing the model's cost estimation accuracy may not obviously improve the optimizer's real-world performance.

6 Conclusions and Future Work

We conduct a comparison and analysis of the performance and mechanisms of the current tree models used in query plan representation in complex workloads under tasks of cost estimation and plan selection. Subsequently, we introduce an innovative tree model BiGG leveraging bidirectional GNN aggregated by GRU to learn query plan representations. Through extensive experiments, our proposed model has shown significant enhancements in accuracy for both cost estimation and plan selection compared to existing models. The next phase of our research will explore applying this model to a complete learning-based query optimizer and further advancing its performance and robustness in plan selection.

References

1. Buterez, D., Janet, J.P., Kiddle, S.J., Oglic, D., Liò, P.: Graph neural networks with adaptive readouts. Adv. Neural. Inf. Process. Syst. **35**, 19746–19758 (2022)
2. Chen, T., Gao, J., Chen, H., Tu, Y.: LOGER: a learned optimizer towards generating efficient and robust query execution plans. Proc. VLDB Endow. **16**(7), 1777–1789 (2023)
3. Cho, K., et al.: Learning phrase representations using RNN encoder-decoder for statistical machine translation. arXiv preprint arXiv:1406.1078 (2014)
4. Ding, B., Das, S., Marcus, R., Wu, W., Chaudhuri, S., Narasayya, V.R.: AI meets AI: leveraging query executions to improve index recommendations. In: Proceedings of the 2019 International Conference on Management of Data, pp. 1241–1258 (2019)
5. Ioannidis, Y.: The history of histograms (abridged). In: Proceedings 2003 VLDB Conference, pp. 19–30. Elsevier (2003)
6. Kamali, A., Kantere, V., Zuzarte, C., Corvinelli, V.: Roq: robust query optimization based on a risk-aware learned cost model. arXiv preprint arXiv:2401.15210 (2024)
7. Kingma, D.P., Ba, J.: Adam: a method for stochastic optimization. arXiv preprint arXiv:1412.6980 (2014)
8. Leis, V., Gubichev, A., Mirchev, A., Boncz, P., Kemper, A., Neumann, T.: How good are query optimizers, really? Proc. VLDB Endow. **9**(3), 204–215 (2015)
9. Li, L., Jamieson, K., et al.: A system for massively parallel hyperparameter tuning. Proc. Mach. Learn. Syst. **2**, 230–246 (2020)
10. Liu, S., Chen, X., Zhao, Y., Chen, J., Zhou, R., Zheng, K.: Efficient learning with pseudo labels for query cost estimation. In: Proceedings of the 31st ACM International Conference on Information & Knowledge Management, pp. 1309–1318 (2022)
11. Marcus, R., Negi, P., Mao, H., Tatbul, N., Alizadeh, M., Kraska, T.: Bao: making learned query optimization practical. In: Proceedings of the 2021 International Conference on Management of Data, pp. 1275–1288 (2021)
12. Marcus, R., et al.: Neo: a learned query optimizer. arXiv preprint arXiv:1904.03711 (2019)
13. Mikolov, T., Chen, K., Corrado, G., Dean, J.: Efficient estimation of word representations in vector space. arXiv preprint arXiv:1301.3781 (2013)
14. Mou, L., Li, G., Zhang, L., Wang, T., Jin, Z.: Convolutional neural networks over tree structures for programming language processing. In: Proceedings of the AAAI Conference on Artificial Intelligence, vol. 30 (2016)
15. Paszke, A., et al.: Pytorch: An imperative style, high-performance deep learning library. Adv. Neural Inf. Process. Syst. **32** (2019)
16. Poess, M., Floyd, C.: New TPC benchmarks for decision support and web commerce. ACM SIGMOD Rec. **29**(4), 64–71 (2000)
17. Poess, M., Nambiar, R.O., Walrath, D.: Why you should run TPC-DS: a workload analysis. In: VLDB, vol. 7, pp. 1138–1149 (2007)

18. Rossi, E., Charpentier, B., Di Giovanni, F., Frasca, F., Günnemann, S., Bronstein, M.: Edge directionality improves learning on heterophilic graphs. arXiv preprint arXiv:2305.10498 (2023)
19. Shi, Y., Huang, Z., Feng, S., Zhong, H., Wang, W., Sun, Y.: Masked label prediction: unified message passing model for semi-supervised classification. arXiv preprint arXiv:2009.03509 (2020)
20. Sun, J., Li, G.: An end-to-end learning-based cost estimator. arXiv preprint arXiv:1906.02560 (2019)
21. Tai, K.S., Socher, R., Manning, C.D.: Improved semantic representations from tree-structured long short-term memory networks. arXiv preprint arXiv:1503.00075 (2015)
22. Vaswani, A., et al.: Attention is all you need. Adv. Neural Inf. Process. Syst. **30** (2017)
23. Yu, X., Li, G., Chai, C., Tang, N.: Reinforcement learning with tree-LSTM for join order selection. In: 2020 IEEE 36th International Conference on Data Engineering (ICDE), pp. 1297–1308. IEEE (2020)
24. Zhao, Y., Cong, G., Shi, J., Miao, C.: QueryFormer: a tree transformer model for query plan representation. Proc. VLDB Endow. **15**(8), 1658–1670 (2022)
25. Zhao, Y., Li, Z., Cong, G.: A comparative study and component analysis of query plan representation techniques in ML4DB studies. Proc. VLDB Endow. **17**(4), 823–835 (2023)
26. Zhou, J., et al.: Graph neural networks: a review of methods and applications. AI Open **1**, 57–81 (2020)

FairMC Fair-Markov Chain Rank Aggregation Methods

Chiara Balestra[1(✉)], Antonio Ferrara[2,3], and Emmanuel Müller[1]

[1] TU Dortmund University, Dortmund, Germany
chiara.balestra@cs.uni-dortmund.de
[2] Centai, Turin, Italy
[3] TU Graz, Graz, Austria

Abstract. Given a set of voters' preferences expressed as rankings, rank aggregation approaches combine them into a unique consensus ranking. Even if some methods guarantee fair representation of single voters, they still overlook unfair biases potentially affecting individuals from marginalized groups in the original rankings.

Rank aggregation is employed in many high-stakes decision-making processes, hence, due to the high societal influence, the development and study of fair rank aggregation approaches is essential. We introduce FairMC, a new fair rank aggregation approach based on Markov Chains to derive consensus rankings. By modifying the transition matrix and enforcing fairness in the transition probabilities across the groups, we obtain fairness in the representation of the items at the ranking level. The resulting ranking assures higher visibility for protected groups while being close to the original rankings.

Keywords: Rank aggregation · Fair rankings · Markov Chains

1 Introduction

Rankings are ubiquitous and arise as the output results of several automatized processes. The necessity of finding a unique consensus ranking has received interest due to its applicability in multiple domains [7,8,12]. Given a set of r preference rankings from multiple sources, rank aggregation methods find a consensus ranking out of a list of potentially numerous rankings. The problem of aggregating rankings is not trivial; the Kemeny optimal ranking problem, i.e., aggregating rankings under the minimal sum of Kendall's Tau distances constraint [10], is NP-hard [2,7]. Various techniques have been proposed to find an approximation of the optimal solution: the well-known Borda counting method [8] but also several other combinatorial methods for Kemeny optimal ranking exist [1,15]. Additional methods exploit the representation of the items and the ranking as nodes and edges of a graph [7].

Given a set of potentially many rankings, rank aggregation aims to obtain a unique consensus ranking. Far more complicated is when the "items" to be ranked are people with emerging fairness-related concerns; the rankers can be biased towards some categories, and rank aggregation methods can propagate or even expand such biases to the

R. Wrembel et al. (Eds.): DaWaK 2024, LNCS 14912, pp. 315–321, 2024.
https://doi.org/10.1007/978-3-031-68323-7_26

consensus rankings. The issue, often referred to as *position bias*, refers to the fact that the users need to explore the rankings fully, and elements not ranked in the top positions suffer from low visibility; this is particularly relevant when it affects differently the various groups of items. The challenge is that the fairness issue is present in both the original and the final rankings; the generated consensus ranking should represent the preferences of single voters and show no or slight bias against protected communities and marginalized groups. The goal is to satisfy some *group fairness* constraints, i.e., aiming at the same treatments in the various groups. We refer to [14] that defines constraints for fairness of exposure in ranking outputs, and [17] that deals with the fair top-k ranking problem. The fairness-aware PageRank [13] comprehends various fair versions of the PageRank algorithm, ensuring a proportional allocation of scores to the different groups. Their Markov Chain considers unweighted graphs and doesn't allow for self-loops, two important ingredients that are instead needed to solve the rank aggregation problem and make fairness-aware PageRank not suited for our task. We finally cite some methods for fair rank aggregation: [5,16] found theoretical bounds for picking the best fair version of the input rankings, [3,4] use post-processing methods to re-rank the output of several other traditional non-fair rank aggregation methods with different fairness metrics, [11] represents an in-processing method.

We propose FairMC, an in-processing method, jointly aggregating rankings and achieving fairness; conversely to [11], FairMC does not assume that the number of input rankings is small. We focus on *group fairness*, i.e., aiming at the same treatments in the various groups, on a fixed number N of candidates and a fixed number of r voter input rankings; each ranker is supposed to provide a full rank of the candidates and each of the candidates belongs or not to a pre-specified, e.g., protected by law, *protected group*. We simplify the multi-class case, where several protected groups must be equally represented, to this binary distinction. Proposing FairMC as a fair rank aggregation method, we integrate group fairness into rank aggregation mechanisms. Particularly interesting is the fact that our contribution does not optimize for any group fairness evaluation metric; the approach builds on the well-known Markov Chains methods for ranking aggregation and modifies the underlying graphs to obtain a fair stationary probability distribution by enforcing group equality in the transition probabilities. Although not directly optimizing for any metric, we show in the experiments that FairMC performs reasonably well in most datasets.

2 Markov Chain Methods for Rank Aggregation

Markov Chain methods base the computation of a consensus ranking on the representation of the rankings as a graph, where the nodes represent the items to be ranked and the edges incorporate the ordering relationships of the input rankings.

Graphs. A directed graph is a pair $G = (\mathcal{N}, \mathcal{E})$ where $\mathcal{N}$ represents the set of the nodes while $\mathcal{E}$ represents the set of the edges. We indicate with $N = |\mathcal{N}|$ the total number of nodes. An edge in a directed graph $e = (v, u)$ is an ordered pair of nodes $v, u \in \mathcal{N}$, and $(v, u) \in \mathcal{E} \not\Rightarrow (u, v) \in \mathcal{E}$. If the edges are weighted, we speak about weighted graphs; weights, designed by $e_{u,v}$, can be assigned to each edge, and they represent the strength of the link between u and v. Mathematically, we assign to each edge a weight $e_{u,v} > 0$; an edge is not part of the graph if and only if its weight is zero, i.e., $(u, v) \notin \mathcal{E} \Leftrightarrow e_{u,v} = 0$. The adjacency matrix $W \in \mathbb{R} \times \mathbb{R}$ whose entries are $W_{u,v} = e_{u,v}$ entirely defines the graph $(\mathcal{N}, \mathcal{E})$. Given a node $v \in \mathcal{N}$, $\mathcal{N}_G(v)$ indicates the set of neighbors of v, i.e., $\mathcal{N}_G(v) = \{u \in \mathcal{N} \mid e_{u,v} > 0\}$; in the case G is a directed graph, we can distinguish between $\mathcal{N}_G^+(v) = \{u \in \mathcal{N} \mid (v, u) \in \mathcal{E}\}$, the *out-neighborhood* and $\mathcal{N}_G^-(v) = \{u \in \mathcal{N} \mid (u, v) \in \mathcal{E}\}$, the *in-neighborhood*.

Markov Chains for Rank Aggregation. We consider a finite set $\Sigma = \{\sigma_1, \ldots, \sigma_r\}$ of ranked lists of the $\mathcal{N}$ elements where each σ_i represents the ranking provided by the ranker i. From the rankings in Σ, Markov Chain methods for rank aggregation construct a directed weighted graph $G = (\mathcal{N}, \mathcal{E})$ whose nodes are the elements in $\mathcal{N}$ and whose edges $\mathcal{E}$ and respective weights represent their pairwise orderings in Σ. The pairwise relationships among elements of $\mathcal{N}$ in the rankings Σ are used to determine the existence and the weight of the edges connecting them, i.e., given $u, v \in \mathcal{N}$, the weight $e_{u,v}$ of the directed edge (u, v) is assigned so that it includes the frequency in which v is ranked higher than u by the input rankings. The Markov process is then defined by the obtained adjacency matrix W. Given a directed weighted graph, it is possible to normalize the edges such that $\sum_{v \in \mathcal{N}_G^+(u)} e_{u,v} = 1$ for all $u \in \mathcal{N}$ (in case, by adding self loops); thus, the adjacency matrix can be interpreted as a probability transition matrix, where $W_{u,v}$ represents the probability for a random walker on the graph to transition from node u to node v. A probability matrix describes the underlying Markov Chain process, and it is called *ergodic* in the case that the probability of ending up in one node is bigger than 0 for each node $u \in \mathcal{N}$. The stationary probability distribution π is the vector satisfying $\pi_u = \sum_{v \in \mathcal{N}_G^-(u)} W_{u,v} \pi_v$ where π_v represents the probability of being in the node v; π can be efficiently computed when the Markov Chain is *ergodic*. From π, the consensus ranking is defined as the underlying ranking obtained using the probabilities of the stationary distribution as importance scores, i.e., the highest probability places the corresponding item in the first position of the consensus ranking. For the construction of the graphs given by the various rank aggregation approaches we refer to the literature [6,7].

Group Fairness. Each ranker assigns an ordering to the elements in $\mathcal{N}$. Rank aggregation methods assume the existence of a *consensus ranking*, i.e., a ranking that equally and fairly considers the opinion of each ranker. The composition of $\mathcal{N}$ is generally a priori unknown. We use the notation $\mathcal{N} = \mathcal{P} \cup (\mathcal{N} \setminus \mathcal{P})$, where $\mathcal{P}$ is the protected category while $\mathcal{N} \setminus \mathcal{P}$ is the set of non-protected items. For the moment, we reduce to study the case of $\mathcal{P} = \mathcal{P}_1 \cup \cdots \cup \mathcal{P}_k$ where $\mathcal{P}_1, \ldots, \mathcal{P}_k$ are k protected categories in $\mathcal{N}$; thus, we have only a binary classification over the items. We indicate with p a protected item, i.e., $p \in \mathcal{P}$, u a non-protected item, i.e., $u \in \mathcal{N} \setminus \mathcal{P}$ and $v \in \mathcal{N}$ for the general individual. Given a ranking σ_r generated by the rth ranker, we write $\sigma_r(v)$ to indicate the ranking position of element v in the ranking σ_r. Each ranking σ_r is an element of S_N, the symmetric group or permutation group of N elements.

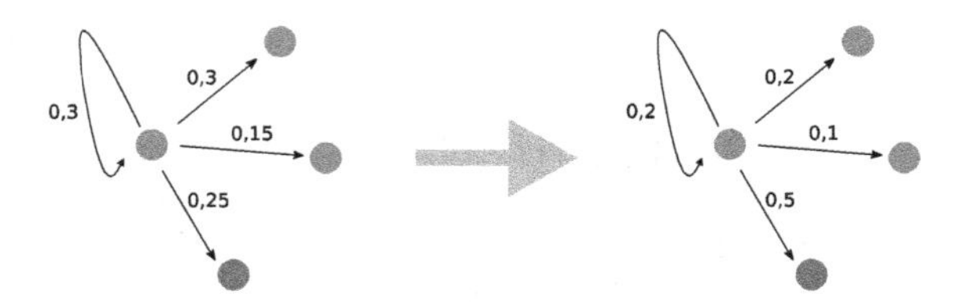

Fig. 1. Toy example for our proposed approach for the node v in the graph.

3 FairMC

Markov Chain methods allow interpreting the search of the consensus ranking in terms of transition probabilities among nodes of a graph and the graph-structured representation has advantages in terms of computational complexity and interpretability. However, as for the other rank aggregation methods, Markov Chain rank aggregation approaches do not consider whether the obtained consensus ranking satisfies fairness requirements. Markov Chain methods rely on the computation of the stationary probability distribution of the Markov process defined on the directed graph of items by an adjacency matrix; we state that it is sufficient to make the adjacency matrix fair in order to obtain a fair consensus ranking. Our claim derives from observing that the transition probability reflects the biases of the rankers, where the probability of transitioning to a protected node or to an unprotected node is not balanced.

Our approach is based on the same assumptions of the Markov Chain methods [6,7]; the proposed modification is applicable to each of the three Markov Chain methods. We use the graph $G = (\mathcal{N}, \mathcal{E})$ derived from Markov Chain and modify it obtaining a new graph $\tilde{G} = (\tilde{\mathcal{N}}, \tilde{\mathcal{E}})$, where $\tilde{\mathcal{N}} = \mathcal{N}$. The edges' weights are reweighted using the following definition

$$\tilde{e}_{u,v} = \begin{cases} e_{u,v} \cdot \frac{|\mathcal{P}|}{N \sum_{w \in \mathcal{P}} e_{u,w}} & \text{if } v \in \mathcal{P} \\ e_{u,v} \cdot \frac{|\mathcal{N} \setminus \mathcal{P}|}{N \sum_{w \in \mathcal{N} \setminus \mathcal{P}} e_{u,w}} & \text{if } v \in \mathcal{N} \setminus \mathcal{P} \end{cases} \tag{1}$$

Furthermore, for each node u for which $\mathcal{N}_G^+(u) \cap \mathcal{P} \neq \emptyset$, we create an edge from u to v with weight $\tilde{e}_{u,v} = \frac{|\mathcal{P}|}{N}$ for each $v \in \mathcal{P}$; instead, for each node u' for which $\mathcal{N}_G^+(u') \cap (\mathcal{N} \setminus \mathcal{P}) \neq \emptyset$, we create an edge from u' to v' with weight $\tilde{e}_{u',v'} = \frac{|\mathcal{N} \setminus \mathcal{P}|}{N}$ for each $v' \in \mathcal{N} \setminus \mathcal{P}$.

Figure 1 illustrates a toy example of the rescaling mechanism. Recall that reweighting the edges' weights is equivalent to redefining the adjacency matrix and that the weights represent the probability of transitioning from one node to another one. The rescaled weights guarantee that

1. the relative importance of the original transition probabilities is respected,
2. the sum of the probabilities to transition to a protected node is equal to the sum of the probabilities to transition to a non-protected, i.e., it holds $\sum_{v \in \mathcal{P}} \tilde{e}_{u,v} = \sum_{v \in \mathcal{N} \setminus \mathcal{P}} \tilde{e}_{u,v}$ for each $u \in \mathcal{N}$. Thus, the transition probabilities are *fair* with respect to the protected group.

In the experimental section, we show that our approach assures fairer consensus rankings without harshly disrupting the Kemeny distance.

4 Experiments

We use *group exposure* [9] and *fairness@k*, i.e., the difference in the percentage of items from the two groups in the top k, to evaluate the fairness of the consensus rankings. We group the methods in FairMC, Markov Chain methods (MC), non-fair aggregation approaches[1], and the EPIRA methods[2]. We use the 30 Mallows datasets (with different dispersion parameters) described in [3]. The code is available on GitHub[3].

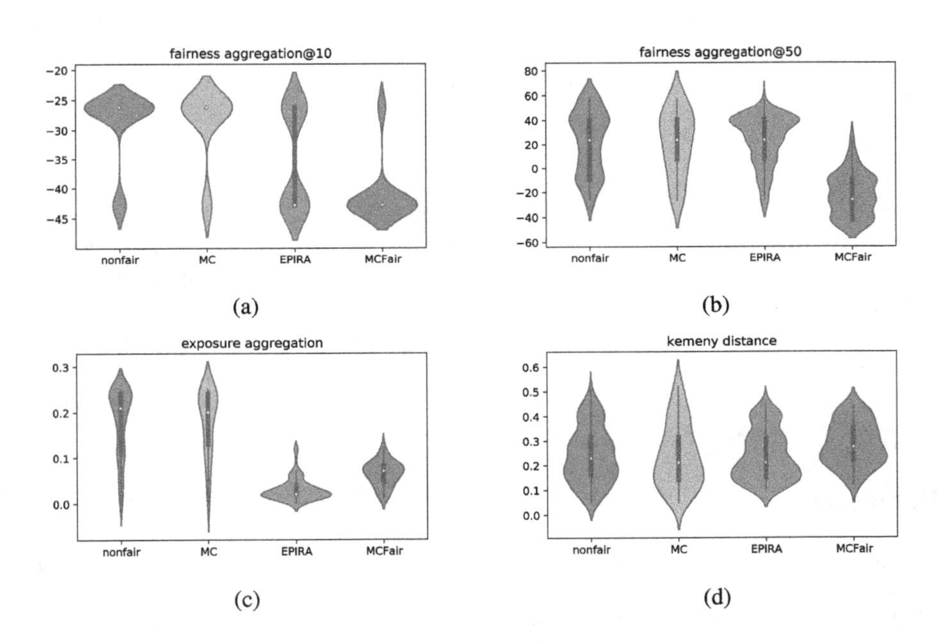

(a) (b) (c) (d)

Fig. 2. Violin plots of summary statistics of performance.

4.1 Performance Evaluation

Figure 2a and 2b show the violin plots of fairness@k of the consensus rankings of the top 10% and 50% of the full ranking. At each level, the non-fair approaches, Markov Chain approaches, and the competing fair rank aggregation approaches tend to obtain

[1] best rank aggregated, Borda aggregation, exponential enhanced Borda, exponential weighting, highest rank, lowest rank, mc4, robust aggregation, round Robin, stability enhanced and stability selection.

[2] EPIK, EPIRA + {Kemeny, Copeland, Copeland noWiG, Schulze, Borda, Maximin Borda} [3].

[3] github.com/chiarabales/fairMC.

consensus rankings favoring the unprotected group; this is particularly visible at 50%, where a clear difference among fair and non-fair methods appears. On the contrary, the FairMC approaches tend to favor the protected group, both at 10 and 50%. In Fig. 2c, we do not expect to perform better than Cachel et al. [3] with respect to the exposure as their methods directly optimize for the exposure of the consensus rankings while we do not. Nevertheless, we can observe how the consensus rankings obtained through the FairMC approaches visibly outperform all non-fair methods in exposing the protected group.

The Kemeny distance in Fig. 2d evaluates the distance between the original rankings and the obtained consensus ranking, and it is not a fairness metric. Higher values of the Kemeny distance indicate that the consensus ranking is distant from the original rankings. We observe a slight tendency to increase the Kemeny distance when adding fairness constraints in the Markov Chain, as expected given the trade-off between optimizing for the Kemeny distance and for the fairness metrics.

5 Conclusions

We propose FairMC, introducing fairness at the level of the nodes in Markov Chain-based rank aggregation methods. Our method achieves extraordinarily good results with respect to the exposure of protected groups and the fairness@k with respect to common rank aggregation approaches without optimizing for any fairness metrics; furthermore, the Kemeny distance with respect to the original rankings remains acceptably low after the introduced modification. We believe in the potential of acting at the level of nodes to achieve fairer consensus rankings and that further development in this direction can achieve unbiased results.

References

1. Ailon, N., Charikar, M., Newman, A.: Aggregating inconsistent information: ranking and clustering. J. ACM (JACM) **55**, 1–27 (2008)
2. Bartholdi, J., Tovey, C.A., Trick, M.A.: Voting schemes for which it can be difficult to tell who won the election. Soc. Choice Welfare **6**, 157–165 (1989)
3. Cachel, K., Rundensteiner, E.: Fairer together: Mitigating disparate exposure in Kemeny rank aggregation. In: FACCT (2023)
4. Cachel, K., Rundensteiner, E., Harrison, L.: Mani-rank: multiple attribute and intersectional group fairness for consensus ranking. In: ICDE (2022)
5. Chakraborty, D., Das, S., Khan, A., Subramanian, A.: Fair rank aggregation. In: NeurIPS (2022)
6. DeConde, R.P., Hawley, S., Falcon, S., Clegg, N., Knudsen, B., Etzioni, R.: Combining results of microarray experiments: a rank aggregation approach. Stat. Appl. Genet. Mol. Biol. **5** (2006)
7. Dwork, C., Kumar, R., Naor, M., Sivakumar, D.: Rank aggregation methods for the web. In: WWW, pp. 613–622 (2001)
8. Emerson, P.: The original Borda count and partial voting. Soc. Choice Welfare **40**, 353–358 (2013)
9. Ferrara, A., Bonchi, F., Fabbri, F., Karimi, F., Wagner, C.: Bias-aware ranking from pairwise comparisons. Data Min. Knowl. Discov. 1–25 (2024)

10. Kendall, M.G.: A new measure of rank correlation. Biometrika **30**, 81–93 (1938)
11. Kuhlman, C., Rundensteiner, E.: Rank aggregation algorithms for fair consensus. VLDB (2020)
12. List, C.: Social choice theory (2013)
13. Pitoura, E., Stefanidis, K., Koutrika, G.: Fairness in rankings and recommendations: an overview. VLDB (2022)
14. Singh, A., Joachims, T.: Fairness of exposure in rankings. In: KDD (2018)
15. van Zuylen, A., Williamson, D.P.: Deterministic algorithms for rank aggregation and other ranking and clustering problems. In: Kaklamanis, C., Skutella, M. (eds.) WAOA 2007. LNCS, vol. 4927, pp. 260–273. Springer, Heidelberg (2008). https://doi.org/10.1007/978-3-540-77918-6_21
16. Wei, D., Islam, M.M., Schieber, B., Basu Roy, S.: Rank aggregation with proportionate fairness. In: ICMD (2022)
17. Zehlike, M., Bonchi, F., Castillo, C., Hajian, S., Megahed, M., Baeza-Yates, R.: FA* IR: a fair top-k ranking algorithm. In: CIKM (2017)

LSiX: A Scheme for Efficient Multiple Continuous Window Aggregation Over Streams

Shun Kawakami[1(✉)], Savong Bou[2], and Toshiyuki Amagasa[2]

[1] Graduate School of Science and Technology, University of Tsukuba, Tsukuba, Japan
s2320583@u.tsukuba.ac.jp

[2] Center for Computational Sciences, University of Tsukuba, Tsukuba, Japan

Abstract. Stream processing engines need to process multiple queries over streams simultaneously, and continuous window aggregation plays a critical role in various applications as a part of data analysis pipelines. However, the system suffers from scalability issues when dealing with massive queries with different window and slide sizes over data streams with high input rates. To address this problem, we propose LSiX (longest-shortest-window-based indexing) to aggregate multiple queries over data streams efficiently. Specifically, we use two arrays based on the longest and shortest windows among all registered queries, and all query results are computed by sharing partial aggregation from the two arrays. This allows for efficient aggregate computation, with a maximum of two operations for each query. The results of our experiment show that LSiX can be at least 3 times faster than the comparative methods, including the state-of-the-art method, MCQA.

Keywords: Sliding window · Aggregation · Streams

1 Introduction

Recently, there has been a growing need for near real-time processing in data stream management systems (DSMS) [2,3] due to the increasing prevalence of continuous data streams in various domains. There has also been a significant rise in the demand for aggregation computation, such as sum, max, and average on time series [4,5]. For instance, equity investors require real-time delivery of their own custom aggregates [10], aggregating the latest data in the data stream.

To apply aggregation over data streams, in many cases, we apply *sliding window aggregation (SWAG)* [1], $Q[w, s]$, that continuously outputs aggregated values within the latest records in the window of size w whenever s new records arrive. Specifically, there are two SWAG types: *time-based* and *count-based*; the former defines a window in terms of time, whereas the latter defines the window according to the number of records in the window.

In most applications, DSMS needs to handle multiple queries over streams, often ranging from tens to hundreds, and must process these queries against

R. Wrembel et al. (Eds.): DaWaK 2024, LNCS 14912, pp. 322–328, 2024.
https://doi.org/10.1007/978-3-031-68323-7_27

streams with very high input rates. To process such streams, there have been many studies on SWAG for a single query [6,7,11]. However those algorithms impractical to deal with thousands of queries simultaneously.

MCQA [9] is a state-of-the-art algorithm for efficient aggregation for multiple queries. However, MCQA's processing time became unstable when the ratio of window size to slide size or the difference between window sizes is large.

To address this problem, this paper proposes the LSiX (longest-shortest-window-based indexing) algorithm to process multiple SWAG queries over data streams efficiently. LSiX uses a query's longest and shortest windows to create two arrays, *Lo* and *Sh*, and shares their (partial) aggregation results among all queries, ensuring stability and significantly reducing processing time. The required time and space of LSiX are respectively $2n + 2 + \lceil \frac{W_{max}}{W_{min}^2} \rceil$ and $W_{max} + \lceil \frac{W_{max}}{W_{min}} \rceil$ where n is the total number of queries, W_{max} and W_{min} are respectively the longest and shortest window sizes. Experiments show that LSiX is at least 3.5 times faster than the state-of-the-art method MCQA and at least 3 times faster than the extended L-BiX [6] for multiple queries.

This paper is organized as follows: Sect. 2 discusses existing approaches, Sect. 3 presents the proposed approach, including experimental evaluations in Sect. 4. Finally, Sect. 5 concludes this paper and mentions future works.

2 Related Work

There have been many methods for SWAG computation. More precisely, the following works focus on processing a single query: L-BiX [6], and FlatFIT [11].

Let us take a closer look at most relevant works. L-BiX [6] is the most efficient method that combines backward aggregation (from latest to oldest record) and forward aggregation (from oldest to latest record). Additionally, FlatFIT [11] uses two arrays storing partial aggregate values and pointers, obtains the index of the required partial aggregate value via pointers, yielding the final result.

In the meantime, multiple queries often need to be processed simultaneously, and the approaches mentioned above cannot deal with multiple queries efficiently because they are optimized for processing a single query. MCQA [9] was proposed to address the mentioned problem. MCQA is an efficient method of aggregating multiple queries by aggregating smaller queries first and then reusing the aggregate values for aggregating larger queries.

Table 1. Required time and space.

Method	Time	Space
L-BiX	$3n$	$\sum_{i=1}^{n}(\frac{W_i}{Sl})$
FlatFIT	$log(n)$	$2^{\lceil log(n) \rceil + 1}$
MCQA	$\sum_{i=2}^{n}(\lfloor \frac{W_i}{W_{i-1}} \rfloor + W_i \% W_{i-1}) + \lfloor \frac{W_1}{Sl} \rfloor + W_1 \% Sl$	$\lceil \frac{W_{max}}{Sl} \rceil \times n$
LSiX	$2n + 2 + \lceil \frac{W_{max}}{W_{min}^2} \rceil$	$W_{max} + \lceil \frac{W_{max}}{W_{min}} \rceil$

The needed time and space are summarized in Table 1. L-BiX's needed time depends only on n, which is large, so LSiX's needed time is significantly smaller, and the space requirement is also smaller. MCQA's needed time strongly depends on the size of two consecutive windows and the space requirement depends on n. In contrast, LSiX ignores other queries and only finds from two arrays, so needed time and space is significantly smaller then that of MCQA

3 Proposed Method: Longest-Shortest-Window-Based Indexing (LSiX)

Given a set of user-defined SWAGs, let W_{max} and W_{min} be the longest and shortest window sizes, respectively. Then, LSiX uses only two arrays to maintain sub-aggregations: Lo of size W_{max} and Sh of size $\lceil \frac{W_{max}}{W_{min}} \rceil$.

Lo maintains all sub-aggregations up to W_{min} ($< W_{max}$). In Lo, aggregation depends on the current index; the index moves to the right each time a record arrives. When the index of Lo has moved by $Wmin$, a new *cycle* of size W_{min} starts, and we maintain the sub-aggregations within the current cycle. Lo basically aggregates a new record and the aggregate values of records that have arrived so far in the current cycle(forward aggregation) and previous indexes of the current cycle maintain raw records. However, only when the end of the current cycle is reached, it updates all current cycle starting with the newest record(backward aggregation). Therefore, Most of Lo maintains such results.

Sh maintains the result of backward aggregation for the aggregated value within each cycle in Lo. The indexes of Sh correspond one-to-one with the cycles of Lo from left to right. Sh is updated only immediately after backward aggregation of Lo, starting from the index corresponding to the current cycle.

Lo combines backward and forward aggregation, making it easier to discard old records and add new ones. Meanwhile, Sh enables large aggregation results to be retained with fewer aggregations and less frequency. When obtaining a query, we primarily utilize a value from Sh. The remaining old part is used from backward aggregation result of Lo, and the remaining new part is used from forward aggregation result of Lo. This allows each query to be obtained in two aggregation operations, making LSiX more stable and requiring fewer aggregation operations than other methods.

Figure 1 shows examples (aggregation function is max). In Fig. 1(a), the latest record is 4, so $Lo[10:12]$ has a forward aggregation ($Lo[10] = 6, Lo[11] = 2, Lo[12] = max([Lo[10], Lo[11], 4])$), and the other has a backward aggregation (e.g. $Lo[4] = 1, Lo[3] = max(Lo[4], 3)$, $Lo[2] = max(Lo[3], 7), Lo[1] = max(Lo[2], 3), Lo[0] = max(Lo[1], 9)$). Sh has a backward aggregation starting from $Sh[1]$ ($Sh[1] = Lo[5], Sh[0] = max(Lo[0], Sh[1]), Sh[3] = max(Lo[15], Sh[0])$). In Fig. 1(b), a new record (3) arrives and Lo's index advances one to the right. Aggregating a new record and $Lo[12]$, which maintains the aggregate values of $Lo[10:12]$, $Lo[13]$ maintains the value with forward aggregation of the current cycle. Then store a raw record (4) in $Lo[12]$. In Fig. 1(c), a new record (1) arrives and Lo's index reaches

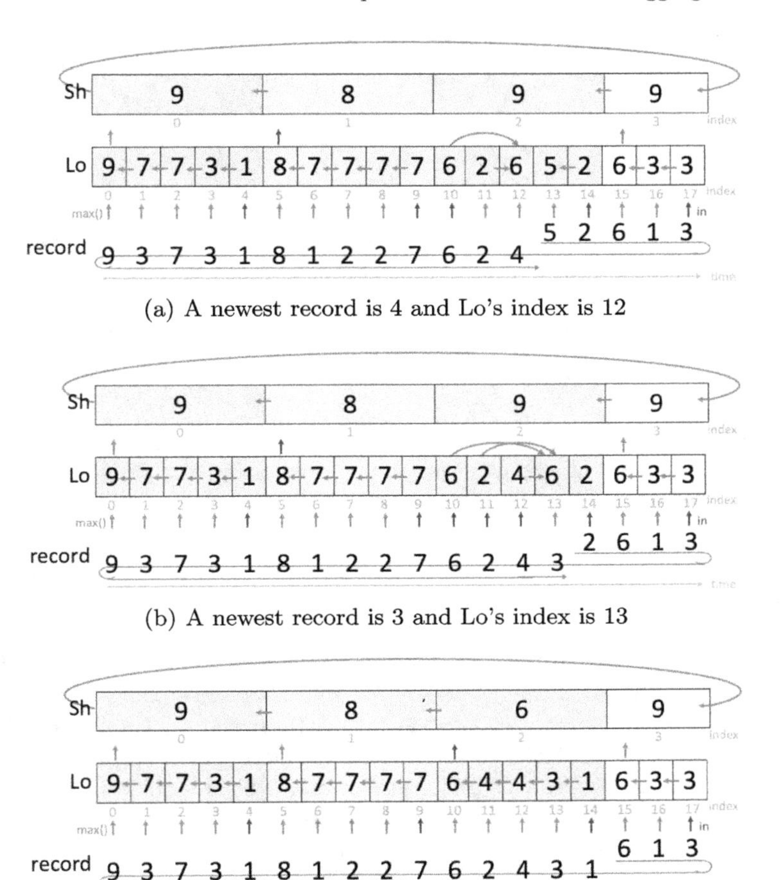

(a) A newest record is 4 and Lo's index is 12

(b) A newest record is 3 and Lo's index is 13

(c) A newest record is 1 and Lo's index is 14

Fig. 1. Example of Lo and Sh with $W_{min} = 5$ and $W_{max} = 18$, slide size = one record. The color of the array indicates the cycle($= W_{min}$). Also, aggregation takes place based on the direction of the red arrows and insertion takes place based on the direction of the blue arrows. (Color figure online)

the end of the current cycle. Therefore, the current cycle of $Lo[10 : 14]$ is updated in backward aggregation ($Lo[14] = 1, Lo[13] = max(Lo[14], 3)$, $Lo[12] = max(Lo[13], Lo[12]), Lo[11] = max(Lo[12], Lo[11])$, $Lo[10] = max(Lo[11], Lo[10])]$). Subsequently, Sh is updated in backward aggregation starting from $Sh[2]$ ($Sh[2] = Lo[10], Sh[1] = max(Lo[5], Sh[2])$, $Sh[0] = max(Lo[0], Sh[1]), Sh[3] = max(Lo[15], Sh[0])$.

The following is an example of obtaining the query. Obtaining $Q[15, 4]$ in Fig. 1(a), we use $Sh[0]$ (result of aggregating $Lo[0 : 9]$) for the most part. The rest old part is used from $Lo[16]$(result of aggregating $Lo[16 : 17]$), and the new part is used from $Lo[12]$(result of aggregating $Lo[10 : 12]$). This query is obtained

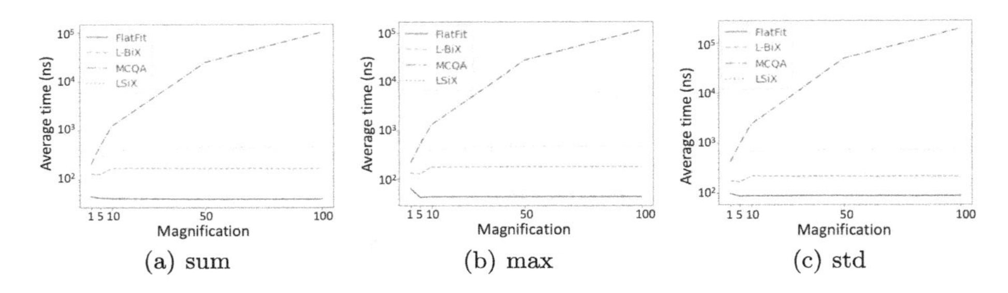

Fig. 2. Results of run time experiment with varying window size for count-based window

by aggregating these three values ($Result = max(Sh[0], Lo[16], Lo[12])$). Other queries is obtained by aggregating up to three values in the same way.

4 Experiment

4.1 Data and Evaluations

To verify the effectiveness of the proposed method, we compared with MCQA, L-BiX and FlatFIT by averaging the processing time per slide.

The DEBS [8] dataset was used for validation: the DEBS12 Grand Challenge Datatset (DEBS) is a standard dataset widely used for query processing load assessment in similar research frameworks. The records contain 51 fields that indicate energy metrics and sensor status, one of which was used in the experiment. The dataset contains about 32 million records.

The query processing load affects SWAG's performance, prompting experiments with varying window sizes. LSiX supports both count-based and time-based windows, but in this experiment only count-based was used. The results were averaged from five independent runs of each algorithm. For the aggregate functions, we employed sum, maximum and standard deviation (std). All experiments were performed on an Apple M1 Pro, 16 GB memory machine.

4.2 Varying the Window Size

In this experiment, the slide size was fixed at 10 and the window size was varied in 5 patterns by multiplying $W = [35, 65, 75, 100, 110, 160, 205, 220, 235, 280]$ by 1, 5, 10, 50, 100. The results for sum, max, std are shown in Fig. 2.

In general, processing time tends to increase as window size increases. In addition, as explained in Sect. 4.2, the processing time of MCQA increases significantly as the window size to slide size ratio increases or the difference between two consecutive window size increases. The processing time of L-BiX and FlatFiT increases slightly as the number of backward aggregation increases with window size. In contrast, the processing time of LSiX decreases as the window size increases. This is because the increase in shortest window size reduces $\lceil \frac{W_{max}}{W_{min}^2} \rceil$, the number of times Sh is performed. Therefore, the difference in processing time becomes more pronounced as the window size increases.

5 Conclusion

This paper has proposed an efficient method for aggregating multiple queries over data stream. The method uses two arrays Lo and Sh based on the longest and shortest window sizes and shares the results to seek all queries efficiently. The arrays Lo and Sh can combine the two forward and backward aggregation methods to keep partial aggregations within a range with a minimum of aggregation operations. As a result, the workload primarily depends on the number of queries, and all queries can be aggregated quickly.

Experimental results show the proposed method significantly outperforms the state-of-the-art methods MCQA and L-BiX in terms of processing time.

Acknowledgements. This paper is based on results obtained from JSPS KAKENHI Grant Numbers JP23K16888 and JP23K24949, JST CREST Grant Number JPMJCR22M2, and "Research and Development Project of the Enhanced Infrastructures for Post-5G Information and Communication Systems" (JPNP20017), commissioned by the New Energy and Industrial Technology Development Organization (NEDO).

References

1. Arasu, A., Babu, S., Widom, J.: The CQL continuous query language: semantic foundations and query execution. VLDB J. **15**(2), 121–142 (2006)
2. Bou, S., Amagasa, T., Kitagawa, H.: Filtering XML streams by XPath and keywords. In: iiWAS, pp. 410–419 (2014)
3. Bou, S., Amagasa, T., Kitagawa, H.: An improved method of keyword search over relational data streams by aggressive candidate network consolidation. In: Hartmann, S., Ma, H. (eds.) DEXA 2016. LNCS, vol. 9827, pp. 336–351. Springer, Cham (2016). https://doi.org/10.1007/978-3-319-44403-1_21
4. Bou, S., Amagasa, T., Kitagawa, H.: Scalable keyword search over relational data streams by aggressive candidate network consolidation. Inf. Syst. **81**, 117–135 (2019)
5. Bou, S., Amagasa, T., Kitagawa, H.: PR-MVI: efficient missing value imputation over data streams by distance likelihood. In: iiWAS, vol. 13635, pp. 338–351 (2022)
6. Bou, S., Kitagawa, H., Amagasa, T.: L-BiX: incremental sliding-window aggregation over data streams using linear bidirectional aggregating indexes. Knowl. Inf. Syst. **62**(8), 3107–3131 (2020)
7. Bou, S., Kitagawa, H., Amagasa, T.: CPiX: real-time analytics over out-of-order data streams by incremental sliding-window aggregation. IEEE Trans. Knowl. Data Eng. **34**(11), 5239–5250 (2022)
8. Jerzak, Z., Heinze, T., Fehr, M.: The DEBS 2012 grand challenge. In: Sixth ACM International Conference on Distributed Event-Based Systems, pp. 393–398 (2012)
9. Liu, W., Zhang, T., Liu, J.: Window-based multiple continuous query algorithm for data streams. J. Supercomput. **75**(9), 5782–5807 (2019)

10. Ma, Y., Mao, R., Lin, Q.: Multi-source aggregated classification for stock price movement prediction. Inf. Fusion **91**, 515–528 (2023)
11. Shein, A.U., Chrysanthis, P.K., Labrinidis, A.: FlatFIT: accelerated incremental sliding-window aggregation for real-time analytics. In: 29th International Conference on Scientific and Statistical Database Management, pp. 5:1–5:12. ACM (2017)

Applications

QPAVE: A Multi-task Question Answering Approach for Fine-Grained Product Attribute Value Extraction

Kassem Sabeh[1(✉)], Mouna Kacimi[2], and Johann Gamper[1]

[1] Free University of Bozen-Bolzano, Bolzano, Italy
{ksabeh,jgamper}@unibz.it
[2] Wonder Technology Srl, Trento, Italy
mouna@wonderflow.ai

Abstract. Extracting product attribute values is essential for e-commerce applications like product search, retrieval, and recommendation. While existing methods excel in extracting values from product descriptions, little attention has been given to extracting values from product profiles, which may contain coarse-grained information consisting of multiple attributes at the same time. In this work, we propose QPAVE, a novel multi-task question answering approach for fine-grained attribute value extraction. QPAVE treats each fine-grained attribute as a question, identifying values from the context of a coarse-grained attribute. To capture dependencies, we employ a hypernetwork to parameterize the decoding layer of the question answering model with the embeddings of the coarse-grained attribute. Additionally, the model predicts the product category in a multi-task training setup to learn category-aware token embeddings. We conducted extensive experiments on a large distantly annotated dataset and a gold human-annotated test set showing superior performance over several state-of-the-art methods. Our code and data are available at https://github.com/kassemsabeh/qpave. We also release our model as a demo at https://bit.ly/3U8rbuI.

Keywords: Attribute Value Extraction · Question Answering · Multi-task Learning

1 Introduction

Product attributes are essential for e-commerce platforms, enabling functions like product search [34], recommendation [26], and question answering [33]. The accuracy of these attributes significantly influences the efficiency of automated product processing, impacting customer shopping experiences. However, these attributes are often noisy, incomplete, and overly general [4,32]. Figure 1 shows an example of an "Electric Shaver" product where some attributes like *Brand* and *Blade Material* are accurate, while others like *Special Feature* are not. We refer to such attributes (e.g., *Special Feature*) as coarse-grained attributes since they

R. Wrembel et al. (Eds.): DaWaK 2024, LNCS 14912, pp. 331–345, 2024.
https://doi.org/10.1007/978-3-031-68323-7_28

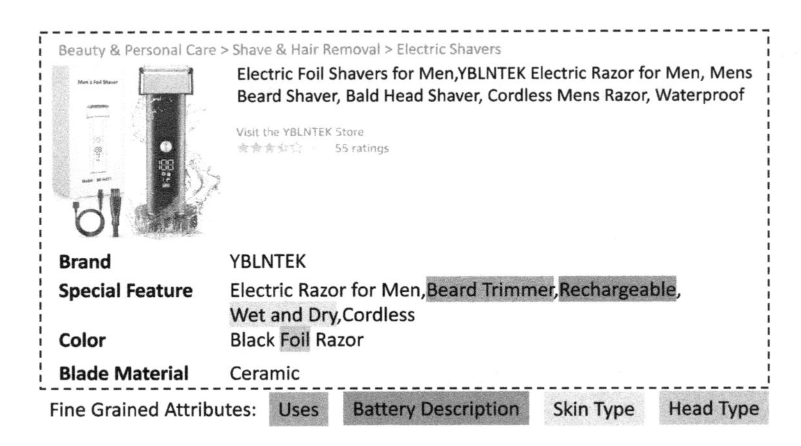

Fig. 1. An example of a product profile with coarse-grained attributes.

contain information about other attribute groups, known in similar products as *Uses*, *Battery Description*, and *Skin Type*. To make product profiles more accurate, we need tailored techniques that extract fine-grained attribute values from existing coarse-grained attributes.

Relevant approaches to this problem focus on attribute value extraction from product descriptions. Early approaches utilize rule-based techniques, which rely on domain-specific knowledge to guide the extraction [11,27]. To tackle the closed-world assumption problem, other approaches adopted sequence tagging based methods [9,29,35] to identify the values from the input text. The most recent approaches formulate the problem as an instance of the question answering (QA) task [28,32] to scale to a large number of attributes and achieve better performance on unseen data.

The above methods have three main limitations: First, they rely solely on product titles and descriptions, which cannot provide all attribute values. For example, in Fig. 1, the attribute *Battery Description* has the value *"Rechargeable"*, but this is only mentioned under the coarse-grained attribute *Special Feature*, and not in the product title. Second, they do not model the dependency between attributes which is crucial for coarse-grained attributes since they typically provide hints about applicable fine-grained attributes. For instance, the coarse-grained attribute *Special Feature* often contains information about the attribute *Uses* in the "Electric Shavers" category. Third, they do not take the product category into account during the extraction which is needed to determine the set of applicable attributes and attribute values. For example, "Vacuum Cleaners" products may use the value "Wet and Dry" for the attribute *Recommended Uses*, while in "Electric Shavers" products, this value could refer to the attribute *Skin Type*.

To address these limitations, we propose **QPAVE**, a multi-task **Q**uestion answering framework for fine-grained **P**roduct **A**ttribute **V**alue **E**xtraction. Inspired by the latest advances in question answering [33], we formulate the fine-grained attribute value extraction problem as a question answering task.

We employ the contextual encoder from BERT [3] as the backbone of our question answering model. To model the dependency on the coarse-grained attribute, we propose to parameterize the decoding layer of the question answering model with the embeddings of the coarse-grained attribute through a hypernetwork [6]. In addition, we predict the product category in a multi-task training setup, which allows us to learn category-specific token embeddings that improve the extraction. Finally, we enhance the generalization performance of the model on unseen attributes by pre-training the encoder on a masked language modelling (mlm) task using an e-commerce corpus.

We summarize the main contributions of the paper as follows: (1) We propose QPAVE, a novel multi-task framework for fine-grained product attribute value extraction. To the best of our knowledge, this is the first paper to explore fine-grained attribute value extraction. (2) We improve the attribute value extraction by incorporating information about the coarse-grained attributes through an adaptive decoder and using multi-task learning to predict the product categories. (3) We introduce a large and novel dataset for end-to-end evaluation of the fine-grained attribute value extraction task. (4) We conduct extensive experiments and demonstrate significant improvements of the proposed approach over several state-of-the-art baselines.

2 Related Work

Initial approaches for attribute value extraction utilized rule-based techniques [5,16,27] which rely on domain-specific seed dictionary and pattern matching rules to guide the extraction. After that, several named entity recognition (NER) methods [13,15,19] were proposed for the attribute value extraction task. However, these approaches carry a closed world assumption and have limited coverage. Dealing with the closed-world assumption, several sequence tagging models have been proposed [8,35]. Zheng et al. [35] propose an end-to-end tagging model using a combination of BiLSTM, CRF, and Attention. Xu et al. [29] scale up these methods by utilizing one set of BIO tags and attribute embeddings. Karamanolakis et al. [9] propose a taxonomy-aware multi-task learning approach that leverages the taxonomy to improve the extraction quality. Yan et al. [30] use a Mixture-of-Experts module and a hypernetwork to build a decoder that is parameterized with attribute embeddings. Recent approaches formulate the attribute value extraction problem as a question answering task [7,22,28]. Wang et al. [28] build a question answering model that jointly encodes both the attribute and product context to scale to a large number of attributes. Yang et al. [32] extend this model to deal with multiple information sources and long sequences by using the ETC [1] model. Most recent techniques adopt generative approaches to solve the attribute value extraction task in an open-world setting [12,23,31]. For example, Shinzato et al. [25] fine-tune a pre-trained T5 model [20] to decode the set of attribute value pairs from the product text. However, most of these approaches extract the value for a given attribute from the product title. Our work differs from previous approaches as it involves extracting fine-grained attribute values from a given context of a coarse-grained attribute.

3 QPAVE

3.1 Problem Definition

Our work deals with two types of attributes that are typically used to describe products:

Coarse-Grained Attribute: is an attribute that combines multiple fine-grained attributes within its value.

Fine-Grained Attribute: is a single, straightforward attribute with only one associated value.

The main goal of *Fine-grained attribute value extraction* is to extract the value for a given fine-grained attribute from the value of a coarse-grained attribute in the product profile. For instance, the value of the coarse-grained attribute *Special Feature* in Fig. 1 is 'Electric Razor for Men,Beard Trimmer...,Cordless', from which we want to extract 'Beard Trimmer' as the value of the fine-grained attribute *Uses*.

To achieve this, we formulate the fine-grained attribute value extraction problem as a question answering task. Given a context (text sequence) and a question, the aim of question answering is to identify the best span in the context that corresponds to the answer. Thus, we treat the value of the coarse-grained attribute as context and turn the fine-grained attribute into a question. More formally, we denote the coarse-grained attribute as β and its context as $X = (x_1, \ldots, x_n)$. We refer to the product category as $c \in C$, where C is a predefined set of categories. We denote the fine-grained target attribute as $A = (a_1, \ldots, a_m)$. The model then extracts the best value $\hat{V}$ from the context X for the target attribute A together with its begin and end indices b and e, respectively, which can be formalized as:

$$\hat{V} = \arg\max_{V} Pr(V \mid X, A, \beta) = \arg\max_{b,e} Pr(w_b^x, w_e^x \mid X, A, \beta) \qquad (1)$$

Data Analysis. To show the importance of fine-grained attribute value extraction, we collected 50,000 product profiles from 9 domains in the Amazon Review Dataset [17]. As a first step, we have identified candidate coarse-grained attributes using distant supervision. Each attribute containing values of other attributes in the same product category was selected as a candidate attribute. This yielded 32,348 (66.8%) products having candidate coarse-grained attributes. These products were assessed by human annotators finding out that 30,145 products have 36,041 coarse-grained attributes. The data analysis clearly shows the need for fine-grained attribute value extraction to improve the quality of the attributes. We provide more details on the data collection in Sect. 4.

3.2 Model Overview

We propose a novel question answering model using a multi-task framework with an adaptive decoder. The overall model architecture is shown in Fig. 2. Essentially, our multi-task model is composed of three key components: the question

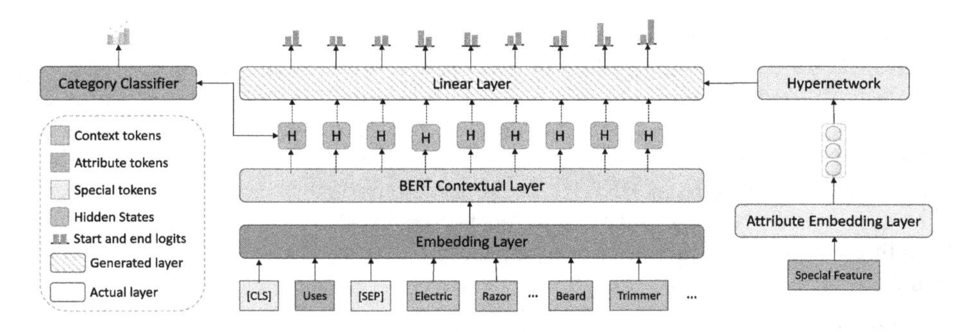

Fig. 2. Overall QPAVE model architecture.

answering (QA), the adaptive decoder (AD), and the category classifier (CC). The question answering component aims to find the best answer span from the context that corresponds to the value. The adaptive decoder is designed to dynamically adapt the extraction model based on the coarse-grained attribute β. The category prediction component is used to incorporate the product categories in the overall model. Furthermore, we pre-train the encoder on a Masked Language Modelling (mlm) task to improve the generalization performance of the model.

3.3 Question Answering

The question answering component is composed of three main layers: an input embedding layer, a contextual encoder layer, and an output layer that produces the final results.

Embedding Layer. In the embedding layer, every word in the context of the coarse-grained attribute and question is mapped into a d-dimensional embedding vector. This embedding representation is obtained by concatenating a word embedding, a segment embedding, and a positional embedding. We add an additional [CLS] token, i.e., classification token, at the beginning of the input sequence to capture the embedding of the whole sequence. We also use a special token [SEP], i.e., separator token, at the end of each input in the sequence to help the model understand the start and end of the context and the question. We initialize the embeddings in this layer from the pre-trained BERT language model and allow them to be learned during training instead of fixing them. We refer to the embedding of an input token x_t as $e_t = e(x_t)$.

Contextual Encoder Layer. The encoder layer captures a contextualized representation for every word in the input sequence. The input sequence to the encoder layer is the concatenation of the question and the context, in addition to the special tokens [CLS] and [SEP]. In this layer, we utilize the contextual encoder with self-attention mechanism developed in BERT [3]. The encoder allows the question (target attribute) A and context X to attend to each other from the bottom layer to the top layer using the self-attention mechanism. Let

$E = (e_1, ..., e_l)$ denote the concatenated embeddings of the question, context and special tokens from the embedding layer. The output from the encoder layer contains a [CLS] embedding, which resembles the entire sequence, and a sequence of contextual embeddings representing the encoded question and context. The [CLS] embedding is essential for the category prediction component of our model. The output encoding from BERT can be represented as $H = BERT(E)$, where $H = (H_1, ..., H_l)$ are the output hidden representations that are computed by encoding the input sequence, where $H_t \in \mathbb{R}^{d_h}$, and d_h is the embedding dimension.

Output Layer. The output layer of the question answering model is a linear layer with a softmax activation function to compute the probabilities for the start and end indices of the answer span. The begin $\hat{b}$ and end $\hat{e}$ indices are given by

$$\begin{aligned} \hat{b} &= \arg\max_t(softmax(W_b H_t)) \\ \hat{e} &= \arg\max_t(softmax(W_e H_t)) \end{aligned} \tag{2}$$

where H_t is the contextual embedding of token x_t in the input context. W_b and W_e are two trainable matrices that map the embeddings to the output logits for both the begin and end indices respectively. To capture the dependency between the begin and end indices, we use the score obtained by adding the start and end logits. First we check for the valid answers, i.e., answers where the begin index is lower than the end index. Then we sort them according to their combined score and choose the best one.

3.4 Adaptive Decoder

One of the key ideas in our proposed approach is to dynamically adapt the parameters of the extraction model to the coarse-grained attribute β. To achieve this, we employ separate decoders that are correlated based on the embeddings of the coarse-grained attribute β. This approach facilitates knowledge-sharing across different attributes while simultaneously providing a distinct and customized decoder parameterization for each attribute β, which enriches our model with attribute-specific knowledge. For fine-grained attribute value extraction, the question answering model takes the context and question as an input, and computes the probabilities of the start and end indices of the answer span. To make the model aware of the coarse-grained attribute β, we need to infuse this attribute information into some components of the question answering architecture. The BERT contextual layer of the QA model encodes the input to obtain a contextualized representation for each token. Based on this representation, the output linear layer then assigns a start and end probability for each token. In our approach, we propose to assign each coarse-grained attribute β a unique and adapted linear decoder whose parameters are generated based on the information of attribute β.

As introduced in Sect. 3.3, the output decoder layer of the question answering model is composed of a linear layer with softmax activation to compute the probabilities. In this work, we parameterize this linear layer with the attribute embeddings $r \in \mathbb{R}^{d_r}$ of the attribute β, where d_r is the embedding dimension. We leverage the flexibility of a hypernetwork [6] to generate the parameters of the linear layer based on the embeddings r. Hypernetworks, introduced by Ha et al. [6], are neural networks tasked with predicting the weights of another neural network, known as the target network. The weights generated by the hypernetwork are used to adjust the parameters of the target network with respect to some specific inputs. In our model, we use the hypernetwork to learn a linear transformation that maps the attribute embeddings to the parameters of the linear layer (W_{linear}) in the question answering model using a sigmoid activation as follows:

$$W_{linear} = \texttt{Reshape}(sigmoid(W_{hyper}\, r + b_{hyper})), \tag{3}$$

where W_{hyper} and b_{hyper} are learnable weights of the hypernetwork, and the `Reshape` operator reshapes a one dimensional vector into a matrix with equal number of elements.

The quality of the attribute embedding r plays a crucial role in deriving the parameters of the linear layer of our model. This embedding should capture the similarity between different attributes and consequently generate similar parameters for similar attributes. In our approach, we propose to use the name of the coarse-grained attribute β as a proxy to parameterize the linear layer for the extraction task. We initialize the attribute embedding layer with GloVe [18] pre-trained word embeddings. Note that the weights of our attribute embedding layer are considered to be learnable parameters and not fixed.

3.5 Category Classifier

We use multi-task learning setup to train the model on the auxiliary task of product category prediction. The aim is to prompt the model to predict the product category based solely on the question and context, encouraging the encoder to learn token embeddings that are distinctive of the categories. As shown in Fig. 2, we introduce a separate classifier $f^{cat}(H)$ to predict the category of the product $c \in C$. As mentioned earlier, the [CLS] token is inserted into the input sequence and attends to all other tokens from the context and question in the contextual encoder layer. The resulting enriched representation H_{cls} of the [CLS] token serves as a global embedding representing the entire input sequence. We pass this global representation to a classification layer to predict the probabilities of the product categories, i.e.,

$$f^{cat}(H) = softmax(W_{cat} H_{cls} + b_{cat}) \tag{4}$$

where w_{cat} and b_{cat} are trainable parameters. We jointly train QPAVE for the question answering and product category prediction tasks as described in Eq. 5:

$$\mathcal{L}_{total} = \lambda \mathcal{L}_{QA} + (1 - \lambda)\mathcal{L}_{CC} \tag{5}$$

where $\lambda \in [0, 1]$ is a tunable hyper-parameter to balance the losses among the two tasks. $\mathcal{L}_{QA}$ and $\mathcal{L}_{CC}$ are the losses for the question answering and category prediction tasks respectively.

Table 1. Statistics of AZ-main dataset.

Attributes	Train	Validation	Test
Brand	27811	3311	3371
Finish	10721	1356	1422
Style	10478	1349	1314
Material	10070	1237	1274
All Attributes	307680	38459	36041

Table 2. Statistics of AZ-zero-shot dataset.

Attributes	Train	Validation	Test
Display Style	0	0	638
Plug Format	0	0	508
Battery Cell Type	0	0	314
Tab Material	0	0	192

3.6 Masked Language Modelling

To handle generalization to new attributes, we pre-train the encoder layer of our model using the masked language model (mlm) [3] task on an e-commerce corpus containing product text obtained from the Amazon Review Dataset [17]. In the mlm task, we mask 15% of the input tokens [3], i.e., we replace them with a special token [MASK], and the objective is to predict the correct token based only on the context. This is achieved by adding a classification layer on top of the BERT encoder, and then computing the probability of each word in the vocabulary using a softmax function:

$$\hat{y} = softmax(W_o H_{mask}) \tag{6}$$

where H_{mask} is the output contextual embedding of the masked word and W_o is the output matrix that maps the output embedding to the output logits for the vocabulary. The mlm task ensures that the decoder learns a robust representation of the input sequence that captures both local and global contextual information of attributes and their values. Pre-training can also improve the generalization ability of the model by encouraging the encoder to effectively handle out-of-vocabulary words. By learning to generate contextualized embeddings for unknown attributes, the model generalizes to new vocabulary of attributes.

4 Experimental Setup

Datasets. Existing datasets [29,32] for attribute value extraction are not tailored towards extracting fine-grained attribute values from coarse-grained attributes. Hence, we build a dataset for the task of fine-grained attribute value extraction to evaluate the effectiveness of our proposed approach. Our dataset is derived from a public product collection - the Amazon Review Dataset [17]. First, we extract the attribute value combinations from the product catalog of each product using BeautifulSoup[1]. Then, we clean the values by removing html tags, whitespaces, and invalid text unicode characters using regex expressions and a few hard-coded rules. We also record the category of the collected product. Following previous work [9,29,30], we obtain the annotations for the fine-grained attribute value pairs using the catalog information of other products

[1] https://www.crummy.com/software/BeautifulSoup/bs4/doc/.

in the same category by distant supervision. The resulting dataset contains over 380k examples with more than 170 unique attributes and 140k unique values. We refer to this dataset as AZ-main where it contains information from 571 unique categories in 9 product domains: 'Appliances', 'Automotive', 'Clothing Shoes & Jewelry', 'Electronics', 'Home and Kitchen', 'Office Products', 'Sports & Outdoors', 'Tools', and 'Grocery & Food'.

We randomly sample 80% of the data for training, 10% for validation, and 10% for testing. The annotations obtained by distant supervision are often noisy and inconsistent, so they can not be considered as golden labels. Therefore, we further manually annotate the test set to ensure the reliability of our evaluation. In the annotation of the dataset, a total of 10 annotators actively participated in the labeling process. To ensure robustness and reliability, each example in the test set was independently labeled by three annotators. We considered a label valid only if at least two of the three annotators agreed on the assignment. This resulted in a test set of 36k instances. To test the performance of the models on individual attributes, we randomly select four frequent attributes (Brand, Finish, Style, Material). We also assess the ability of the models in scaling up to an arbitrary number of attributes by evaluating the models on all the attributes in the dataset. See Table 1 for statistical details of attributes in each AZ-main dataset split.

To further assess the models ability to generalize to new attributes, we split the AZ-main dataset into separate training and test subsets with distinct attributes. Specifically, four attributes, namely Display Style, Plug Format, Battery Cell Type, and Tab Material, are chosen and exclusively included in the test set (Table 2), ensuring they are not seen during training. This subset is named AZ-zero-shot, created for evaluating zero-shot extraction performance.

Implementation Details. All models are implemented using Pytorch[2] and trained on NVIDIA Tesla V100 GPUs. We use the validation set of our dataset to select the optimal hyper-parameters for all models, while we report our final results on the test set. During training, optimization is performed using Adam [10] with $1e^{-5}$ initial learning rate. For a fair comparison, all models are initialized with 768-dimension BERT-base [3]. We choose $d_r = 100$ as the dimension for the attribute embedding. The hyper-parameter λ is set to 0.8. The batch size is set to 64.

Baselines. We compare our proposed model with a series of strong baselines: *Dict* [24] is a simple dictionary matching by returning the most frequent seen value in the context. *BERT-MLC* [2] is a BERT-base classification model that predicts all the values of the fine-grained attributes from the representation of the given input. *SUOpenTag* [29] is a sequence tagging model which encodes the product text and the query fine-grained attribute separately with BERT. *AVEQA* [28] and *MAVEQA* [32], which are state-of-the-art question answering models for product attribute value extraction. We also use *QPAVE (AD)*, an ablation of our model without multi-task learning and mlm pre-training, for a fair comparison with the other baselines.

[2] https://pytorch.org/.

Table 3. Performance comparison on the AZ-main dataset (best in boldface and second best underlined).

Models	Brand			Finish			Style			Material			All Attributes		
	P	R	F_1	P	R	F_1	P	R	F_1	P	R	F_1	P	R	F_1
DICT	15.67	12.69	14.02	58.57	33.18	42.36	19.17	22.37	20.65	36.40	30.69	33.30	28.82	26.27	27.49
BERT-MLC	16.76	11.55	13.67	77.61	35.52	48.73	33.63	27.18	30.07	63.62	32.74	43.23	36.71	28.29	31.96
SUOpenTag	76.37	71.28	73.73	76.25	48.87	59.57	55.40	45.81	50.15	72.78	52.43	60.95	65.17	59.77	64.53
AVEQA	91.82	90.44	91.13	86.23	71.51	78.19	75.03	67.65	71.15	86.36	74.48	79.98	87.66	80.94	84.17
MAVEQA	92.06	89.91	90.97	82.42	70.97	76.28	76.03	66.43	70.91	82.37	72.34	77.03	86.97	79.18	82.89
QPAVE (AD)	91.90	90.51	91.23	87.24	83.54	85.34	79.09	75.27	77.13	87.22	80.13	81.86	89.96	83.85	86.80
QPAVE	**93.75**	**92.64**	**93.19**	**88.62**	**86.56**	**87.58**	**82.30**	**77.17**	**79.65**	**89.46**	**81.16**	**85.11**	**90.83**	**85.91**	**88.30**

Evaluation Metrics. Following previous works [28,32], we use Precision (P), Recall (R), and F1 score (F_1) as the evaluation metrics for each attribute. These metrics are computed based on the number of true positives (TP), false positives (FP), and false negatives (FN). We follow the Exact Match [21] criteria in which the full sequence of extracted values needs to be correct. Each experiment is repeated 5 times and we report the average scores.

$$P = \frac{TP}{TP + FP} R = \frac{TP}{TP + FN} F_1 = 2 \times \frac{P \times R}{P + R} \tag{7}$$

5 Experimental Results

5.1 Results on All and Selected Attributes

In the first experiment, we compare the performance of QPAVE with the baseline models on the attributes in the AZ-main dataset. As shown in Table 3, QPAVE consistently outperforms the other compared models across all selected attributes. The overall improvement in F_1 score is up to 8.5% compared to AVEQA. One key observation is that without multi-task learning and mlm pre-training, our QPAVE (AD) method outperforms all baselines and achieves state-of-the-art performance on the AZ-main dataset. This result demonstrates that the adaptive decoder allows for joint modelling of different coarse-grained attributes, which facilitates knowledge sharing and improves the overall performance. There are several key observations in Table 3. First, Dict and BERT-MLC perform very poorly compared to all other baselines, as they make a closed-world assumption and can not discover new attributes outside the training data. As a result, they suffer from low recall. SUOpenTag does not make a close-world assumption and hence has a higher recall. Second, inline with previous works [32], we also observe that AVEQA and MAVEQA achieve similar results. This is because our input sequences are short, and thus AVEQA and MAVEQA are both equally effective. Third, QPAVE significantly outperforms both SUOpentag and AVEQA by 23.77% and 4.13% in F_1 score, respectively. While QPAVE (AD) also shows comparable performance and outperforms all other baselines, QPAVE outperforms QPAVE (AD) by 1.5%. This is mainly because by learning to predict the product category, QPAVE learns category aware contextual embeddings that are more effective for the extraction task.

5.2 Results of Discovering New Attributes

Table 4. Zero-shot extraction results on AZ-zero-shot dataset.

Models	Display Style			Plug Format			Battery Cell Type			Tab Material		
	P	R	F_1	P	R	F_1	P	R	F_1	P	R	F_1
SUOpenTag	88.91	47.26	61.71	89.25	28.57	43.29	85.75	37.45	52.13	93.13	47.10	62.56
AVEQA	96.35	65.41	77.92	91.07	45.86	61.01	85.73	**41.83**	**56.23**	91.26	61.94	73.79
MAVEQA	95.76	62.38	75.55	89.64	47.17	61.76	81.16	40.23	53.80	89.23	61.32	72.68
QPAVE (AD)	94.60	65.14	77.15	90.01	46.87	61.64	**86.23**	40.02	54.66	92.40	63.42	75.21
QPAVE	**96.74**	**74.10**	**83.92**	**92.27**	**55.39**	**69.22**	82.85	**41.83**	55.60	**93.68**	**70.32**	**80.34**

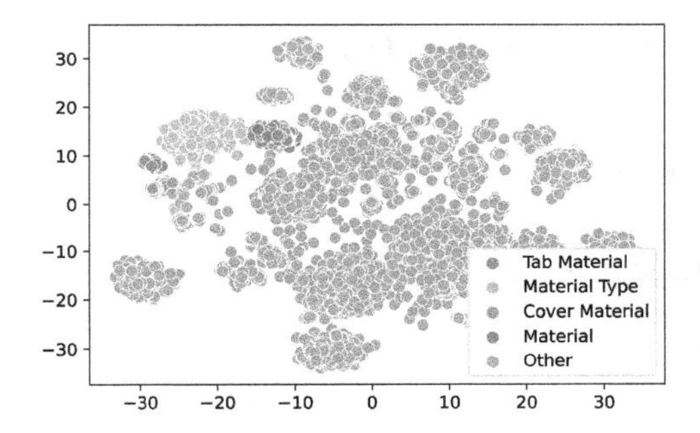

Fig. 3. Embeddings projections of semantically related new and existing attributes.

We conduct zero-shot extraction experiments to examine the generalization ability of our model on unseen attributes. We exclude Dict and BERT-MLC from this experiment as they can not handle new attributes. It can be seen in Table 4 that our model achieves better performance than all the baselines on the four new attributes, where the gains in recall are up to 8.7%. This is because the mlm pre-training task encourages the encoder to learn contextual embeddings for unknown attributes. Moreover, the category classification component also boosts the zero-shot extraction performance by exploiting the similarities across categories of products.

There are several interesting observations in Table 4. First, as there are no examples from the zero-shot attributes in our training set, it is natural that the overall performance of all models is worse as compared to the results on the AZ-main dataset. Second, the performance of QPAVE depends on the attribute. We can observe that the F_1 score for the attribute *Tab Material* (80.34%) is much higher than the attribute *Battery Cell Type* (55.60%). We believe that some attributes in the training set are semantically related to some of the unseen attributes and thus, they provide hints that guide the extraction. To investigate this, we map the embeddings of the zero-shot and existing attributes to a two-dimensional space using t-SNE [14] as indicated in Fig. 3. We can observe that the attributes *Material Type* and *Cover Material*, which are present in the training set, are semantically related to the zero-shot attribute *Tab Material*, which boosts the extraction results.

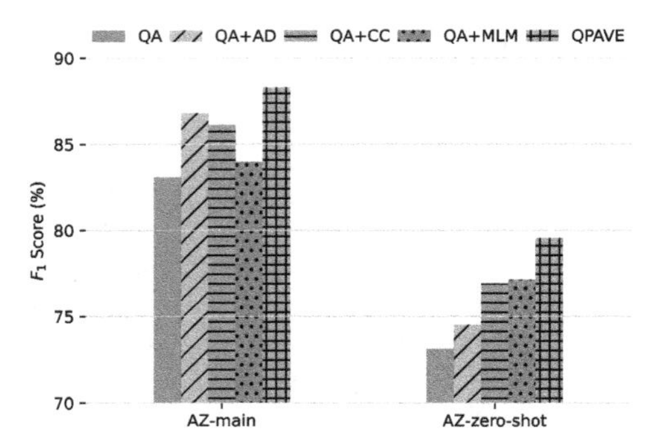

Fig. 4. F_1 (%) results of ablation study on AZ-main and AZ-zero-shot datasets.

5.3 Ablation Study

To understand the importance of the different components in our model, we conduct an ablation study by testing the individual impact of each component. This resulted in three variant models: question answering (QA), question answering + adaptive decoder (QA+AD), and question answering + category classifier (QA+CC). Additionally, we train another model, QA+MLM, to investigate the impact of the mlm pre-training on the performance. The F_1 scores on both AZ-main and AZ-zero-shot datasets are illustrated in Fig. 4. The results show that the adaptive decoder (AD) notably enhances the performance on AZ-main from 83.08 to 86.80 as compared to the basic QA model. This observation validates that modelling the dependency on the coarse-grained attributes is crucial for fine-grained attribute value extraction. Another interesting observation is that mlm pre-training significantly improves zero-shot extraction by up to 4.0% over the basic QA model, with a modest 0.90% increase on AZ-main. This is inline with our expectations, as the main role of mlm pre-training is to improve model generalization on unseen attributes.

Moreover, it is also clear that the category classifier (CC) significantly enhances performance on both AZ-main (by 3.04%) and AZ-zero-shot (by 4.0%) datasets. We believe that the category classifier allows the model to learn category enriched embeddings that distinguish targeted attribute values from other deflecting values in the context. To further confirm this hypothesis, we choose one of the four selected attributes in the AZ-main-dataset, namely, *Brand*, and we map its hidden states, to a two-dimensional space using t-SNE [14]. Figure 5 shows the projections of the embeddings of the attribute *Brand* for the QA+CC model. Each point is a distinct example in the dataset, and the four colors represent the product category of the example. We plot the top four categories of the attribute *Brand* in the AZ-main dataset (blue: Parts, orange: Body & Trim, green: Shocks & Suspension, red: Keychains). From Fig. 5, we can observe that the embeddings of the attribute *Brand*, generated by the QA+CC model, are clustered according to the categories. This indicates that related categories have similar embeddings for the same attribute. On the other hand, the attributes gen-

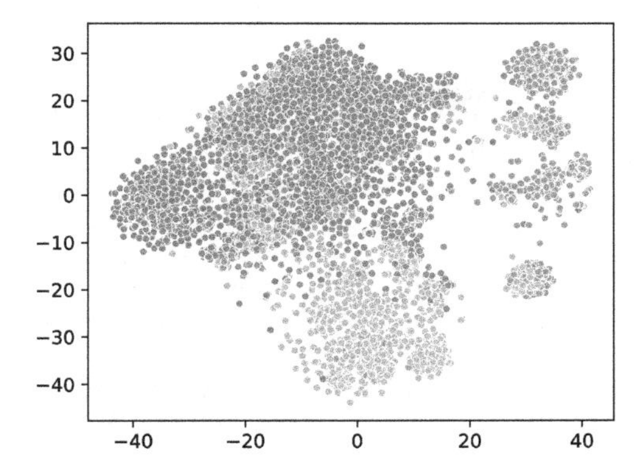

Fig. 5. T-SNE visualization of the attribute *Brand*.

erated by the QA model, not shown here due to space limitations, are scattered and do not encode any category information. This validates that the category classifier allows the model to learn category specific embeddings that are effective for the extraction task. Finally, the experimental results in Fig. 4 show that QPAVE, which benefits from all the components, achieves the best performance on AZ-main and AZ-zero-shot datasets.

6 Conclusion

This paper presents a novel mutli-task approach for fine-grained attribute value extraction. We build a question answering model that treats the fine-grained attribute as question and extracts its value from the context of the coarse-grained attributes. The contextual encoder from BERT is deployed to encode the question and the context. We employ a hypernetwork to dynamically adapt the extraction model based on the coarse-grained attribute. We infuse category information into the overall model by learning to predict the product categories in a multi-task training setup. A masked language modelling task is introduced to improve the model generalization ability. We also introduce a novel dataset for fine-grained attribute value extraction. Our experiments demonstrate the effectiveness and robustness of the proposed approach. One limitation of our work is that the model expects both the target fine-grained attribute as input along the context. As future work, we plan to explore generative approaches that could identify both the target fine-grained attribute and its corresponding value from the input text.

References

1. Ainslie, J., et al.: ETC: encoding long and structured inputs in transformers. In: EMNLP, pp. 268–284 (2020)
2. Chen, W.T., Xia, Y., Shinzato, K.: Extreme multi-label classification with label masking for product attribute value extraction. In: Proceedings of ECNLP 5, pp. 134–140 (2022)
3. Devlin, J., Chang, M.W., Lee, K., Toutanova, K.: BERT: pre-training of deep bidirectional transformers for language understanding. arXiv preprint arXiv:1810.04805 (2018)
4. Dong, X.L., et al.: AutoKnow: self-driving knowledge collection for products of thousands of types. In: Proceedings of ACM SIGKDD, pp. 2724–2734 (2020)
5. Gopalakrishnan, V., Iyengar, S.P., Madaan, A., Rastogi, R., Sengamedu, S.: Matching product titles using web-based enrichment. In: Proceedings of ACM CIKM, pp. 605–614 (2012)
6. Ha, D., Dai, A.M., Le, Q.V.: Hypernetworks. In: International Conference on Learning Representations. OpenReview.net, Toulon (2017)
7. Hu, M., Xiao, J., Liu, Y., Guo, W., Fan, X.: Fusing attribute type features for attribute value extraction from product via question answering. In: Proceedings of ACM MLNLP, pp. 179–184 (2022)
8. Huang, Z., Xu, W., Yu, K.: Bidirectional LSTM-CRF models for sequence tagging (2015)
9. Karamanolakis, G., Ma, J., Dong, X.L.: TXtract: taxonomy-aware knowledge extraction for thousands of product categories. arXiv preprint arXiv:2004.13852 (2020)
10. Kingma, D.P., Ba, J.: Adam: a method for stochastic optimization. arXiv preprint arXiv:1412.6980 (2014)
11. Kozareva, Z., Li, Q., Zhai, K., Guo, W.: Recognizing salient entities in shopping queries. In: Proceedings of ACL, pp. 107–111 (2016)
12. Li, Y., Xue, B., Zhang, R., Zou, L.: AtTGen: attribute tree generation for real-world attribute joint extraction. In: Proceedings of ACL, pp. 2139–2152 (2023)
13. Ling, X., Weld, D.: Fine-grained entity recognition. In: Proceedings of AAAI, vol. 26, pp. 94–100 (2012)
14. Van der Maaten, L., Hinton, G.: Visualizing data using t-SNE. J. Mach. Learn. Res. **9**(11) (2008)
15. More, A.: Attribute extraction from product titles in ecommerce. arXiv preprint arXiv:1608.04670 (2016)
16. Nadeau, D., Sekine, S.: A survey of named entity recognition and classification. Lingvisticae Invest. **30**(1), 3–26 (2007)
17. Ni, J., Li, J., McAuley, J.: Justifying recommendations using distantly-labeled reviews and fine-grained aspects. In: Proceedings of EMNLP-IJCNLP, pp. 188–197 (2019)
18. Pennington, J., Socher, R., Manning, C.D.: GloVe: global vectors for word representation. In: Proceedings of EMNLP, pp. 1532–1543 (2014)
19. Putthividhya, D., Hu, J.: Bootstrapped named entity recognition for product attribute extraction. In: Proceedings of EMNLP, pp. 1557–1567 (2011)
20. Raffel, C., et al.: Exploring the limits of transfer learning with a unified text-to-text transformer. J. Mach. Learn. Res. **21**(1), 5485–5551 (2020)
21. Rajpurkar, P., Zhang, J., Lopyrev, K., Liang, P.: SQuAD: 100,000+ questions for machine comprehension of text. In: Proceedings of EMNLP, pp. 2383–2392 (2016)

22. Sabeh, K., Kacimi, M., Gamper, J.: CAVE: correcting attribute values in e-commerce profiles. In: Proceedings of ACM CIKM, pp. 4965–4969 (2022)
23. Sabeh, K., Kacimi, M., Gamper, J.: GAVI: a category-aware generative approach for brand value identification. In: Proceedings of ICNLSP, pp. 110–119. Association for Computational Linguistics (2023)
24. Shinzato, K., Yoshinaga, N., Xia, Y., Chen, W.T.: Simple and effective knowledge-driven query expansion for QA-based product attribute extraction. In: Proceedings of ACL, pp. 227–234 (2022)
25. Shinzato, K., Yoshinaga, N., Xia, Y., Chen, W.T.: A unified generative approach to product attribute-value identification. In: Findings of ACL, pp. 6599–6612 (2023)
26. Tan, J., Xu, S., Ge, Y., Li, Y., Chen, X., Zhang, Y.: Counterfactual explainable recommendation. In: Proceedings of ACM CIKM, pp. 1784–1793 (2021)
27. Vandic, D., Van Dam, J.W., Frasincar, F.: Faceted product search powered by the semantic web. Decis. Support Syst. **53**(3), 425–437 (2012)
28. Wang, Q., et al.: Learning to extract attribute value from product via question answering: a multi-task approach. In: Proceedings of ACM CIKM, pp. 47–55 (2020)
29. Xu, H., Wang, W., Mao, X., Jiang, X., Lan, M.: Scaling up open tagging from tens to thousands: Comprehension empowered attribute value extraction from product title. In: Proceedings of ACL, pp. 5214–5223 (2019)
30. Yan, J., Zalmout, N., Liang, Y., Grant, C., Ren, X., Dong, X.L.: AdaTag: multi-attribute value extraction from product profiles with adaptive decoding. In: Proceedings of ACL-IJCNLP, pp. 4694–4705 (2021)
31. Yang, L., et al.: MixPAVE: mix-prompt tuning for few-shot product attribute value extraction. In: Findings of ACL 2023, pp. 9978–9991 (2023)
32. Yang, L., et al.: MAVE: a product dataset for multi-source attribute value extraction. In: Proceedings of ACM WSDM, pp. 1256–1265 (2022)
33. Zhang, W., Deng, Y., Lam, W.: Answer ranking for product-related questions via multiple semantic relations modeling. In: Proceedings of ACM SIGIR, pp. 569–578 (2020)
34. Zhao, J., Chen, H., Yin, D.: A dynamic product-aware learning model for e-commerce query intent understanding. In: Proceedings of the ACM CIKM, pp. 1843–1852 (2019)
35. Zheng, G., Mukherjee, S., Dong, X.L., Li, F.: OpenTag: open attribute value extraction from product profiles. In: Proceedings of ACM SIGKDD, pp. 1049–1058 (2018)

Open-Source Drift Detection Tools in Action: Insights from Two Use Cases

Rieke Müller[1], Mohamed Abdelaal[1(✉)], and Davor Stjelja[2]

[1] Software AG, Darmstadt, Germany
{rieke.mueller,mohamed.abdelaal}@softwareag.com
[2] Granlund Oy, Helsinki, Finland
Davor.Stjelja@granlund.fi

Abstract. Data drifts pose a critical challenge in the lifecycle of machine learning (ML) models, affecting their performance and reliability. In response to this challenge, we present a benchmark study, called D3Bench (Source code is available at https://github.com/mohamedyd/D3Bench), which evaluates the efficacy of open-source drift detection tools. D3Bench examines the capabilities of Evidently AI, NannyML, and Alibi-Detect, leveraging real-world data from two smart building use cases. We prioritize assessing the functional suitability of these tools to identify and analyze data drifts. Our findings reveal that Evidently AI stands out for its general data drift detection, whereas NannyML excels at pinpointing the precise timing of shifts and evaluating their consequent effects on predictive accuracy.

1 Introduction

In this paper, we present a benchmark, denoted as D3Bench[1,2], that rigorously evaluates three leading open-source tools designed for the detection of data drifts, employing a range of established criteria. To conduct this assessment, we measure both functional and non-functional performance attributes. Our examination encompasses several key aspects: functional suitability, integration capability, compatibility with various data types, user-friendliness, and performance metrics including time efficiency and resource consumption. This holistic approach ensures a comprehensive analysis, providing valuable insights into the efficacy of each tool. D3Bench examines two distinct use cases, each using a univariate time series data set with building data from building management experts. These datasets comprise authentic data, where detecting data drifts poses a substantial challenge due to the data's intricate nature. The first use case delves into a dataset that encapsulates a concept shift within the dataset. Unlike the predictable nature of seasonal trends, concept shifts represent a change in the underlying relationships between data attributes, often resulting in a more elusive form of data drift that can be particularly challenging to detect and quantify.

[1] D3Bench is an abbreviation of **D**ata **D**rift **D**etection **Bench**mark.
[2] An extended version of the paper is available at https://arxiv.org/abs/2404.18673.

R. Wrembel et al. (Eds.): DaWaK 2024, LNCS 14912, pp. 346–352, 2024.
https://doi.org/10.1007/978-3-031-68323-7_29

Transitioning to the second use case, we encounter a different kind of complexity: a data drift characterized by seasonal trends. The cyclical nature of this data offers a unique opportunity to explore how data drifts can manifest over time, influenced by predictable, recurring patterns. Since these datasets involve predictions for a single variable, D3Bench focuses exclusively on univariate methods for detecting data drifts. Together, these two use cases provide a comprehensive view of the nuanced and multifaceted nature of data drifts, enabling a thorough comparison of the efficacy of our detection tools in real-world scenarios.

2 Architecture

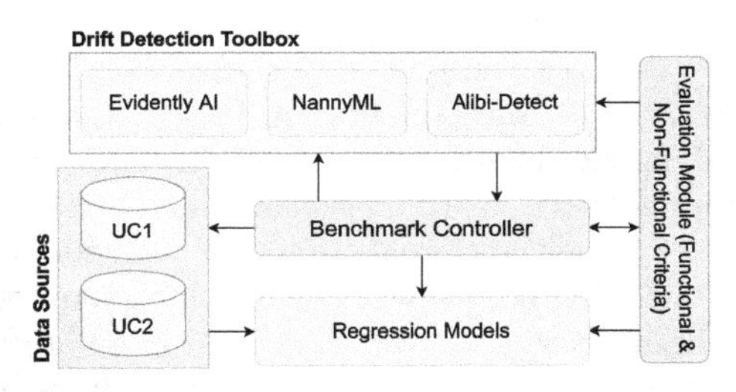

Fig. 1. Architecture of D3Bench

D3Bench is designed to assess individual drift values, which in turn inform the functional suitability of the system. Additionally, it evaluates non-functional criteria such as runtime, CPU runtime, and memory consumption. At the core of D3Bench lies the benchmark controller, the orchestrator that manages the various components of the benchmarking process. This controller is responsible for loading the datasets, which are derived from two distinct use cases, and for carrying out the necessary preprocessing steps. It systematically partitions the dataset into training and test sets, setting the stage for a thorough comparison of the drift detection tools based on pre-selected benchmarks. Within the scope of D3Bench, we focus on three widely recognized tools: *Alibi-Detect*[3], *NannyML*[4], and *Evidently AI*[5]. These tools, each with their unique methodologies, are applied independently to the dataset to ensure a comprehensive evaluation. The outcomes of the benchmarking process are documented. Results are not only displayed in the console for immediate review but are also stored in a CSV file, facilitating subsequent in-depth analysis. This methodical approach ensures that D3Bench provides a robust framework for benchmarking drift detection tools, yielding insights that are both actionable and accessible (Fig. 1).

In D3Bench, we present two real-world use cases within the field of building management. Each case involves the analysis of time series data that exhibits data drift over time. The first use case (UC1: Occupancy Detection) examines the temporal pattern of room occupancy, distinguishing between states of being occupied by one or more individuals and being unoccupied. The dataset encompasses 46,555 time series entries, spanning from March 30, 2021, at 12:12 AM to July 11, 2021, at 11:57 PM, with data recorded at 3-minute intervals. The dataset includes air quality parameters of a room as variables. It measures the

[3] https://github.com/SeldonIO/alibi-detect.
[4] https://www.nannyml.com/.
[5] https://www.evidentlyai.com/.

room temperature (`temperature`) and the current CO2 level of the room (`co2`). A Deep Learning classification model predicts based on these variables whether a room is occupied (value: 1) or not (value: 0). At the same time, the ground truth is available that describes the actual reality of whether the room is occupied or not. The classification model has been trained on data from a room in another building and improved by transfer learning with a sample from the room under investigation from March 22, 2021, to March 26, 2021.

The second use case (UC2: Energy Consumption Prediction) describes the temporal evolution of thermal energy usage across a network of 36 university buildings, offering a comprehensive dataset encompassing 1,556,915 individual time series entries. These entries log thermal energy usage, quantified in megawatt-hours (MWh), and span from the early hours of January 1, 2019, at 01:00 AM, to the end of the recorded period on June 1, 2022, at 12:00 AM. Situated in Tampere, Finland, these buildings are integral components of the local district heating system. To monitor thermal energy usage, each building is equipped with a primary heat meter. These meters are either advanced smart meters, which offer real-time data and remote monitoring capabilities, or traditional analog metering devices, which require manual reading. These meters log the daily energy consumption and estimate the hourly usage by computing the difference between the cumulative readings of consecutive hours.

3 Comparative Analysis

Results of UC1. Table 1 summarizes the estimated drift values from the three open-source tools in UC1. Below, we present the results of each tool. For Evidently AI, the Epps-Singleton test does not yield results for the variables considered, so the remaining eleven methods are examined. Among these, the majority of the methods (7 out of 11) do not detect a data drift in the `occupancy` variable. However, a data drift in the variables `CO2` (detected by 10 out of 11 methods) and `temperature` (detected by all 11 methods) is reliably identified. Evidently AI's report provides a detailed analysis for both input and target variables. NannyML identifies data drift in the `temperature` variable using three of four methods for over 50% of the chunks (cf. Table 1). For the `CO2` variable, two methods detect drift in more than 50% of the chunks. Meanwhile, no drift is detected in the target variable. In NannyML, both the Jensen-Shannon and Hellinger distances accurately detect data drift in the input and target variables. Conversely, the Kolmogorov-Smirnov test and Wasserstein distance fail to identify shifts in the input variables for most data segments, yielding inaccurate assessments. However, they accurately confirm the absence of drift in the target variable.

Table 1. UC1 results using three open-source tools, indicating correct decisions with ✗ for Occupancy and ✓ for other variables.

Tool	Evidently AI						
Variable	Occupancy		CO2		Temperature		
Methods	D-value	Drift?	D-value	Drift?	D-value	Drift?	Group
Kolmogorov-Smirnov Test	0,86	✗	0,0	✓	0,0	✓	1
Wasserstein Distance	0,02	✗	0,27	✓	1,35	✓	1
Kullback-Leibler Divergence	$2,76e^{-4}$	✗	0,09	✗	1,11	✓	3
Population Stability Index	$5,6e^{-4}$	✗	0,18	✓	1,90	✓	1
Jensen-Shannon Distance	0,459	✗	0,15	✓	0,43	✓	1
Anderson Darling Test	$1,19e^{-3}$	✓	$1,00e^{-3}$	✓	$1,00e^{-3}$	✓	2
CramÃĩr-von-Mises Test	$2,53e^{-7}$	✓	$7,08e^{-8}$	✓	$2,87e^{-7}$	✓	2
Hellinger Distance	$8,36e^{-3}$	✗	0,15	✓	0,45	✓	1
Mann-Whitney U-Rank Test	$7,92e^{-3}$	✓	0,0	✓	0,0	✓	2
Energy Distance	$8,25e^{-3}$	✗	1,45	✓	0,41	✓	1
Epps-Singleton Test	–	–	–	–	–	–	–
T-Test	0,01	✓	$5,99e^{-129}$	✓	0,0	✓	2
Spot-The-Difference Test							

Tool	NannyML						
Variable	Occupancy		CO2		Temperature		
Methods	D-value	Drift?	D-value	Drift?	D-value	Drift?	Group
Kolmogorov-Smirnov Test	0,01	✗ (0%)	0,30	✗ (20%)	0,62	✗ (20%)	3
Wasserstein Distance	0,01	✗ (0%)	9,31	✗ (10%)	0,23	✓ (50%)	3
Kullback-Leibler Divergence							
Population Stability Index							
Jensen-Shannon Distance	0,03	✗ (0%)	0,33	✓ (100%)	0,63	✓ (100%)	1
Anderson Darling Test							
CramÃĩr-von-Mises Test							
Hellinger Distance	0,02	✗ (0%)	0,29	✓ (100%)	0,60	✓ (100%)	1
Mann-Whitney U-Rank Test							
Energy Distance							
Epps-Singleton Test							
T-Test							
Spot-The-Difference Test							

Tool	Alibi-Detect						
Variable	Occupancy		CO2		Temperature		
Methods	D-value	Drift?	D-value	Drift?	D-value	Drift?	Group
Kolmogorov-Smirnov Test	0,86	✗	0,0	✓	0,0	✓	1
Wasserstein Distance							
Kullback-Leibler Divergence							
Population Stability Index							
Jensen-Shannon Distance							
Anderson Darling Test							
CramÃĩr-von-Mises Test	$2,53e^{-7}$	✓	$2,87e^{-7}$	✓	$7,08e^{-8}$	✓	2
Hellinger Distance							
Mann-Whitney U-Rank Test							
Energy Distance							
Epps-Singleton Test							
T-Test							
Spot-The-Difference Test	$3,63e^{-118}$	NA	$3,63e^{-118}$	NA	$3,63e^{-118}$	NA	NA

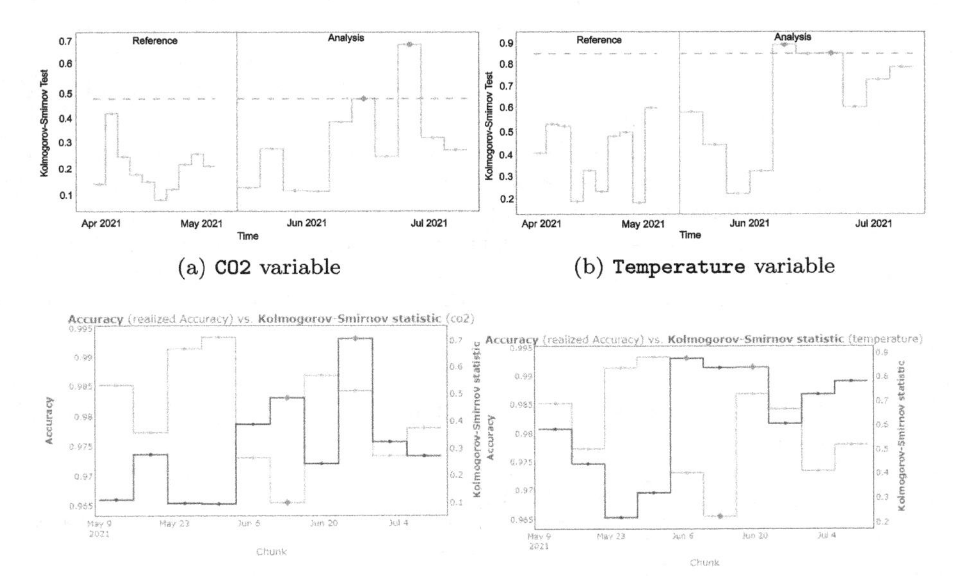

(a) CO2 variable

(b) Temperature variable

(c) Predictive accuracy vs. KST test of CO2

(d) Accuracy vs. KST test of Temperature

Fig. 2. Reports generated by NannyML for UC1

The NannyML report facilitates the detection of drift at specific times within the test data. As an example, the Kolmogorov-Smirnov test reveals drift in 20% of the chunks for each input variable. Figure 2a displays the CO2 variable's test outcomes, with alarms activated for June and July 2021. Similarly, Fig. 2b illustrates the `temperature` variable's test results, also triggering alarms during the same period. These findings prompt an investigation into any changes in the prediction model's performance. NannyML offers a feature to assess the model's accuracy. Figure 2c compares the model's accuracy (in light blue) with the CO2 variable's test results (in purple).

Table 1 presents the outcomes of three Alibi-Detect drift detection methods. The spot-the-difference test identifies drift within the entire dataset rather than isolating specific variables, hence it indicates a data drift but does not distinguish the impact on the two variables. Conversely, both the Kolmogorov-Smirnov and Cramér-von-Mises tests detect drift in the input variable, yet their results are inconsistent when applied to the target variable. Moreover, the table indicates that only the Kolmogorov-Smirnov test accurately detects data drifts across all variables (Group One). The Cramér-von-Mises test incorrectly flags the target variable, classifying it in Group Two. The spot-the-difference test's outcome is uncategorized (N/A) since it does not provide variable-specific results. However, it correctly identifies a data drift, confirming that $P_{tr}(y, x) \neq P_{tst}(y, x)$.

Results of UC2. Table 2 summarizes the performance of three open-source detection tools in UC2. Notably, Evidently AI identifies data drifts in both the target and input variables using nine out of twelve methods tested (Group 1). Group 2 contains only the Energy Distance method, which measures the distance between two random vectors. In this method, a calculated distance below 0.1 indicates no detected drift in the target variable's training and test distributions. The Mann-Whitney U Rank Test, Group 3's sole method, identifies no data drift in the input variable. With a p-value of one, it suggests a 100% probability of a Type I error, far exceeding the 5% significance level, leading to the acceptance of the null hypothesis of identical cumulative distribution functions. Finally, Group 4, featuring only the Kullback-Leibler Divergence, does not detect a data drift in any variable. The divergence value returned by Evidently AI is below the threshold of 0.1.

Table 2. UC2 results using three open-source tools, indicating correct decisions with ✓ for all variables.

Tool	Evidently AI					NannyML					Alibi-Detect				
Variable	consumption		temp_outside			consumption		temp_outside			consumption		temp_outside		
Methods	D-Value	Drift?	D-Value	Drift?	Group	D-Value	Drift?	D-Value	Drift?	Group	D-Value	Drift?	D-Value	Drift?	Group
Kolmogorov-Smirnov Test	$2,51e^{-35}$	✓	$3,37e^{-48}$	✓	1	0,44	✗ (0%)	0,416	✗ (0%)	4	$2,51e^{-35}$	✓	0,0	✓	1
Wasserstein Distance	0,26	✓	0,19	✓	1	0,13	✗ (20%)	7,31	✗ (0%)	4					
Kullback-Leibler Divergence	0,07	✗	0,08	✗	4										
Population Stability Index	0,24	✓	0,22	✓	1										
Jensen-Shannon Distance	0,15	✓	0,15	✓	1	0,51	✓ (100%)	0,48	✓ (100%)	1					
Anderson Darling Test	0,00	✓	0,00	✓	1										
CramÃĺr-von-Mises Test	$6,43e^{-9}$	✓	$1,98e^{-10}$	✓	1						$6,43e^{-9}$	✓	$1,98e^{-10}$	✓	1
Hellinger Distance	0,15	✓	0,15	✓	1	0,47	✓ (100%)	0,44	✓ (100%)	1					
Mann-Whitney U-Rank Test	$1,26e^{-16}$	✓	1.00	✗	3										
Energy Distance	0,06	✗	0,42	✓	2										
Epps-Singleton Test	$2,82e^{-223}$	✓	$4,22e^{-234}$	✓	1										
T-Test	$8,84e^{-50}$	✓	$2,60e^{-16}$	✓	1										
Spot-The-Difference Test											0,07	NA	0.07	NA	NA

Table 2 presents the detection capabilities of Alibi-Detect, with two out of three methods identifying a data drift in both the target and input variables. As listed in the table, the spot-the-difference test fails to detect a data drift. This test was unable to calculate a test variable for each variable, leading to a single drift value computed for the entire dataset, reflected identically in both variable columns. The Kolmogorov-Smirnov and Cramér-von-Mises tests accurately identify data drifts in both variables (group one). The spot-the-difference test's outcome cannot be classified within these groups (effectively, NA) since it provides no variable-specific results. Ultimately, it does not detect a data drift. Alibi-Detect does not provide visualizations or reports; hence, no results report is included.

Runtime and RAM Consumption: Figure 3 compares the runtime and RAM usage of three tools, both with and without report generation. Alibi-Detect, which does not produce reports, has metrics only for the latter case. The presented values represent the means from five trials for each tool. For UC1, Fig. 3a depicts that Alibi-Detect's runtime is substantially longer than the other tools, exceeding NannyML's by approximately 17,840% and Evidently AI's by about 11,366%. Conversely, Evidently AI's runtime is approximately 57% greater than NannyML's. In terms of RAM usage, NannyML requires approximately 1.5% more RAM than Evidently AI, while Alibi-Detect is the most efficient, using about 0.5% less RAM than Evidently AI and around 2% less than NannyML. For Evidently AI, running the tool for detection along with report generation increases the runtime by about 57%, while the RAM usage slightly decreases by approximately 0.2%. In the case of NannyML, the runtime sees a substantial increase of approximately 2330% when adding report generation, with virtually no change in RAM usage.

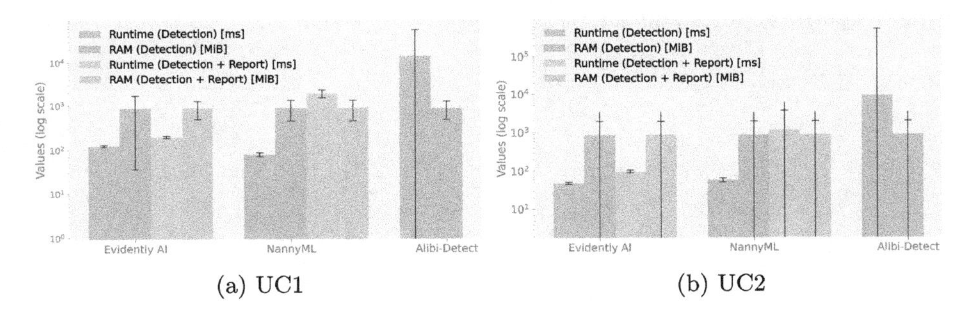

(a) UC1 (b) UC2

Fig. 3. Runtime and RAM consumption of the compared tools

For UC2, Fig. 3b shows that Evidently AI and NannyML have comparable detection runtimes at 48.26 ms and 57.92 ms, respectively, with NannyML being approximately 20% slower. Alibi-Detect's runtime is significantly longer, at 9219.69 ms, making it roughly 19,000% slower than Evidently AI. The reason for this is the Spot-The-Difference test, as it trains a classifier that differentiates between test and training data. Moreover, Alibi-Detect's RAM usage for detection is slightly higher, at about 2.6% more than Evidently AI's and 2.7% more than NannyML's, while Evidently AI's RAM consumption is marginally higher, by approximately 0.3%, compared to NannyML. The figure shows that Alibi-Detect has a much larger variability in both runtime and RAM usage during detection compared to Evidently AI and NannyML. Meanwhile, Evidently AI and NannyML show more consistency in runtime but have a considerable variance in RAM usage, suggesting variability in their memory consumption.

When comparing the runtime for detection alone to detection with report generation, Evidently AI shows an increase of over 103%, while NannyML exhibits a dramatic increase, almost 1876%. For RAM usage, Evidently AI sees a slight increase of about 2.6% when generating reports, whereas NannyML shows a marginal decrease of approximately 0.3% in RAM consumption in the report generation scenario.

4 Conclusion

In this study, we conducted a comparative analysis of three open-source tools—Evidently AI, NannyML, and Alibi-Detect—focused on detecting data drifts utilizing datasets from smart building applications. The tools were assessed based on a spectrum of functional and non-functional criteria using real-world time series data. Our findings highlighted that Evidently AI excels at identifying broad data drifts. NannyML, on the other hand, stands out for its precision in pinpointing the specific timing of drifts and evaluating their impact on the accuracy of predictions. The practical deployment of these tools hinges on their functional suitability, and in this regard, Evidently AI emerged as the preferred option. Its rapid computational capabilities, along with its adeptness at generating insightful reports and seamless integration with existing ML tools.

A Study on Database Intrusion Detection Based on Query Execution Plans

Tadeusz Morzy and Maciej Zakrzewicz(✉)

Institute of Computing Science, Poznan University of Technology, Poznań, Poland
{maciej.zakrzewicz,tadeusz.morzy}@put.poznan.pl

Abstract. Database Intrusion Detection (DID) is critical to systems that do not fully apply the Principle of Least Privilege and users are able to use their privileges to compromise data privacy. Existing DID tools focus on analyzing SQL texts and they miss the cases of query equivalence and do not reveal data sources hidden behind views, synonyms, aliases. We propose to use Query Execution Plans (QEPs) to detect misuse queries and we describe a QEP-based DID System.

1 Introduction

Database Intrusion Detection focuses on inspecting database workload to detect and prevent execution of misuse/anomalous SQL queries, like SQL injection attacks, malicious insider activity, application errors. Database Intrusion Detection Systems (DIDS) are an additional layer of protection between users and data, enriching existing access control mechanisms. Most DIDS rely on query syntax analysis, which is very difficult given the expressiveness of SQL – false alarms may be generated for refactored but still valid queries, while intruders may make their queries "look safe" and bypass the DIDS. In this paper we propose a DIDS that analyzes Query Execution Plans (QEPs) to detect anomalous queries. We show that several types of anomalies can be easily detected by comparing QEP trees to a reference set representing normal workload.

2 QEP-Based Detection of Anomalous SQL Queries

In a perfect world, a database user should only see data that relate to the job role, business unit, location, etc. – the Principle of Least Privilege (POLP). Many database mechanisms support POLP: discretionary access control, mandatory access control, role-based access control, row level access control. But following POLP is not easy for database administrators – generic user groups, application-level accounts, complex applications – result in granting more privileges than needed and malicious users may use those extra privileges to compromise data privacy. *Anomalous queries* are queries that try to access data not intended to be accessed by a particular user while technically accessible to the user due to imperfect POLP policy. They can be also interpreted as "queries never seen

R. Wrembel et al. (Eds.): DaWaK 2024, LNCS 14912, pp. 353–358, 2024.
https://doi.org/10.1007/978-3-031-68323-7_30

before". Examples include: reading tables not to be read by the user, reading table columns or records not to be read by the user. Existing DID tools focus on SQL text, either inspecting queries to detect forbidden keywords, names, or comparing syntactical feature vectors with pre-learned profiles of past user queries. Not all anomalies can be detected in this way. Malicious users can write their queries using tampered views, synonyms, tricky joins or subqueries to bypass black list rules of a text-based DIDS.

Notice that semantics of a query is represented not only by its text, but also by its QEP – an ordered set of data access steps and operations. A QEP is represented by a tree whose leaves are tables/indexes, internal nodes are algebraic operations, edges represent data flow. The root is the final operator producing the query answer [8]. QEPs show features not visible in the query's text. They expose tables participating in views, tables hidden behind synonyms, table and column aliases, actual joins and cross joins, unnested subqueries.

Table 1. Costs of node edit operations in QEP trees (reference QEP → new QEP)

Node edit operation	Cost	Explanation of the arbitrary value chosen
SORT node insert/delete	0	Sorting does not compromise data privacy
SELECTION node insert	0	Narrowing data range does not compromise data privacy
SELECTION node delete	10	Widening data range may compromise data privacy
PROJECTION node insert	0	Narrowing data range does not compromise data privacy
PROJECTION node delete	10	Widening data range may compromise data privacy
AGGREGATION node insert	0	Generalization does not compromise data privacy
AGGREGATION node delete	10	Drilling down may compromise data privacy
TABLE node delete	0	Skipping a table does not compromise data privacy
TABLE node insert	100	Adding another table may compromise data privacy
INDEX node insert	0	Narrowing data range does not compromise data privacy
INDEX node delete	10	Widening data range may compromise data privacy
All other node edit operations	1	

Our idea of using QEPs for anomalous query detection is to compare a QEP of a new query with QEPs of reference queries (normal workload). If the new QEP is identical/close enough to one of the reference QEPs, then the query is considered typical, otherwise an alert is raised. To measure similarity of two QEPs, we use Tree Edit Distance (TED) (minimal-cost sequence of node edit operations that transform one tree into another [9,14]), with different arbitrary unit costs of node edits for different relational operations (Table 1). A query is considered anomalous when:

$$\min_{r \in R} TED(r, q) > \delta \qquad (1)$$

where R is the reference QEP set, q – QEP being verified, δ – an arbitrary threshold. See the example in Fig. 1. The left query operates on a different table than the right reference query. It results in a different QEP tree: "table access orders" replaced with "table access salaries". The distance between the QEPs is 200 (node delete: 100, node insert: 100), which might signal an alert. Table 2 gives feature comparison details for our proposed QEP-based method vs. the popular SQL-text-based approaches.

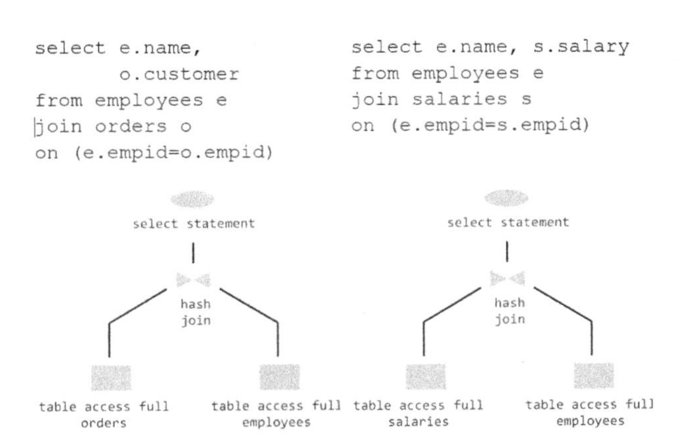

Fig. 1. Sample SQL queries and their QEPs

To verify a new query, reference QEPs are necessary. They should represent typical/legal set of database operations. They can be pre-learned by e.g. running a legitimate application and collecting all QEPs, or can be continuously extracted from a database cursor cache, as a moving window that spans last n QEPs. The reference QEPs can describe generic database workload or be user-specific. Verification of a new query must be fast, not to hurt performance. Calculating TEDs is fast for a pair of QEP trees but not necessarily fast for thousands of QEP trees. We introduce two types of preliminary filters to quickly accept or reject a trivial case QEP: (1) *Hash Value Filter* detects that hashes of two QEPs are identical, (2) *Table Names Filter* detects non-identical sets of source tables, signaling a heavy tree edit operation, exceeding the arbitrary δ.

Table 2. SQL-text-based vs. QEP-based intrusion detection

Anomalous query	Text based ID	QEP based ID

skipping selection predicates
`select * from t where c=:b1;`
`vs. select * from t;`
	yes	yes

querying different data tables
`select * from t;`
`vs. select * from s;`
	yes	yes

querying different columns
`select a, b from t;`
`vs. select b, c from t;`
	yes	no

querying a table through a tampered view/synonym
`select * from s;`
`vs. drop view s; create view s as select * from t;`
`select * from s;`
	no	yes

using different literals/bind variables
`select * from t where c=1;`
`vs. select * from t where c=2;`
	practically not	generally not (selectivity may influence QEP)

disabling join conditions
`select * from s join t on (s.c = t.c);`
`vs. select * from s join t on (s.c = s.c);`
	no	yes

using tampered subqueries
`select * from (select * from s);`
`vs. select * from (select * from t);`
	no	yes

3 Experimental Evaluation

We implemented a prototype QEP-based DIDS for Oracle Database 18c. QEPs were extracted using native DBMS features, TED implementation was based on APTED [15]. Two forms of experiments were conducted: (1) accuracy evaluation by analyzing queries run by students performing hands-on practices for their database fundamentals course, where incorrect queries were considered anomalies and the reference QEPs have been defined by a trainer, (2) performance evaluation by measuring TED calculation time for realistic-size query sets. Table 3 shows the accuracy experiments' results. Anomalous join queries and aggregation queries were easily detected. Simple queries containing "incorrect" selections or projections were more difficult to detect as column names were not exhibited in QEPs. Therefore, alerts were mostly based on QEP side effects like choosing different join algorithms or index access methods due to different column cardinalities. Figure 2 shows TED calculation time for varying sets of QEP trees and for two tree sizes (2.1 GHz Intel Core i7-4600U CPU), with- and without the preliminary filtering phases. We observed sub-second exe-

Table 3. Accuracy experiment results ($\delta = 9$)

SQL syntax area	Num of queries	True positives	True negatives	False positives	False negatives	Accuracy
selection and projection	968	78	780	43	67	89%
joins	449	31	395	6	17	95%
aggregation	358	47	289	11	11	94%

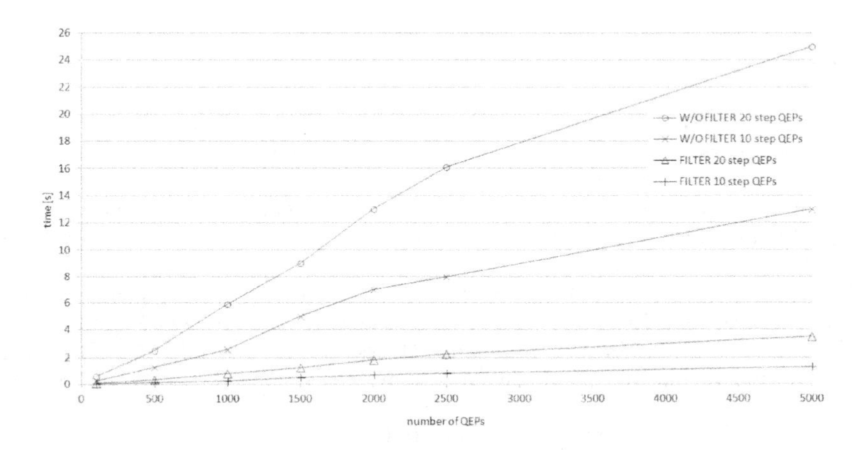

Fig. 2. TED calculation times

cution times for 2,500 trees of 10 nodes on average and 1,000 trees of 20 nodes on average.

4 Related Work

DEMIDS [2] derived user profiles from audit logs and used them to compute query anomaly score. In [15] queries were tokenized and then converted into feature vectors to be verified against pre-learned profiles. [1] used analysis and profiling of an application in order to create a representation of its interaction. [3] used two multi-layer perceptrons. SQL query feature model was used in [10], and a random forest and PCA detected anomalies. [5] described an SQL Injection Attack detection framework which leveraged knowledge from pre-deployment testing of web applications. Other interesting approaches have been described in [4,6,7,11,13].

5 Conclusions and Future Work

We have addressed a question that, to the best of our knowledge [12], has not been answered before: can analysis of QEPs help detect database intrusions? We developed a model and a research prototype of an QEP-based DIDS system. Our solution offers significant advantages when compared to popular text-based

DIDSs: (1) actual data sources - like tables hidden behind SQL views, synonyms, aliases and subqueries - are revealed, (2) the subject of analysis is what a query does, and not how it is formulated. It does, however, suffer from the problem of not being able to check what actual data items are accessed by database users. Our experimental results are promising. Although our test group members were not real malicious users, we were able to demonstrate that "incorrect" database interactions can be efficiently detected by analyzing QEPs.

References

1. Bossi, L., Bertino, E., Hussain, S.R.: A system for profiling and monitoring database access patterns by application programs for anomaly detection. IEEE Trans. Software Eng. **43**(5), 415–431 (2017)
2. Chung C.Y., Gertz M., Levitt K., DEMIDS: a misuse detection system for database systems. In: Third Annual IFIP TC-11 WG 11.5 Working Conference on Integrity and Internal Control in Information Systems, pp. 159–178 (1999)
3. Choi, S.-G., Cho, S.-B.: Adaptive database intrusion detection using evolutionary reinforcement learning. In: Pérez García, H., Alfonso-Cendón, J., Sánchez González, L., Quintián, H., Corchado, E. (eds.) SOCO/CISIS/ICEUTE -2017. AISC, vol. 649, pp. 547–556. Springer, Cham (2018). https://doi.org/10.1007/978-3-319-67180-2_53
4. Creech, G., Hu, J.: A semantic approach to host-based intrusion detection systems using contiguous and discontiguous system call patterns. IEEE Trans. Comput. **63**(4), 807–819 (2014)
5. Dharam, R., Shiva, S.-G.: Runtime monitoring framework for SQL injection attacks. Int. J. Eng. Technol. **6**(5), 392–401 (2014)
6. Hu, Y., Panda, B.: Identification of malicious transactions in database systems. In: Seventh International Database Engineering and Applications Symposium (2003)
7. Hashemi, S., Yang, Y., Zabihzadeh, D., Kangavari, M.R.: Detecting intrusion transactions in databases using data item dependencies and anomaly analysis. Exp. Syst. **25**(5), 460–473 (2008)
8. Ioannidis Y.E.: Query optimization. ACM Comput. Surv. (CSUR) Surv. **28**(1) (1996)
9. Pawlik, M., Augsten, N.: Tree edit distance: Robust and memory-efficient. Inf. Syst. **56**, 157–173 (2016)
10. Ronao, C.A., Cho, S.-B.: Anomalous query access detection in RBAC-administered databases with random forest and PCA. Inf. Sci. **369**(10), 238–250 (2016)
11. Rappel, N., Stein, N., Flath, Ch.M.: Dynamic intrusion detection in database systems: a machine-learning approach. In: ICIS 2016 Proceedings
12. Santos, R.J., Bernardino, J., Vieira, M.: Approaches and challenges in database intrusion detection. ACM SIGMOD Rec. **43**(3), 36–47 (2014)
13. Sallam, A., Bertino, E.: Detection of temporal data ex-filtration threats to relational databases. In: IEEE Conference on Collaboration and Internet Computing, pp. 146–155 (2018)
14. Tai, K.-C.:The Tree-to-tree correction problem. J. ACM (1979)
15. Valeur, F., Mutz, D., Vigna, G.: A learning-based approach to the detection of SQL attacks. In: Julisch, K., Kruegel, C. (eds.) DIMVA 2005. LNCS, vol. 3548, pp. 123–140. Springer, Heidelberg (2005). https://doi.org/10.1007/11506881_8

Visual Transformers Meet Convolutional Neural Networks: Providing Context for Convolution Layers in Semantic Segmentation of Remote Sensing Photovoltaic Imaging

Jose Alejandro Libreros[1(✉)], Muhammad Hamza Shafiq[1,2], Edwin Gamboa[1], Martin Cleven[2], and Matthias Hirth[1]

[1] TU Ilmenau, Ilmenau, Germany
{jose.libreros,edwin.gamboa,matthias.hirth}@tu-ilmenau.de
[2] E.ON Digital Technology GmbH, Hannover, Germany
{muhammad-hamza.shafik,martin.cleven}@eon.com

Abstract. The energy potential of Photovoltaic Panel (PVPs) decreases due to obstructions. Identification of rooftops and relevant objects to optimize PVPs capacity is usually done using aerial imagery. However, the manual process is tedious, costly, and time-consuming involving subjective expert judgment. While machine learning-based segmentation algorithms, e.g., Convolutional Neural Network (CNNs), show high accuracy while segmenting images with extensive training data, performance decreases in the context of PVPs, due to lack of labelled data for panels and obstacles. This data scarcity results from costs of acquiring and annotating high-resolution aerial imagery implying non-robust pattern learning.

This paper introduces the application of Visual Transformer (ViTs) to segment relevant objects for identifying PVPs potential using aerial images. ViTs addresses the need for models with inductive bias, allowing direct image learning rather than feature representations. Consequently, ViT requires less training data to achieve optimal performance, with advantages in scenarios with limited training data available. In our study, we compare segmentation capabilities of a ViT and a CNNs using same small size, low resolution aerial images as training data.

Our results suggest better performance of ViTs compared to CNNs with the same training and test dataset for the PVP context, and slightly better results when using a hybrid CNN-ViT model. In this last case, it is clear that adding global information learned by ViTs to local information learned by CNNs yields better performance using a small amount of low-cost, low-resolution reduced data.

Keywords: Visual Transformers (ViTs) · Convolutional Neural Networks (CNNs) · Photovoltaic panels

R. Wrembel et al. (Eds.): DaWaK 2024, LNCS 14912, pp. 359–366, 2024.
https://doi.org/10.1007/978-3-031-68323-7_31

1 Introduction

Photovoltaic Panel (PVP) are one promising ways to decarbonize cities. However, some areas on rooftops are not feasible for generating electricity from PVPs due to obstacles on the roofs or the presence of trees that cover the rooftops [11]. This poses challenges in identifying available rooftop areas for installing PVPs.

Aerial imagery provides a cost-effective and non-invasive way to delineate suitable rooftop zones. Yet, segmenting obstacles, rooftops, and trees becomes a challenge that requires both human and algorithmic involvement. While manual approaches can be tedious and subjective, algorithmic methods offer efficiency and accuracy, benefiting from advances in computer vision techniques, particularly those based on machine learning.

In the past, Convolutional Neural Network (CNNs) were widely considered one of the best supervised machine learning techniques for pixel-wise classification in semantic segmentation tasks. However, challenges remain with CNNs, especially concerning dataset size and feature detection. Recently, the new paradigm Visual Transformer (ViTs) has emerged for segmentations, by appling a self-attention mechanism directly on the image, not on intermediate convolutional operations, as CNNs. Therefore, we evaluate the two models' performance in detecting available rooftops, trees, obstacles, and backgrounds using an aerial images. In addition, we test whether obtaining information from the context is more accurate for identifying small obstacles on rooftops and solar panels. The appraoch offers the opportunity that even small existing datasets could be handled with automated segmentation models that may be scalable to other scenarios.

Our study uses a dataset taken from OpenNRW data with about 2902 images, with a defined Ground Truth (GT). We used two CNN-based models, one ViT-based model, and a hybrid model with ViT as the backbone of a CNN scheme. We found that ViT significantly outperformed the semantic segmentation performance of CNN-based models, but also we confirmed that the performance of CNN-based models can be improved by using information from ViT in a hybrid model. This remainder of the paper is organized as follows. Section 2 presents the related work on CNN and ViT, then we present the materials and methods in Sect. 3. The results are presented in Sect. 4, and the discussion is given in Sect. 5.

2 Related Work

Previous research shows that multispectral information of aerial image is not linearly separable, e.g., overlapping obstacles in this context, thus limiting generalization to other types of scenarios [10]. It has been reflected in lack of performance of classical techniques, such as mathematically based [15] and unsupervised methods [9].

Supervised learning methods, especially CNNs, have shown high performance in semantic segmentation compared to the previously mentioned approaches.

CNNs take advantage of deep, multilayered neural networks with numerous parameters, which enhance its ability to represent information and learn from data, as they are nonlinear models to represent the features, allowing classification of nonlinear multispectral information. CNNs have been applied in various domains, such as remote sensing images [12], with reported limitations. Zhang et al. [16], conducted a study using remote sensing images of an Asian city as input data for three CNNs: SegNet, DeepLab, and the proposed HSCNN model. They reported limitations of DeepLab and SegNet, suggesting the need to use features of pixels and the spatial relation between them for segmentation. There are other reports of performances of less than 10% when segmenting buildings [5]. In general, there seems to be intolerance to a multiclass response when little data is available. Also, many data-unbalanced classes or a too-small training data set might cause a lack of performance on semantic segmentation.

Recently, ViTs have become relevant for retrieving global information for images. ViT is an adaptation of the transformer architecture used in texts, as proposed by Dosovitskiy et al. [6]. A ViT works by using a standard transformation encoder of a linear projection of flattened patches of the image, making better use of background information and relying less on high-frequency texture attributes [7]. Thus, ViTs could help obtain contextual information that helps achieve well-defined object fragmentations. However, current works have not tested CNNs and ViTs with aerial imagery and larger datasets yet, where spatial resolution is relevant to obtain information about multiple objects, which is very different from other domains.

This paper presents the first approach using ViTs for segmenting areal images to identify possible locations for PVPs. Through visual inspection of the results, we compared the capacity of having ViT vs. CNN and vs. ViT + CNN through experiments on some models with input data. We show that ViTs outperforms CNN when segmenting obstacles for the PVPs. We also found that a hybrid model of ViT + CNN segments PVP more accurately than ViT and by far more accurately thanCNN. This is very important when considering the intention of calculating power capacity by exploring the installed capacity. Thus, reducing the inductive bias of ViT without losing the original image content may imply better performance even with cheap low-resolution data. That is, our model may be scale-invariant and domain-adaptive. Paying more attention to information about an object's interior implies a better differentiation of objects, i.e., a better performance in identifying edges.

3 Materials and Methods

3.1 Dataset

We trained the models using Open NRW data from [1]. The dataset consists of 2902 aerial images with a spatial resolution of 10 cm. There are four classes: tree, existing photovoltaic panel, obstacle, and background. A group of annotators, supervised by E.ON Digitial Technology GmbH, provided the dataset labels for calculating the available rooftop area for solar panel installation. The set of labels is used here as GT.

3.2 Semantic Segmentation Models

We perform the experiments using two state-of-art CNNs, SegNet and DeepLabv3, a ViT (SegFormer) model and a pyramid hybrid model with CNN - Feature Pyramid Network (FPN) - ViT (UperNet-Swin).

For the **SegNet** model, we directly used the structure proposed by Badrinarayanan et al. [2] consisting of an encoder, a decoder, and a classifier. The encoder uses the first 13 convolutional layers of the VGG16 network, each with its corresponding decoder layer, for a total of 13 decoder layers. The last decoder generates a multichannel feature map as input to the classifier, which outputs a probability vector of length K – the number of classes. The final predicted category corresponds to the class with the maximum probability at each pixel. In terms of training, SegNet can be trained end-to-end using Stochastic Gradient Descent.

We used the **DeepLabv3** model [4] with a ResNet101 backbone. The network uses dilated convolutions after a variation of the pooling proposed by ResNet101. Dilated convolutions attempt to include more information without increasing the kernel size by itself. Finally, the model uses a fully connected conditional random field model, which refines the segmentation result to better capture the segmentation boundaries [3]. From SegNet and DeepLabv3 instance models, we compute the class weights based on the dataset before training, which avoids not initializing with a random neural network.

Proposed by Xie et al. [14], the Semantic Segmentation Framework is a Multilayer Perceptron (MLP) decoder. **SegFormer** has two attractive features: 1) it includes a novel hierarchically structured transformer encoder that outputs multiscale features. It does not require positional coding, thus avoiding interpolation of positional codes, which reduces performance when the test resolution differs from the training resolution. 2) It avoids complex decoders. The proposed MLP decoder aggregates information from different layers, thus combining both local and global attention to produce powerful representations.

We use also **UperNet-Swin** [13], an implementation of Upernet, which uses the Swin Transformer as a backbone for learning. Unified Perceptual Parsing Network (UperNet) is a framework built as a FPN, which is a generic feature extractor that exploits multi-level features to obtain multi-level representations of the scene in an inherent and pyramidal hierarchy of a deep CNN [8]. For Swin Transformer backbone and Segformer training, we use a pre-trained network (transfer learning).

3.3 Training Phase and Performance Evaluation

We have divided the dataset into training data (2579 images) and test data (323 images). For each current epoch, we analyze whether the score has not improved in a neighborhood of 5 epochs, which acts as a stopping condition to define a checkpoint and, thus, the maximum number of epochs for the current model. A combined qualitative and quantitative analysis is shown to get a better understanding of the behavior of each model. We report the median Intersection over Union (mIoU) and the loss per model as a standard metric to measure the performance of semantic image segmentation and the loss during the testing phase. Additionally, we report the mIoU per class for each model, and the confusion matrix for the PVP class. This allows us to understand what happens during certain phases of the models and how information is retained or lost, which is a critical factor in explaining the quantitative results.

4 Results

This section presents the results and an analysis of our proposed experimental setup, which is the basis for the discussion in Sect. 5.

SegNet and DeepLabv3 have taken 246 and 116 min respectively, and 62 and 17 epocs respectively, to reach convergence. On the other hand, SegFormer and UperNet-Swin have taken 44 and 33 epochs respectively, and 377 and 476 min respectively. SegNet reached its best performance with 62 epochs but has a higher loss. The training time for ViT based models is higher than for CNN models. We believe that it can be explained by the time required by ViT-based models to get more information from the context per epoch.

Table 1 shows the loss during the test phase and the mIoU per model with all images, ranging from 0 to 1. From the mIoU vs. loss using the ViT-based models (SegFormer and UpperNet-Swin), we see that less loss does not necessarily mean more mIoU. This is useful for analyzing the other evaluation metrics and for visual inspection of the output, which we will discuss in Sect. 5. From the table, we can also conclude that the obstacle class is difficult to segment using CNN-based approaches. Furthermore, we see significant improvements for each class when using ViT-based models, especially for obstacle and PVPs. This is important because the size ranges for obstacles and PVPs are significantly small with respect to background and trees, even in the training data set. This leads us to believe that the information captured from the context is crucial when it comes to encoding the largest amount of information available for small objects,

Table 1. Median IoU and loss during training phase per model, and the confusion matrix of pixels with GT (BG: background, OB: obstacle, PVP and TR: tree)

Model	Loss	mIoU					Conf. Matrix PVP %			
		Train	BG	OB	PVP	TR	BG	OB	PVP	TR
SegNet	0.65	0.26	0.64	0.08	0.21	0.48	06	44	50	00
DeepLabv3	0.21	0.35	0.82	0.22	0.26	0.58	05	07	88	01
SegFormer	0.23	0.55	0.92	0.44	0.56	0.65	17	03	80	00
UperNet-Swin	0.17	0.51	0.86	0.33	0.58	0.61	03	05	91	01

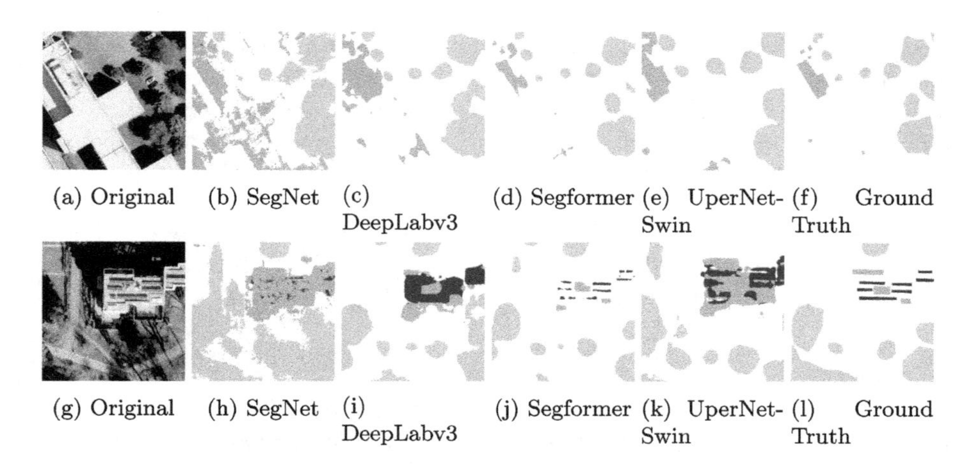

Fig. 1. Semantic segmentation output of each tested model architecture with some test images.

which is particularly necessary in remote sensing imaging. Although segmenting large extensions of crops is not the same thing, our set of images belongs to cities with a very specific target. Table 1 also includes the confusion matrix for the PVPs class. Specifically, it shows the distribution of detected percentages for each class, with PVPs as the GT. Note that for the specific class PVPs segmentation, the hybrid model ViT-CNN UperNet-Swin has the best performance, followed by the CNN-based DeepLabv3.

Figure 1 shows the output of the segmentation models for some critical images. By visual inspection of Fig. 1 (b)–(h), and compared to the other rows of Fig. 1, SegNet fails to discriminate between color variations, where it is particularly visible with lower zones of blue color (PVPs) but *obstacle*. For example, having pixels with blue colors, SegFormer classifies them as PVPs.

5 Discussion

Our results suggests that, given the same amount of training data, two possible alternatives to explore are either having a stratified set of models or including more information. SegNet has the worst results when examining the confusion matrix shown in Fig. 1 and the visual results shown in Fig. 1. The confusion matrix shows false positives PVPs, where in reality, those pixels belong to other classes, especially background. DeepLabv3 has reduced such false negative and false positive rates concerning its CNN counterpart SegNet. With SegNet, we observe better results in the identification of PVPs (see Table 1), but at the cost of having more unwanted pixels labeled as obstacles. Also, there are still limitations in differentiating spatial distributions when using CNNs to perform semantic segmentation for aerial images. In contrast, SegFormer has the lowest proportion of false pixel-based classifications.

In this application case, with particular attention to PVPs, we focus on examining the number of true and false positives and negatives by condensing these measures into more elaborate metrics as the mIoU. Analyzing the results from visual inspection of the output is challenging, as there are clear highlights and differences worth analyzing. In fact, there is more presence of pixels misclassified as background and tree than the GT. CNNs could achieve better performance by modifying their architecture to adopt the behavior of transformers. In our case, we extended the CNN by adding a ViT as a backbone of a CNN, which is the case of UperNet-Swin. In addition, as a comparison, SegFormer (and the Swin-based backbone of UperNet) has been shown to be highly accurate for segmenting aerial images to detect not only large areas, such as trees or large roofs but also small areas, such as obstacles or existing PVPs with high refinement. The presence of small objects on aerial images with high variations in the objects' region will decrease the homogeneity. This indicates poor contextual information. The results of DeepLabv3 and SegNet suggest that they are less noise-resistant.

An additional factor for the lack of performance could be missclassifications in the GT data. A deep study of the ability of these models to discard wrong segmentations in training could be analyzed in the future. But even here, we salvage the capacity of a model using contexts like ViT or a hybrid model CNN - ViT to avoid such wrong predictions in some training images. Part of our contribution is to remark that ViT is more tolerant to outliers, requiring smaller amounts of training data set for industrial applications. Nevertheless, this paper shows that in so few cases, introducing two different architectures improves the mIoU score.

6 Conclusion

In this paper, we analyzed the behavior of CNNs and ViTs, identifying PVP as an unexplored application case. Visual inspection and quantitative results allowed us to formalize generalized pitfalls of the CNNs and how the ViTs provide further pattern recognition. Our results showed that the ViTs have high performance with a significantly lower amount of training data, which means more generalization capacity of the ViTs. We also analyzed how the models could meet in a hybrid model. We demonstrated that a hybrid model is more suitable for some challenging classes with small objects in the images. It shows open possibilities for future research in this direction.

References

1. Solarpotenzialkataster von NRW und deutschlandweit, https://open.nrw/open-data/showroom/solarpotenzialkataster-von-nrw-und-deutschlandweit
2. Badrinarayanan, V., Kendall, A., Cipolla, R.: SegNet: a deep convolutional encoder-decoder architecture for image segmentation. IEEE Trans. Pattern Anal. Mach. Intell. **39**(12), 2481–2495 (2017). https://doi.org/10.1109/TPAMI.2016.2644615

3. Chen, L.C., Papandreou, G., Kokkinos, I., Murphy, K., Yuille, A.L.: DeepLab: semantic image segmentation with deep convolutional nets, atrous convolution, and fully connected CRFs. IEEE Trans. Pattern Anal. Mach. Intell. **40**(4), 834–848 (2018). https://doi.org/10.1109/TPAMI.2017.2699184
4. Chen, L.C., Papandreou, G., Schroff, F., Adam, H.: Rethinking atrous convolution for semantic image segmentation (2017). https://doi.org/10.48550/arXiv.1706.05587
5. Ding, L., et al.: Looking outside the window: wide-context transformer for the semantic segmentation of high-resolution remote sensing images. IEEE Trans. Geosci. Remote Sens. **60**, 1–13 (2022). https://doi.org/10.1109/TGRS.2022.3168697
6. Dosovitskiy, A., et al.: An image is worth 16x16 words: transformers for image recognition at scale (2021). https://doi.org/10.48550/arXiv.2010.11929
7. Ghiasi, A., et al.: What do vision transformers learn? A visual exploration (2022). https://doi.org/10.48550/arXiv.2212.06727
8. Lin, T.Y., Dollar, P., Girshick, R., He, K., Hariharan, B., Belongie, S.: Feature pyramid networks for object detection, pp. 2117–2125 (2017)
9. Lv, Z., Hu, Y., Zhong, H., Wu, J., Li, B., Zhao, H.: Parallel K-means clustering of remote sensing images based on mapreduce. In: Wang, F.L., Gong, Z., Luo, X., Lei, J. (eds.) WISM 2010. LNCS, vol. 6318, pp. 162–170. Springer, Heidelberg (2010). https://doi.org/10.1007/978-3-642-16515-3_21
10. Mountrakis, G., Im, J., Ogole, C.: Support vector machines in remote sensing: A review. ISPRS J. Photogrammetry Remote Sens. **66**(3), 247–259 (2011). https://doi.org/10.1016/j.isprsjprs.2010.11.001
11. Ren, H., Xu, C., Ma, Z., Sun, Y.: A novel 3D-geographic information system and deep learning integrated approach for high-accuracy building rooftop solar energy potential characterization of high-density cities. Appl. Energy **306**, 117985 (2022). https://doi.org/10.1016/j.apenergy.2021.117985
12. Wang, D., Zhang, J., Du, B., Xia, G.S., Tao, D.: An empirical study of remote sensing pretraining. IEEE Trans. Geosci. Remote Sens. 1–1 (2022). https://doi.org/10.1109/TGRS.2022.3176603
13. Xiao, T., Liu, Y., Zhou, B., Jiang, Y., Sun, J.: Unified perceptual parsing for scene understanding. In: Proceedings of the European conference on computer vision (ECCV), pp. 418–434 (2018)
14. Xie, E., Wang, W., Yu, Z., Anandkumar, A., Alvarez, J.M., Luo, P.: SegFormer: simple and efficient design for semantic segmentation with transformers. In: Advances in Neural Information Processing Systems, vol. 34, pp. 12077–12090. Curran Associates, Inc. (2021)
15. Yuan, J., Wang, D., Li, R.: Remote sensing image segmentation by combining spectral and texture features. IEEE Trans. Geosci. Remote Sens. **52**(1), 16–24 (2014). https://doi.org/10.1109/TGRS.2012.2234755
16. Zhang, C., et al.: Convolutional neural network-based remote sensing images segmentation method for extracting winter wheat spatial distribution. Appl. Sci. **8**(10), 1981 (2018). https://doi.org/10.3390/app8101981

Data Quality and Applications

NADA: NMF-Based Anomaly Detection in Adjacency-Matrices for Industrial Machine Log-Files

Sabrina Luftensteiner(✉), Patrick Praher, and Nicole Schwarz

Software Competence Center Hagenberg, Softwarepark 32a,
4232 Hagenberg im Muehlkreis, Austria
sabrina.luftensteiner@scch.at

Abstract. In the manufacturing and process industries, the standard approach to detecting faults or anomalies in equipment is to use condition monitoring methods. The disadvantages of these approaches are the need for expensive sensor hardware and extensive knowledge of mechanical and electronic properties. Since these challenges are not always easy to overcome, we demonstrate a novel approach that allows us to monitor the behavior of industrial equipment based on regular logs with a moderate to low number of log entries. We propose a new approach for detecting anomalies and deviations in log files using the relative or power adjacency matrix of transitions based on non-negative matrix factorization. The experiments section discusses the application to a typical manufacturing process consisting of multiple processing steps and demonstrates the ability to detect regular and irregular machine behavior.

Keywords: Log Files · Anomaly Detection · Process Mining

1 Introduction

In manufacturing, machines are often programmed to repeat specific workflows, with logs documenting both expected behavior and undesired incidents such as errors that indicate system weaknesses [1,2]. These programs usually have a linear workflow and are logged constantly. Manual inspection of these logs is tedious and time-consuming [1], necessitating automated solutions. Process Mining automates the extraction of process knowledge from event logs, making it well-suited for manufacturing due to its process-oriented nature [3,4]. Machine logs detail activities like cutting, coating, and bending, as well as errors from

The research reported in this paper has been funded by the Federal Ministry for Climate Action, Environment, Energy, Mobility, Innovation and Technology (BMK), the Federal Ministry for Labour and Economy (BMAW), and the State of Upper Austria in the frame of the SCCH competence center INTEGRATE in the COMET - Competence Centers for Excellent Technologies Programme managed by Austrian Research Promotion Agency FFG.

R. Wrembel et al. (Eds.): DaWaK 2024, LNCS 14912, pp. 369–374, 2024.
https://doi.org/10.1007/978-3-031-68323-7_32

faulty executions or worn parts. Regular analysis of these logs using Process Mining helps ensure workflows behave as expected, highlighting the need for efficient methods to compare Process Mining graphs and check conformance with base process models [1,5].

The paper is structured as follows: it begins with a review of related work on anomaly detection in logs, conformance checking, and process deviance and variants, discussing various approaches and their drawbacks. Next, it introduces our approach, including event-log preparation and the application of NADA. The experiments section details the setup, workflow structure, types of anomalies, NADA settings, and scenarios. Results are presented textually and visually using different comparison methods. The paper concludes with a discussion of advantages and disadvantages and an overview. The main contribution is a method for detecting anomalies or deviations in processes using Process Mining and event-logs, based on residuals calculation via non-negative matrix factorization. Additionally, it provides an overview of existing approaches to anomaly detection in process mining logs highlighting issues relevant to our use-cases.

2 Related Work

This section covers an insight into related work for the comparison of event-logs to find behavior changes or anomalies in processes.

Anomaly detection approaches have been developed for both log files, recorded on machines, and event-logs, which represent further processed and enriched log files [16]. Various approaches in different application areas haven been proposed, see for e.g. [6] or [2]. The proposed methods are highly sensitive to deviations in process executions and struggle with processes that frequently deviate without a stable base, such as irregular and often occurring error messages. Additionally, these methods often have difficulty identifying frequently occurring anomalies.

Another interesting topic within this area of research is *Conformance Checking. Conformance Checking* aims at detecting inconsistencies between a process model and its corresponding execution log and their quantification by the formation of metrics [7]. Various approaches for conformance checking are available, including but not limited to [7–11]. The conformance checking approaches seem quite interesting, but they require a stable process model as a base to provide valid results. In manufacturing, this may often be not the case as processes are adapted or unpredictable errors are logged, which complicates the automatic creation of a good process model without human intervention.

Process Deviance and Variants are another interesting subtopic regarding the inspection of process workflows and their conformance. Approaches for the generalization and identification of variants are proposed by Li et al. [12], Bolt et al. [13] and Nguyen et al. [14,15]. Process deviance and variants provide suitable approaches to handle variants and the generation of a reference model. Still, human intervention is needed to generate a meaningful base for further usage and to define normal behavior of a process. Furthermore, human input is needed to define positive and negative influences of deviations on the process.

3 Method

3.1 Preparation of Event-Logs

The data consists of three features: CaseID, Activity, and Timestamp. This simplified event-log excludes resources and other irrelevant information. CaseID identifies a process execution, Activity provides a brief description of each step, and Timestamp keeps the data ordered. Users can adjust the CaseID granularity (e.g., per execution, hour, or day) based on their needs, which can help in identifying specific patterns, such as error frequency within certain time spans.

The first step regarding data preparation is the extraction of an adjacency matrix, which represents the process graph derived from the event-logs using Process Mining. We use relative values in the matrix, based on the total number of edge transitions between nodes, to ensure comparability across different scenarios, such as varying execution runtimes.

3.2 NADA Using Event-Logs

The event-logs from Sect. 3.1 are used as base for further processing. *Non-negative matrix factorization* (NMF) is used to factorize a source matrix V into two separate matrices W and H with the property that all three matrices contain only non-negative values. V is the product of the matrices H and W, where W represents the base matrix and H the coefficients matrix. By multiplying W and H, an approximated matrix is generated, which is later used to calculate the residuals and further metrics, such as the standard deviation σ.

As in our case more than one matrix is used for the extraction of a common base matrix W, an adapted method called *Joint non-negative matrix factorization* (JNMF) [17] is used. One common base matrix W is extracted and one individual coefficient matrix H is generated for each adjacency matrix. The reason for multiple matrices arises as the event-logs are gathered over a specific time-span, e.g. a day, and are compared afterwards to see if major behavior changes appeared. After the extraction of the base matrix W, the individual coefficient matrices $H_1 ... H_K$ and the approximated matrices are generated, it is possible to calculate the residuals and determine which of the matrices are better represented in combination with base matrix W. For matrices representing normal behavior, we would expect easier reconstruction and therefore a small residuals. Discrepancies may arise if the behavior of the process is changing. Similar matrices provide similar residual results. The process of calculating the residuals can be summarized as follows:

- Transformation of event-logs to relevant adjacency matrices
- Extraction of base matrix W and coefficient matrices $H_1 ... H_K$ using JNMF
- Calculation of residuals between each approximated and corresponding adjacency matrix
- Comparison of residuals

4 Experiments

4.1 Experimental Setup and NADA Settings

The experimental setup includes six scenarios, ranging from no anomalies to highly variable days with many anomalies. Each scenario represents one day, consisting of 12 program executions, and is used to build the adjacency matrix through Process Mining. An optimal process execution is a linear workflow with 9 steps, including *Start*, *Stop*, and 7 intermediate steps like cutting or bending in manufacturing. To simulate issues, error messages (*Error 1* to *Error 3*) are introduced at varying rates in the scenarios, reflecting common manufacturing disturbances, such as misaligned material. Overall, the steps, further called activities, include the start, intermediate process steps, and the stop. The special use-cases will be represented by additional activities, repeating activities, interrupted process executions and diverging activities.

Since a single adjacency matrix cannot fully represent the workflow, multiple matrices are used to derive a common base matrix via JNMF. Our experiments use 20 matrices from the first 20 days of executions. We test two approaches: one with anomalies during the training phase and one without. In real-world scenarios, we recommend using non-anomalous days for training to improve accuracy. Introducing anomalies in training allows comparisons to supervised methods like K-Nearest Neighbors (KNN). Our approach may be viewed as semi-supervised or unsupervised.

4.2 Scenarios Description

The experimental scenarios are divided by error amount, error appearance, and anomaly amount. Error amount indicates the probability of errors per execution. Error appearance refers to which errors appear in normal versus anomalous days. Anomaly amount denotes the probability of anomalies. Table 1 contains the parameters of the scenarios.

Table 1. Specification of Scenarios. The appearance of errors is defined (N - tolerated during normal execution [considered okay by the operator], A - during anomalous execution) as well as the probability of errors and anomalous executions and if training samples contain anomalous days (AT).

UC	Error 1	Error 2	Error 3	% Err.	% Anom.	AT
UC1	N, A	A	A	0.1	0.1	F
UC2	N, A	A	A	0.5	0.1	F
UC3	N, A	N, A	N, A	0.5	0.1	F
UC4	N, A	N, A	N, A	0.5	0.5	F
UC5	N, A	A	A	0.1	0.1	T
UC6	N, A	N, A	N, A	0.1	0.3	T

4.3 Results

The scenarios are divided into two groups: (I) those without anomalies during training, and (II) those with varying anomaly levels during training. We chose these settings to mimic real-world industrial situations where certain anomalies are considered tolerable by machine operators. In scenarios without anomalies during training, we compare our method (NADA) with the commonly used approach, PCA (AB-PCA), which utilizes adjacency matrices as input. Scenarios with anomalies during training involve our approach (NADA), AB-PCA, and K-Nearest Neighbors (KNN), also utilizing adjacency matrices (AB-KNN). AB-KNN is applicable only in scenarios with anomalies due to training reasons. Mean plus/minus two Standard deviations are used to define the anomalous threshold.

Figure 1 contains further evaluations of our scenarios, where sensitivity/recall, specificity, precision and AUC are calculated.

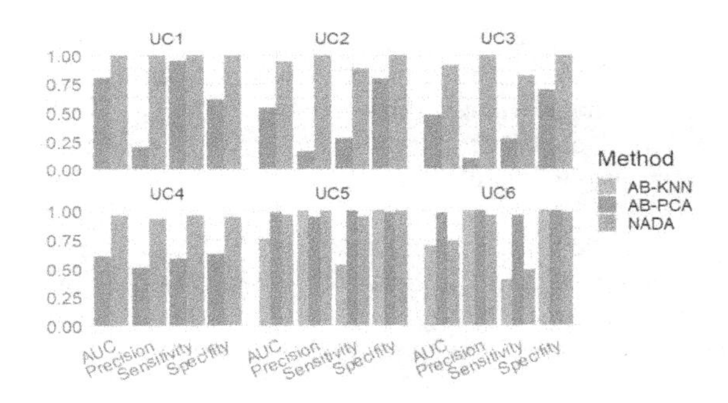

Fig. 1. Sensitivity, Specificity, Precision and AUC results of all scenarios.

5 Discussion and Conclusion

We addressed the problem of detecting anomalies or deviations in log files from industrial equipment for mostly linear programs. Based on Process Mining techniques on event-logs, we developed a method to provide indications for anomalies in the machines behavior. Our approach NADA covers the calculations of residuals using non-negative matrix factorization to determine how similar program executions are based on adjacency-matrices. In the experiments section, we provided a specific setup consisting of a basic program and possible deviations to demonstrate the impact on the methods and their effectiveness. Our method allows us to identify irregular machine behavior although the training data is limited in volume and contains irregularities like occasional warnings which often resembles real world industrial situations.

Regarding future work, we are interested in the integration of possibilities for active learning regarding continuously changing executions.

References

1. Frei, A., Rennhard, M.: Fine particles, thin films and exchange anisotropy. In: 2008 Third International Conference on Availability, Reliability and Security, pp. 610–617 (2008)
2. Breier, J., Branišová, J.: Anomaly detection from log files using data mining techniques. Inf. Sci. Appl. 449–457 (2015)
3. Corallo, A., Lazoi, M., Striani, F.: Process mining and industrial applications: a systematic literature review. Knowl. Process. Manag. **27**, 225–233 (2020)
4. Son, S., et al.: Process mining for manufacturing process analysis: a case study. In: Proceeding of 2nd Asia Pacific Conference on Business Process Management, Brisbane, Australia (2014)
5. Rozinat, A., Van der Aalst, W.: Conformance checking of processes based on monitoring real behavior. Inf. Syst. **33**, 64–95 (2008)
6. Sahlabadi, M., Muniyandi, R., Shukur, Z.: Detecting abnormal behavior in social network websites by using a process mining technique. J. Comput. Sci. **10**(3), 393 (2014)
7. Dunzer, S., Stierle, M., Matzner, M., Baier, S.: Conformance checking: a state-of-the-art literature review. In: Proceedings of the 11th International Conference on subject-oriented Business Process Management, pp. 1–10 (2019)
8. Adriansyah, A., van Dongen, B., van der Aalst, W.: Towards robust conformance checking. In: International Conference on Business Process Management, pp. 122–133 (2010)
9. Adriansyah, A., van Dongen, B., van der Aalst, W.: Conformance checking using cost-based fitness analysis. In: 2011 IEEE 15th International Enterprise Distributed Object Computing Conference, pp. 55–64 (2011)
10. Munoz-Gama, J., Carmona, J., van der Aalst, W.: Hierarchical conformance checking of process models based on event logs. In: International Conference on Applications and Theory of Petri Nets and Concurrency, pp. 291–310 (2013)
11. Myers, D., Suriad, S., Radke, K., Foo, E.: Anomaly detection for industrial control systems using process mining. Comput. Secur. **78**, 103–125 (2018)
12. Li, C., Reichert, M., Wombacher, A.: Mining business process variants: challenges, scenarios, algorithms. Data Knowl. Eng. **70**, 409–434 (2011)
13. Bolt, A., van der Aalst, W., De Leoni, M.: Finding process variants in event logs. In: OTM Confederated International Conferences, pp. 45–52 (2017)
14. Nguyen, H., Dumas, M., La Rosa, M., Maggi, F., Suriadi, S.: Mining business process deviance: a quest for accuracy. In: OTM Confederated International Conferences, pp. 436–445 (2014)
15. Nguyen, H., Dumas, M., La Rosa, M., Maggi, F., Suriadi, S.: Business process deviance mining: review and evaluation. arXiv preprint arXiv:1608.08252 (2016)
16. Van Der Aalst, W.: Process mining: overview and opportunities. ACM Trans. Manage. Inf. Syst. (TMIS) **3**, 1–17 (2012)
17. Zhang, S., Liu, C., Li, W., Shen, H., Laird, P., Zhou, X.: Discovery of multi-dimensional modules by integrative analysis of cancer genomic data. Nucleic Acids Res. **40**, 9379–9391 (2012)

Trusting Fair Data: Leveraging Quality in Fairness-Driven Data Removal Techniques

Manh Khoi Duong(✉) and Stefan Conrad

Heinrich Heine University, Universitätsstraße 1, 40225 Düsseldorf, Germany
{manh.khoi.duong,stefan.conrad}@hhu.de

Abstract. In this paper, we explore bias mitigation techniques that use undersampling to create a fair representation of sociodemographic groups in training sets. These techniques aim to make machine learning models trained on these datasets fair. However, such approaches may exclude relevant data, reducing trustworthiness and potentially harming models' performances. To address this, we propose two criteria in addition to fairness: (1) group coverage and (2) minimal data loss. Group coverage ensures that no group gets entirely removed, while minimal data loss aims to retain as many data points as possible. By proposing a multi-objective optimization approach, we find Pareto-optimal solutions that balance these objectives. This allows users to make informed decisions about the trade-off between fairness and data quality. Our method is distributed as a Python package via PyPI under the name `FairDo` (https://github.com/mkduong-ai/fairdo).

Keywords: Fairness · Bias mitigation · Data quality · Coverage · AI Act

1 Introduction and Related Work

Our work takes its motivation from the AI Act [6] that aims to regulate AI systems in the European Union. Specifically, we deal with the following excerpt from the current corrigendum:

> *"[...] data sets should also have the appropriate statistical properties, including as regards the persons or groups of persons in relation to whom the high-risk AI system is intended to be used, with specific attention to the mitigation of possible biases in the data sets [...]"* (Recital 67)

Some bias mitigation techniques involve the removal of data points from the training set [4,12] that cause discrimination. By doing so, the machine learning model is trained on a fair subset and its predictions are expected to be fair as well. While these techniques tackle the root cause of the problem, they are deemed problematic. One particular problem is the *removal of groups* as a whole, i.e., lack

R. Wrembel et al. (Eds.): DaWaK 2024, LNCS 14912, pp. 375–380, 2024.
https://doi.org/10.1007/978-3-031-68323-7_33

of *coverage*. For instance, removing non-privileged groups can be considered fair by various fairness metrics, yet it is equivalent to underreporting discrimination. Another problem is the *amount of data removed*. When removing too many data points, the resulting dataset may not accurately represent the original data and data quality is compromised.

`AIF360` [1] and `Fairlearn` [2] are popular Python packages that include bias mitigation techniques. However, the included pre-processing methods are unable to deal with *non-binary groups* or transform data in an *uninterpretable* way by editing features or labels. `FairDo` [4] does not come with these drawbacks and we enhance this package to tackle the mentioned problems.

2 Enhancing Trust in Fair Data

To enhance the trustworthiness of fair datasets, we introduce two criteria in addition to fairness. We will dedicate a subsection to each criterion.

2.1 Fairness

We proposed a framework [4] for removing discriminating data points from a given dataset $\mathcal{D} = \{d_i\}_{i=1}^n$. We stated the problem as finding a subset $\mathcal{D}_{\text{fair}} \subseteq \mathcal{D}$, which minimizes discrimination. The combinatorial optimization problem is formulated as:

$$\min_{\mathcal{D}_{\text{fair}} \subseteq \mathcal{D}} \quad \psi(\mathcal{D}_{\text{fair}}), \tag{1}$$

where ψ is a fairness metric that measures discrimination in a dataset. To deal with non-binary groups, we use maximal statistical disparity $\psi_{\text{SDP-max}}$ [4,13]:

$$\psi_{\text{SDP-max}}(\mathcal{D}) = \max_{i,j \in g} |P(Y=1 \mid Z=i) - P(Y=1 \mid Z=j)|. \tag{2}$$

It represents the maximum difference in the probability of receiving the positive outcome $Y = 1$ between any two groups $Z = i$ and $Z = j$ from the set of groups g. Because finding the exact optimal solution for Eq. (1) is an NP-hard problem if ψ is treated as a black-box, we employed genetic algorithms in our prior work [4].

2.2 Coverage

The subset that satisfies coverage is always preferred over the subset that does not. However, if both subsets satisfy coverage, we want to compare them based on the fairness metric ψ. Thus, a penalty is only applied if a group is missing. We denote $\hat{\psi}$ as the penalized version of ψ that enforces this preference.

We derive the penalty for $\psi_{\text{SDP-max}}$ as follows: The highest disparity possible for $\psi_{\text{SDP-max}}$ is 1. Let $|g_m|$ be the number of missing groups and $\epsilon > 0$, then we penalize $\psi_{\text{SDP-max}}$ as follows to enforce preferring coverage over non-coverage:

$$\hat{\psi}_{\text{SDP-max}}(\mathcal{D}) = \max(\psi_{\text{SDP-max}}(\mathcal{D}), [|g_m| > 0] \cdot (1+\epsilon)), \tag{3}$$

where $[|g_m| > 0]$ is an indicator function, which returns 1 if $|g_m| > 0$ and 0 otherwise. Setting ϵ to any positive value ensures that the penalty is higher than the maximum discrimination score.

2.3 Data Loss

By data loss, we refer to the similarity between the fair subset and its original. An efficient measure is the relative amount of data removed. With $\mathcal{D}_{\text{fair}} \subseteq \mathcal{D}$, it is given by:

$$\mathcal{L}(\mathcal{D}, \mathcal{D}_{\text{fair}}) = 1 - \frac{|\mathcal{D}_{\text{fair}}|}{|\mathcal{D}|}. \tag{4}$$

3 Optimization Objectives

We have three objectives to optimize for: *fairness*, *coverage*, and *data loss*. Two of the objectives, fairness and coverage, can be combined into a single objective, as shown in Eq. (3). The third objective, data loss, can be treated in multiple ways, which we will discuss in the following.

3.1 Multi-Objective Optimization

If there are no preferences provided regarding fairness and data loss, we have to treat the problem as a multi-objective optimization problem. The aim is to minimize both discrimination $\hat{\psi}$ and data loss $\mathcal{L}$:

$$\min_{\mathcal{D}_{\text{fair}} \subseteq \mathcal{D}} \quad (\hat{\psi}(\mathcal{D}_{\text{fair}}), \mathcal{L}(\mathcal{D}, \mathcal{D}_{\text{fair}})). \tag{5}$$

Solvers for multi-objective optimization problems aim to find a Pareto front, which is a set of solutions. We used our own modified version of the NSGA-II algorithm [3] to solve this problem.

3.2 Single-Objective Optimization

If the importance of fairness and data loss is known beforehand, the problem can be transformed into a single-objective optimization problem by using the weighted sum of the objectives:

$$\min_{\mathcal{D}_{\text{fair}} \subseteq \mathcal{D}} \quad \frac{\alpha \hat{\psi}(\mathcal{D}_{\text{fair}})}{\beta} + (1 - \alpha)\mathcal{L}(\mathcal{D}, \mathcal{D}_{\text{fair}}), \tag{6}$$

where $\alpha \in [0, 1]$ is a weighting factor that determines the importance of fairness over data loss and β is a normalization factor to make both objectives comparable. There are two meaningful choices for β: We either care about the *absolute* or *relative discrimination* improved by the fair subset. For absolute discrimination, β can be set to the theoretical maximum value of $\hat{\psi}$. In the case of $\hat{\psi}_{\text{SDP-max}}$, we set $\beta = 1 + \epsilon$. For relative discrimination, β can be set to the initial discrimination, i.e., $\beta = \hat{\psi}(\mathcal{D})$. This setting enforces lowering discrimination, even if the original dataset already has a low discrimination score.

4 Evaluation

To evaluate the proposed framework, we conducted two experiments guided by following research questions:

- **RQ1** Which genetic operators for the NSGA-II algorithm are best suited to mitigate bias in datasets?
- **RQ2** What is the impact of pre-processed datasets on the fairness and performance of machine learning models compared to unprocessed data?

Every experiment was performed using the same set of objectives and datasets. The objectives are $\hat{\psi}_{\text{SDP-max}}$ (Definition 3) with $\epsilon = 0.01$ and $\mathcal{L}$ (Definition 4). We experimented on the Adult [9], Bank [11], and Compas [10] datasets, which are widely used in the fairness literature. To ensure the reliability of our results, all experiments were repeated for 10 trials.

4.1 Hyperparameter Optimization

We used grid search to go through all configurations of genetic operators for the NSGA-II algorithm solving the optimization problem in Eq. (5). Specifically, we tested two methods for each genetic operator: *random* and *variable* for the initializer, *elitist* and *binary tournament* for the selection, *1-point* and *uniform* for the crossover, *bit flip* and *shuffle* for the mutation [5,8]. For all selection methods, we set the number of parents to two. For bit flip, the mutation rate was set to 5%. For each operator combination, we evaluated the resulting Pareto front using the *hypervolume indicator* (HV) [7] with a reference point of (1, 1). A higher hypervolume indicates more coverage of the solution space and hence a better Pareto front. A value of 1 indicates the theoretically best possible solution and a value of 0 indicates the worst. The population size and the number of generations were set to 100 and 200, respectively.

To answer **RQ1**, we observed the best results with variable initializer, elitist selection, 1-point crossover, and bit flip mutation across all datasets with HV values of 0.87, 0.89, and 0.90 for Adult, Bank, and Compas, respectively. The hypervolume indicator with the random initializer was significantly lower, around 0.50, compared to the variable initializer across all experiments. The latter is our own proposed method for this work that initializes the population with a set of solutions that vary in the number of data points removed. We note that all results were very consistent, as the maximum standard deviation was 0.01, indicating the stability of NSGA-II in solving the stated optimization problem.

4.2 Bias Mitigation and Classification Performance

To answer **RQ2**, we employ following pipeline: We split each dataset into training and testing sets using an 80-20 split. Next, we applied both the single- and multi-objective optimization approaches to mitigate bias in the training data with varying normalization factors $\beta \in \{(1+\epsilon), \hat{\psi}(\mathcal{D})\}$. We set the population size to 200 and the number of generations to 400 for both solvers.

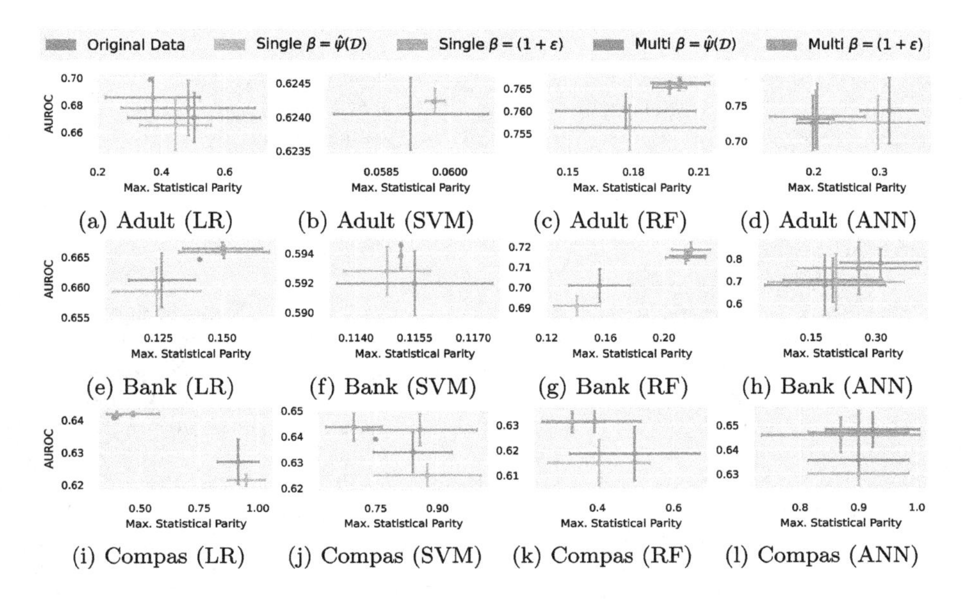

(a) Adult (LR) (b) Adult (SVM) (c) Adult (RF) (d) Adult (ANN)

(e) Bank (LR) (f) Bank (SVM) (g) Bank (RF) (h) Bank (ANN)

(i) Compas (LR) (j) Compas (SVM) (k) Compas (RF) (l) Compas (ANN)

Fig. 1. Results on the test set using two approaches with varying β parameters. x-axis and y-axis represent discrimination and performance metrics, respectively.

Because the multi-objective approach returns a Pareto front of solutions $\mathcal{PF}$ when solving with NSGA-II, we selected a single solution based on β with: $\arg\min_{\mathcal{D}_{\text{fair}} \in \mathcal{PF}} \frac{\hat{\psi}_{\text{SDP-max}}(\mathcal{D}_{\text{fair}})}{\beta} + \mathcal{L}(\mathcal{D}, \mathcal{D}_{\text{fair}})$. For the single objective, we used the genetic algorithm from our prior work [4] with the same genetic operators as the NSGA-II algorithm. An α-value of 0.5 was set to equally weigh the objectives. Next, we trained *Logistic Regression* (LR), *Support Vector Machines* (SVM), *Random Forest* (RF), and *Artificial Neural Networks* (ANNs) and evaluated their predictions on the untouched test set for fairness with $\psi_{\text{SDP-max}}$ and performance with the area under the ROC curve (AUROC). Higher AUROC values indicate better performance.

The classifiers' results are displayed in Fig. 1. We use error bars to display the mean and standard deviation of the respective metrics. We do not observe an overall clear trend in the results or deviation from the original datasets. This emphasizes that the pre-processed datasets can indeed be used reliably as training data. When comparing the classifiers, SVM seems to be least affected by the pre-processed datasets, mostly apparent in Fig. 1b and Fig. 1f.

5 Conclusion

To improve dataset fairness, it is common to remove data points that cause discrimination. However, this does not guarantee that the dataset remains reliable. Hence, we introduced two criteria, *coverage* and *data loss*, that a fair subset must fulfill to preserve its quality. For this purpose, we formulated multiple optimization problems that deal with these criteria. We evaluated our methodology on machine learning models using real-world datasets. The models' fairness and performances were not affected by the data removal process. This implies that our methods produce reliable datasets.

References

1. Bellamy, R.K.E., et al.: AI fairness 360: an extensible toolkit for detecting, understanding, and mitigating unwanted algorithmic bias. CoRR abs/1810.01943 (2018)
2. Bird, S., et al.: Fairlearn: a toolkit for assessing and improving fairness in AI. Technical report MSR-TR-2020-32, Microsoft (2020). https://www.microsoft.com/en-us/research/publication/fairlearn-a-toolkit-for-assessing-and-improving-fairness-in-ai/
3. Deb, K., Pratap, A., Agarwal, S., Meyarivan, T.: A fast and elitist multiobjective genetic algorithm: NSGA-II. IEEE Trans. Evol. Comput. **6**(2), 182–197 (2002)
4. Duong, M.K., Conrad, S.: Towards fairness and privacy: a novel data pre-processing optimization framework for non-binary protected attributes. In: Benavides-Prado, D., Erfani, S., Fournier-Viger, P., Boo, Y.L., Koh, Y.S. (eds.) AusDM 2023. CCIS, vol. 1943, pp. 105–120. Springer, Singapore (2024). https://doi.org/10.1007/978-981-99-8696-5_8
5. Eiben, A.E., Smith, J.E.: Introduction to Evolutionary Computing, 2nd edn. Springer, Cham (2015). https://doi.org/10.1007/978-3-662-44874-8
6. European Commission: Artificial Intelligence Act, Corrigendum (2024). https://www.europarl.europa.eu/doceo/document/TA-9-2024-0138-FNL-COR01_EN.pdf. Accessed 17 May 2024
7. Fonseca, C., Paquete, L., Lopez-Ibanez, M.: An improved dimension-sweep algorithm for the hypervolume indicator. In: 2006 IEEE International Conference on Evolutionary Computation, pp. 1157–1163 (2006). https://doi.org/10.1109/CEC.2006.1688440
8. Goldberg, D.E.: Genetic Algorithms in Search, Optimization and Machine Learning, 1st edn. Addison-Wesley Longman Publishing Co., Inc., USA (1989)
9. Kohavi, R.: Scaling up the accuracy of naive-bayes classifiers: a decision-tree hybrid. In: KDD 1996, pp. 202–207. AAAI Press (1996)
10. Larson, J., Angwin, J., Mattu, S., Kirchner, L.: Machine bias (2016). https://www.propublica.org/article/machine-bias-risk-assessments-in-criminal-sentencing
11. Moro, S., Cortez, P., Rita, P.: A data-driven approach to predict the success of bank telemarketing. Decis. Support Syst. **62**, 22–31 (2014)
12. Verma, S., Ernst, M.D., Just, R.: Removing biased data to improve fairness and accuracy. CoRR abs/2102.03054 (2021)
13. Žliobaitė, I.: Measuring discrimination in algorithmic decision making. Data Min. Knowl. Disc. **31**, 1060–1089 (2017)

"The Absence of Evidence is Not the Evidence of Absence": Fact Verification via Information Retrieval-Based In-Context Learning

Payel Santra[1(✉)], Madhusudan Ghosh[1], Debasis Ganguly[2], Partha Basuchowdhuri[1], and Sudip Kumar Naskar[3]

[1] Indian Association for the Cultivation of Science, Kolkata, India
payel.iacs@gmail.com, partha.basuchowdhuri@iacs.res.in
[2] University of Glasgow, Glasgow, UK
debasis.ganguly@glasgow.ac.uk
[3] Jadavpur University, Jadavpur, Kolkata, India

Abstract. Fact verification, the process of estimating the truth of a particular claim, plays an important role to automatically filter out misinformation and fake news in this age of information overload. Existing approaches of fact verification involve a supervised approach that relies on the existence of manually assessed resources in the form of claims and their truth or falsity. However, constructing such datasets requires significant manual effort, and, hence suffers from lack of scalability. Instead, in this study we demonstrate that an unsupervised non-parametric approach of using 0-shot or k-shot in-context learning turns out to yield results that, in terms of effectiveness measures, are close to that of a standard supervised method that involves fine-tuning a pre-trained transformer. In particular, we demonstrate that sentences extracted from webpages or Wikipedia, similar to the claim that is to be verified, turn out to be useful prompts towards effectively guiding a large language model-based decoder for this particular task.

Keywords: Fact verification · Prompt learning · In-context learning

1 Introduction

The steep rise of fake news, often spread by some harmful motives, underscores the need to develop skills in verifying information's accuracy. This area has seen considerable advancements, especially with the inception of the Fact Extraction and Verification (FEVER) task [8, 17]. Fact-checking typically involves finding potential evidence, using both unsupervised methods [9] and dense retrievers [16]. Techniques such as the Kernel Graph Attention Network (KGAT) [11] and semantic-level graphs [24] have also been applied to assess the validity of claims. In our work, we use the FEVER dataset [17], where the objective is to verify the validity of a claim by retrieving evidence for or against the claim from Wikipedia. Verifying claims with necessary evidence involves manually identifying relevant evidence for each claim, similar to relevance assessments in information retrieval [1]. This process is labor-intensive and prone to biases like

R. Wrembel et al. (Eds.): DaWaK 2024, LNCS 14912, pp. 381–387, 2024.
https://doi.org/10.1007/978-3-031-68323-7_34

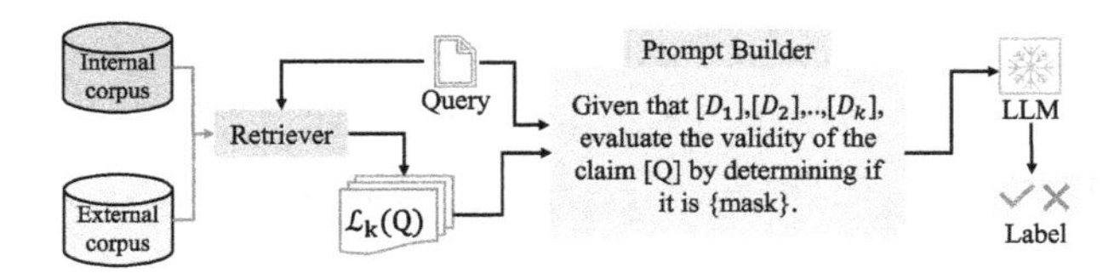

Fig. 1. Schematic of the proposed ICL-based framework for claim verification.

pooling and exposure [21]. We posit that simplifying fact-checking tasks by bypassing evidence collection enables using in-context learning without utilizing the existing training dataset of facts and their veracity. This approach also eliminates the need for manually evaluated evidence relevance for training [2] or evaluation [13].

Our Contributions. In our paper, we enhance claim verification by using in-context learning (ICL) with a large language model (LLM), where we provide context from related sentences in external sources like the Web and Wikipedia. This approach doesn't need labeled datasets or ground-truth tagged evidence, making it more efficient.

2 In-Context Learning for Claim Validity

Brief Review of Supervised Approach. We aim to map an input claim $\mathbf{x}$ and its evidence $e(\mathbf{x})$ to a binary outcome by optimizing the model parameters θ. This process involves identifying a set of potentially relevant text as evidence through an information retrieval (IR) model. For each evidence $e \in M_k(\mathbf{x})$, the parameterized posterior likelihood $P(y = 1|(\mathbf{x}, e); \theta)$ is computed, where $M_k(\mathbf{x})$ includes the top k evident texts related to claim $\mathbf{x}$.

Introduction to In-Context Learning. We introduce a claim validation approach using ICL based framework, different from supervised learning, the function f does not involve training on labeled data. Instead, claim validity is assessed using the input $\mathbf{x}$, the LLM's decoder parameters ϕ_{LLM}, and a prompt with potential k examples $\mathcal{P}_k(\mathbf{x})$. Formally, $P(y = 1|\mathbf{x}) = f(\mathbf{x}, \mathcal{P}_k(\mathbf{x}); \phi_{\text{LLM}})$. The LLM decoder's output, a word sequence, determines class likelihoods by mapping to word sets representing each class, essential for our binary claim verification.

An IR Approach to Obtain Prompts for Claim Verification. To extract text units for a claim $\mathbf{x}$, we perform a keyword search using $\mathbf{x}$ on a sparse index, focusing on sentence-level granularity to align with the claim's length. We employ an IR-based model, LM-Dirichlet [22] which we used in our experiments similar to [20], yielding the top-k sentences $M_k(\mathbf{x})$, which are then incorporated into a prompt template $\mathcal{P}_k(\mathbf{x})$.

Specifically, as choices for the document collections and indexes, we employed the following in three different ways: first, an **in-domain data of labeled examples** from a training dataset for claim verification task, where in addition to a sentence, there is an indication of its label. Generally speaking, the neighborhood set $M_k(\mathbf{x})$ is comprised of pairs of the form $\{(\mathbf{x}', y(\mathbf{x}'))\}$, where $\mathbf{x}'$ is a sentence from training dataset that is semantically similar to the claim $\mathbf{x}$, and $y(\mathbf{x}')$ is its truthfulness or falsity indicator

Table 1. Templates for generating the prompts from a set of k claims similar to the current claim $\mathbf{x}$ (retrieved from a collection), i.e., $M_k(\mathbf{x}) \mapsto \mathcal{P}_k(\mathbf{x})$.

Corpus	Labeled	Prompt template
Internal	Yes	Given that $\langle \mathbf{x_1} \rangle$is $\langle y(\mathbf{x}_1) \rangle, \ldots$ and $\langle \mathbf{x}_k \rangle$ is $\langle y(\mathbf{x}_k) \rangle$ evaluate the validity of the claim $\langle \mathbf{x} \rangle$ by determining if it is $\langle$MASK$\rangle$
External	No	Given that $\langle \mathbf{x_1} \rangle, \langle \mathbf{x_2} \rangle, ..\ \langle \mathbf{x_k} \rangle$ evaluate the validity of the claim $\langle \mathbf{x} \rangle$ by determining if it is $\langle$MASK$\rangle$

ground-truth. Second, we index sentences from the **MS MARCO passage collection**, an extensive database of over 8M web page excerpts. Thirdly, we also index sentences from a **Wikipedia collection** based on the 2020 dump, both do not have any truthfulness indicator label information. Figure 1 presents a schematic depiction of our proposed setup.

Template-Based Prompt Construction. Given instances with and without labels (from FEVER, MS MARCO, and Wiki datasets), we employed distinct templates for mapping similar instance sets $M_k(\mathbf{x})$ to prompts $\mathcal{P}_k(\mathbf{x})$, as detailed in Table 1.

3 Evaluation

Datasets. For claim verification task, we used the FEVER dataset with $\sim$ 185K claims from 5.4M Wikipedia articles [17]. Claims labeled 'not enough information' were excluded from both training and evaluation sets to ignore ambiguity. For ICL experiments, MS MARCO [12] and Wikipedia [19] collections were employed as the external collections (see Sect. 2).

Methods Investigated. To evaluate the effectiveness of the proposed in-context learning (ICL) based framework, we employ a number of standard supervised baselines. Since the prediction in ICL depends on a number of similar data instances, we employ the traditional non-parametric model **K-NN** [7] as a standard baseline. Specifically, the nearest neighbors for each test instance are computed over the training set of the FEVER dataset over a sparse bag-of-words representation of the claims. Next, as parametric supervised approaches, we first employ **Bi-LSTM**, where static word embeddings (specifically, GloVe [14] of 300 dimensions in our experiments) are fed into a Bi-LSTM recurrent network to learn a parameterized representation of the sequence of words to the truth indicator classes [5,10,25]. We use FLAN-T5 (pre-trained with causal masked language model and 4 other different tasks, namely text generation, language translation, sentiment analysis, and text classification) as the LLM realisation for our ICL-based approaches. Consequently, for a fair comparison, we employ the standard task-specific fine-tuning of FLAN-T5 parameters (on the FEVER training dataset) as the other supervised baseline. We call this approach **FLAN-T5-FT** (fine-tuned version of FLAN-T5).

Table 2. Comparisons between our ICL-based methods vs. supervised baselines on the FEVER dataset. All few-shot results used $k = 3$ examples.

Metric	Supervised Models			In-Context Learning						
	K-NN	Bi-LSTM	FLAN-T5-FT	ICL_0	ICL-ID-RP	ICL-MSM-RP	ICL-WK-RP	ICL-ID	ICL-MSM	ICL-WK
Prec	0.6869	0.7784	**0.8144**	0.7709	0.6454	0.6747	0.6493	0.6461	0.7574	**0.7839**
Recall	0.6142	0.7209	**0.8003**	0.6749	0.6445	0.6031	0.6159	0.6219	0.7574	**0.7818**
F1	0.6485	0.7485	**0.8073**	0.7197	0.6450	0.6369	0.6322	0.6337	0.7574	**0.7829**

Table 3. Examples of top-most similar sentences retrieved from the three different collections for a sample claim.

Claim	Retrieved evidences
Amy Winehouse died on 23 July 2011	FEVER (In-domain): Katy Perry died in 2011
	MS MARCO: Winehouse died of alcohol poisoning on 23
	Wikipedia: After Amy Winehouse's untimely death, on 23 July 2011 from alcohol poisoning at the age of 27, the foundation...

An objective of our experiments is to empirically verify if the information from other related claims either from the in-domain training set or from the external collections indeed proves useful for ICL. Hence, as an ablation of ICL, we employ the **random prompting** baseline, where $M_k(\mathbf{x}) \sim \mathcal{C}$, i.e., the neighborhood is a uniform sample of sentences of size k from the collection, the rest of the process of prompt generation being identical to that proposed in Sect. 2 (we denote these ablations by adding the suffix 'RP'). As another ablation of ICL (denoted as ICL_0), we employ the zero-shot setup.

To denote the collection used in our proposed ICL-based approach, we include the strings 'ID', 'MSM' and 'WK' to denote in-domain FEVER training data, MS MARCO and Wiki respectively, thus leading to the following three variants **ICL-ID**, **ICL-MSM** and **ICL-WK**[1]

Implementation Details. Experiments were conducted with **FLAN-T5$_{\text{large}}$** [3], chosen after preliminary tests with various LLMs like, GPT2 [15], OPT [23], Electra [4], GPT-J [18], and GPT-Neo [6], etc. FLAN-T5 was superior for this task.

Results. Table 2 presents a comparison of methods investigated for claim verification. Fine-tuning an LLM (FLAN-T5-FT) yields the best performance, requiring a labeled training set. Conversely, the unsupervised zero-shot ICL_0 approach, with no additional information achieves competitive results, highlighting ICL's potential for this task. The use of random prompts reduces effectiveness compared to zero-shot, highlighting the need for relevant claim selection. In-domain results ('ID') are lower than zero-shot,

[1] Implementation: https://github.com/payelsantra/ICL-WK.

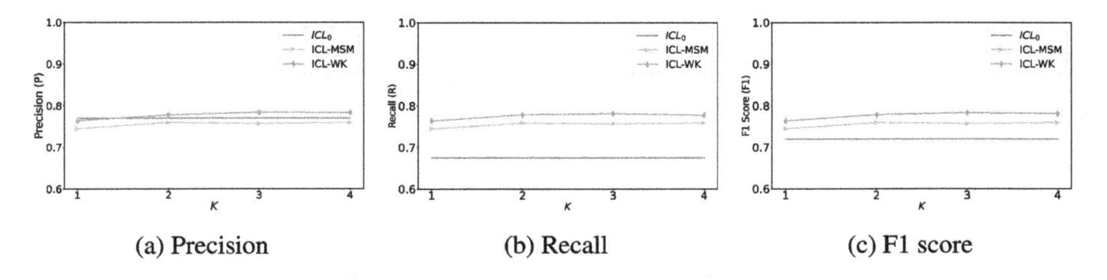

(a) Precision (b) Recall (c) F1 score

Fig. 2. Sensitivity of claim verification effectiveness to the number of sentences (k) provided as context to an LLM.

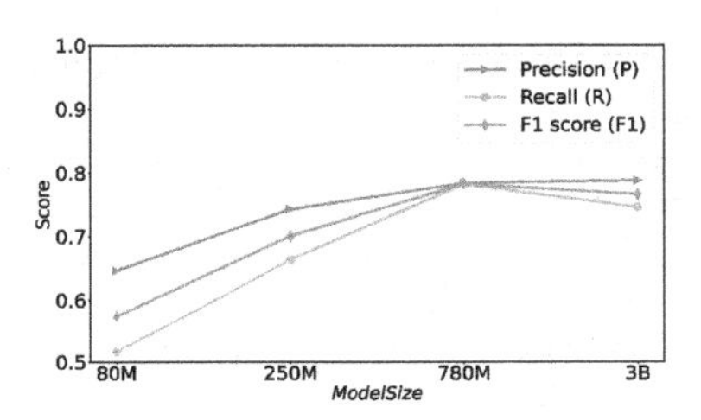

Fig. 3. Performance comparison of Flan-T5 models with different parameter sizes.

likely due to lower average cosine similarities ($\overline{\mathbf{x} \cdot \mathbf{x}'}$) in the FEVER training set. Manual inspection shows that FEVER's training set may contain non-pertinent contexts, e.g., an unrelated claim about Katy Perry for Amy Winehouse's death claim (Table 3). Sentences from MS MARCO and Wikipedia, with higher cosine similarities to the test claim, enhance relevance and improve ICL_0 results. Wikipedia is notably effective, with its context boosting validation, achieving an F1 score just 2.5% lower than the supervised approach FLAN-T5-FT. Table 3 shows MS MARCO and Wikipedia providing pertinent details on Amy Winehouse's death, thus aiding in prediction accuracy. We can observe from Fig. 2 that our ICL-WK model consistently outperforms (when $k = 3$) its corresponding counterparts i.e., baseline ICL_0 and ICL-MSM for different choices of k in terms of Precision, Recall and F-score.

Ablation Study. We now select the best performing ICL approach (i.e., ICL-WK) and carry out experiments to compare the performance obtained via the use of different model variants (involving different parameter sizes). Figure 3 presents a relative comparison in performance across different Flan-T5 model versions. Additionally, it can be observed from Fig. 3 that the Flan-T5-large (780M) model outperforms the other versions.

4 Conclusions and Future Work

This paper demonstrated that a non-parametric approach of ICL that does not require labeled data (hence unsupervised) is able to achieve comparable results with that of a supervised model (FLAN-T5) fine-tuned on labeled examples of claim validations. In future, in addition to retrieving related claims, we would explore the effects of automatically generating evidence on prediction. We would also like to investigate a data-driven adaptation of the prompts.

References

1. Al-Maskari, A., Sanderson, M., Clough, P.: Relevance judgments between TREC and non-TREC assessors. In: SIGIR (2008)
2. Chen, J., et al.: LOREN: logic-regularized reasoning for interpretable fact verification. In: AAAI (2022)
3. Chung, H.W., et al.: Scaling instruction-finetuned language models. arXiv preprint arXiv:2210.11416 (2022)
4. Clark, K., Luong, M.-T., Le, Q.V., Manning, C.D.: ELECTRA: pre-training text encoders as discriminators rather than generators. arXiv preprint arXiv:2003.10555 (2020)
5. Deng, J., Cheng, L., Wang, Z.: Attention-based biLSTM fused CNN with gating mechanism model for Chinese long text classification. Comput. Speech Lang. (2021)
6. Gao, L., et al.: The pile: an 800GB dataset of diverse text for language modeling. arXiv preprint arXiv:2101.00027 (2020)
7. Guo, G., Wang, H., Bell, D., Bi, Y., Greer, K.: KNN model-based approach in classification. In: Meersman, R., Tari, Z., Schmidt, D.C. (eds.) OTM 2003. LNCS, vol. 2888, pp. 986–996. Springer, Heidelberg (2003). https://doi.org/10.1007/978-3-540-39964-3_62
8. Hanselowski, A., et al.: UKP-Athene: multi-sentence textual entailment for claim verification. In: Proceedings of the FEVER Workshop (2018)
9. Jiang, K., Pradeep, R., Lin, J.: Exploring listwise evidence reasoning with T5 for fact verification. In: Proceedings of ACL 2021 (2021)
10. Liu, G., Guo, J.: Bidirectional LSTM with attention mechanism and convolutional layer for text classification. Neurocomputing **337**, 325–338 (2019)
11. Liu, Z., Xiong, C., Sun, M., Liu, Z.: Fine-grained fact verification with kernel graph attention network. In: ACL (2020)
12. Nguyen, T., et al.: MS marco: a human generated machine reading comprehension dataset (2016)
13. Pan,L., Chen, W., Xiong, W., Kan, M.-Y., Wang, W.Y.: Zero-shot fact verification by claim generation. arXiv preprint arXiv:2105.14682 (2021)
14. Pennington, J., Socher, R., Manning, C.D.: GloVe: global vectors for word representation. In: EMNLP 2014 (2014)
15. Radford, A., et al.: Language models are unsupervised multitask learners. OpenAI blog (2019)
16. Subramanian, S., Lee, K.: Hierarchical evidence set modeling for automated fact extraction and verification. In: EMNLP 2020 (2020)
17. Thorne, J., Vlachos, A., Christodoulopoulos, C., Mittal, A.: FEVER: a large-scale dataset for fact extraction and VERification. In: NAACL (2018)
18. Wang, B., Komatsuzaki, A.: GPT-J-6B: a 6 billion parameter autoregressive language model (2021, 2022)
19. WikiDump. wikidump (2020). https://dumps.wikimedia.org/enwiki/20230620/enwiki-20230620-pages-articles-multistream.xml.bz2
20. Wu, Z., et al.: OpenICL: an open-source framework for in-context learning. arXiv preprint arXiv:2303.02913 (2023)
21. Yang, S., Shu, K., Wang, S., Renjie, G., Fan, W., Liu, H.: Unsupervised fake news detection on social media: a generative approach. In: AAA (2019)

22. Zhai, C., Lafferty, J.: A study of smoothing methods for language models applied to information retrieval. ACM Trans. Inf. Syst. (2004)
23. Zhang, S., et al.: OPT: open pre-trained transformer language models. arXiv preprint arXiv:2205.01068 (2022)
24. Zhong, W., et al.: Reasoning over semantic-level graph for fact checking. In: ACL (2020)
25. Zhou, P., Qi, Z., Zheng, S., Xu, J., Bao, H., Xu, B.: Text classification improved by integrating bidirectional LSTM with two-dimensional max pooling. In: COLING 2016 (2016)

Discovering Relationships Among Properties in Wikidata Knowledge Graph

Emetis Niazmand[1,2(✉)] and Maria-Esther Vidal[1,2,3]

[1] TIB Leibniz Information Centre for Science and Technology, Hannover, Germany
{Emetis.Niazmand,Maria.Vidal}@tib.eu
[2] Leibniz University of Hannover, Hanover, Germany
[3] L3S Research Center, Hanover, Germany

Abstract. Wikidata, a community-maintained knowledge graph (KG), may integrate different entities and properties with the same meaning. Contributors can add new properties similar in meaning to other properties in the KG. Detecting relationships among these properties plays a crucial role in interoperability and downstream tasks. We tackle the problem of discovering relationships among properties regarding their instances in Wikidata. Although our approach is *knowledge graph-agnostic*, it can be applied to any KGs. Our approach resorts to Class-based Relationship Discovery (CRD) to capture the most important characteristics of the properties. We evaluate our approach over Wikidata and show the benefits of exploiting statements annotated with qualifiers, references, and ranks. We empirically study the distribution and frequency of relationships among predicates in six domains to provide evidence that KGs enclose relationships that define the same real-world properties.

Keywords: Knowledge Graphs · Wikidata · Knowledge Discovery

1 Introduction

Community-maintained knowledge graphs, such as Wikidata [15], have the potential to be incomplete due to the decentralized nature of their development and maintenance [2].[1] This incompleteness can be enhanced by discovering relationships among properties; found at two levels: inside a knowledge graph (a.k.a. *Intra-KG Relationships*) or in other knowledge graphs (a.k.a. *Inter-KG Relationships*). These relationships can be discovered as similar properties based on different Natural Language Processing (NLP) methods such as Information Retrieval (IR) [13], Embeddings [12], or Association Rule Mining [1]. One of the reasons for opting to use Wikidata to discover relationships among properties is its substantial capacity to interlink classes and properties with various Linked Open Data (LOD) ecosystems, despite some limitations in interconnections [6]. Wikidata contains almost 11,746 properties with descriptions[2].

[1] SPARQL/WIKIDATA Qualifiers, References and Ranks.
[2] Wikidata:Database reports/List of properties/all (Feb 2024).

R. Wrembel et al. (Eds.): DaWaK 2024, LNCS 14912, pp. 388–394, 2024.
https://doi.org/10.1007/978-3-031-68323-7_35

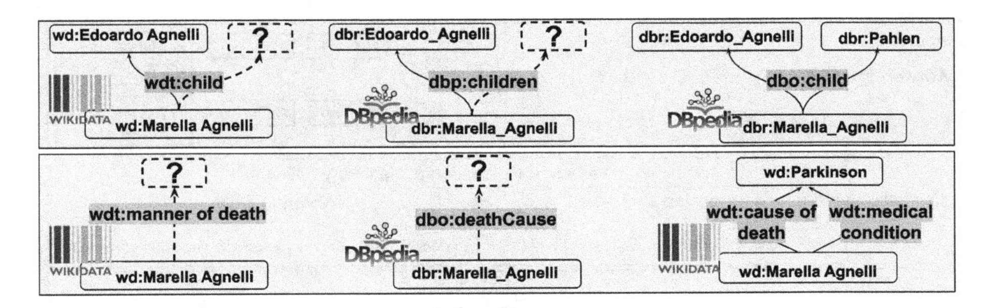

Fig. 1. An example demonstrating discovering relationships among properties in Wikidata and DBpedia [3] to associate various resources and entities (Nov 2022).

Problem Statement and Proposed Solution. We tackle the problem of discovering relationships among predicates in Wikidata. The relationships demonstrate how the predicates are related to each other regarding their corresponding entities and instances. A given predicate may have a relationship with one or more predicates. We describe a Class-based Relationship Discovery (CRD) model which captures relationships among predicates at the level of classes. Our approach considers class-based descriptions of data sources-an abstract description of entities and properties that belong to the same semantic type and their characteristics-to find a set of relationships for a given predicate. In the CRD, diverse metrics are used to discover relationships; they are calculated over all properties of a class and quantify the degree of distance and overlap.

Contributions: This paper makes the following contributions: a) An approach using CRD to discover relationships among properties within the same KG, as *Intra-KG Relationships*, or across different KGs, referred to as *Inter-KG Relationships*, b) An empirical evaluation across six Wikidata domains revealed that within the *Person* domain, relationships are distributed among *86.66%* of predicates, whereas in the *Drug* domain, the distribution stands at *42.39%*.

2 Preliminaries and Motivation

Knowledge Graphs (KGs) [9]. A Knowledge graph $KG = (V, E, L)$ is defined as a directed edge-labeled graph, where V is a set of nodes represented as entities; E is a set of edges, $E \subseteq V \times L \times V$; and L corresponds to a set of labels.

Motivating Example. Consider the example illustrated in Fig. 1. The resource *Marella Agnelli* can be associated with only one object via predicates `wdt:child`, i.e., (`wdt:P40`) in Wikidata or `dbo:children` in DBpedia. Following the open-world assumption (OWA), KGs might lack completeness, implying absent relationships are unknown, not false. Discovering relationships among predicates enhances downstream tasks. For instance, the predicate `dbo:child` can associate the resource *Marella Agnelli* with more objects than predicates `dbp:children`, or `wdt:child`, indicating a relationship between these predicates.

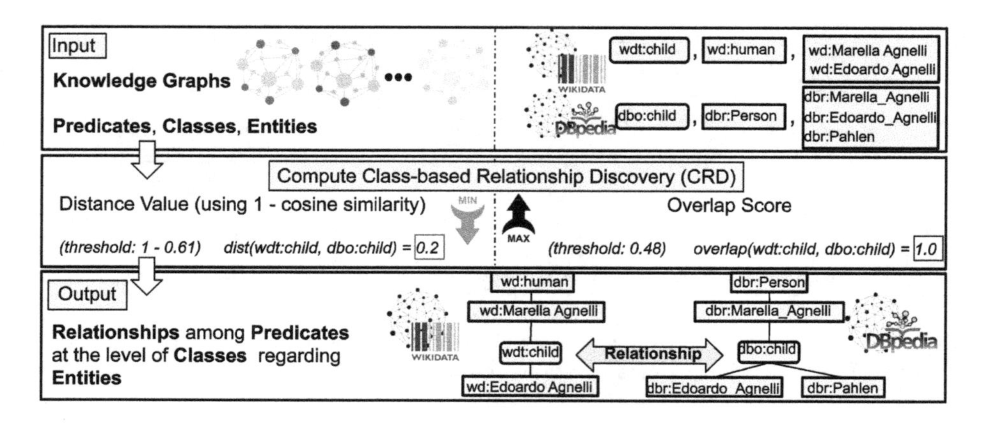

Fig. 2. Running Example: Relationships among predicates inside a knowledge graph or across knowledge graphs are discovered by applying CRD concept.

In another scenario, the resource *Marella Agnelli* cannot be associated with any objects via predicates `wdt:manner of death`, i.e., (`wdt:P1196`) in Wikidata or `dbo:deathCause` in DBpedia. However, two other predicates in Wikidata, `wdt:cause of death`, i.e., (`wdt:P509`) and `wdt:medical condition`, i.e., (`wdt:P1050`), associate the resource *Marella Agnelli* with the object *Parkinson.*

3 Discovering Relationships Among Properties

Problem Statement. Given a set $K = \{KG_1, KG_2, ..., KG_n\}$ of knowledge graphs where $\{KG_i = (V_i, E_i, L_i) | KG_i \in K\}$, and predicate p in L_i. Let C be a class such that $(?resource, wdt : instanceof, C) \in E_i$ in Wikidata or $(?resource, rdf : type, C) \in E_i$ in DBpedia. We tackle the problem of discovering a set of relationships L' belonging to the same class C among the set of KGs. A relationship p' in L' is discovered with respect to its distance value and an overlap score, so that $min(distance_p, distance_{p'}) \equiv max(overlap_p, overlap_{p'})$.

Proposed Solution. We present a method to uncover relationships among KGs in K for predicate p in L_i from KG_i. Our method resorts to Class-based Relationship Discovery (CRD) to discover *Intra-KG* and *Inter-KG Relationships* (Fig. 2). This approach takes into account various KGs, such as Wikidata and DBpedia, as input sources. Entities and properties belonging to each class are extracted using SPARQL queries. The resulting output illuminates the relationships among predicates, demonstrating how predicates are related to each other concerning their associated entities. In CRD, two metrics are employed: distance value and overlap score (Algorithm 1). To compute the thresholds, we employ the*95th* percentile represents the point where only *5%* of the values exceed the assigned threshold, falling into the *95th* percentile category.

Distance Value. We measure distance between predicates using RDF2Vec [12], which outperforms other embedding methods over KGs [11]. Cosine similarity,

Algorithm 1: Discovery Relationships Among Predicates Algorithm

Input : Set of knowledge graphs K; set of predicates L; predicate p in L with domain C and predicate p' with similar domain; Thresholds ð, τ
Output: Set of *Intra-KG Relationships* L' and *Inter-KG Relationships* L''

```
1  L' ← {}, L'' ← {}, D ← {};
2  foreach p' in L from KG' ∈ K do
3      convert p and p' to vectors V^p, V^p';
4      if dist(V^p, V^p') ≤ (1 − ð) then
5          D ← D.add(p');
6          foreach p' in D do
7              evaluation of p over KG ([[p]]^KG) and p' over KG' ([[p']]^KG');
8              if overlap(p, p', KG, KG') ≥ τ then
9                  if KG = KG' then
10                     L' ← L'.add(p');
                   else
11                     L'' ← L''.add(p');
12 return L', L''
```

ranging from *0.0* to *1.0*, is used to compute embedding distances. Higher values indicate greater similarity and closer distances in an abstract space. After converting predicates to vectors using RDF2Vec, the cosine similarity calculates the distance between two property vectors, V^{p} and $V^{\mathrm{p}'}$.

$$dist(V^{\mathrm{p}}, V^{\mathrm{p}'}) = 1 - cos(\theta) = 1 - \frac{V^{\mathrm{p}}.V^{\mathrm{p}'}}{|V^{\mathrm{p}}||V^{\mathrm{p}'}|} \tag{1}$$

The *95th* percentile of distance is at a threshold ð of *0.61*, meaning only *5%* of similarity values exceed *0.61*, while *95%* are below it. Thus, properties with a distance value under *0.39* (*1* - *0.61*) are considered closer in distance.

Overlap Score. We measure the overlap of instances of predicates within or across KGs, ranging from *0.0* to *1.0*. Values closer to *1.0* indicate a high number of shared instances between predicates. Certain predicates in Wikidata, like `wdt:place of birth`, i.e., (`wdt:P19`) and `wdt:place of death`, i.e., (wdt:P20), exhibit high instance overlap. Despite high overlap, they have different statements, such as title. Hence, the predicate `wdt:place of birth` is not considered a relationship for the predicate `wdt:place of death`. The use of statements (qualifier references) in Wikidata distinguishes it from other knowledge graphs in order to discover relationships among predicates [8]. If predicates p and p' share the same knowledge graph ($KG = KG'$), the relationships are *Intra-KG Relationships*, otherwise they are *Inter-KG Relationships.* The expression $[[\mathrm{p}]]^{\mathrm{KG}}$ denotes the evaluation of an input SPARQL query to retrieve the instances associated with predicate p over KG, respecting a specific resource.

$$Overlap(p, p', KG, KG') = \frac{|[[p]]^{\mathrm{KG}} \cap [[p']]^{\mathrm{KG'}}|}{min(|[[p]]^{\mathrm{KG}}|, |[[p']]^{\mathrm{KG'}}|)} \tag{2}$$

The *95th* percentile overlap score threshold is *0.48*, indicating properties with scores above *0.48* have significant commonality.

4 Empirical Evaluation

We empirically study the application of CRDs as a solid approach for discovering relationships among properties inside the same KG or within other KGs. The empirical evaluation aims to formulate the following research questions: **RQ1)** Are there any relationships among predicates in Wikidata? **RQ2)** How frequent are relationships among predicates in Wikidata?

Table 1. The table describes the distribution of relationships among properties per domain in Wikidata. The values in **bold** depict the highest value for relationships among predicates. *The limit has been applied to retrieve the predicates.

Wikidata's Domain	#properties	%total relationships	#frequent relationships	%frequent relationships
Person	3,000*	**86.66%**	**3**	46.38%
Music	683	73.02%	**3**	44.6%
History	195	41%	2	56.25%
Film	1,000*	72.1%	**3**	**71.7%**
Sport	678	44.98%	2	51.47%
Drug	467	42.39%	1	60.6%

Benchmark. We conducted our evaluation on a large cross-domain KG, Wikidata. Predicates and associated entities and classes extracted from the domains, *Person*, *Music*, *History*, *Film*, *Sport*, and *Drug* by executing SPARQL queries.

Metrics. In our experiment, we used two metrics: distance to measure similarity or dissimilarity between predicates, and overlap score to determine the overlap of instances associated with predicates to reveal their relationships.

Implementation. Experiments were run on a Windows 10 machine with an Intel i7-9850H 2.6 GHz CPU and 16 GB 1333 MHz DDR3 RAM. We implemented our approach in Python 3.7.5. We use the TIB SPARQL endpoint of Wikidata[3].

Experimental Results. The distribution and frequency of relationships in Table 1 indicate that more predicates in domains *Person*, *Music*, and *Film* can have relationships to other predicates. The similarity metrics and overlap scores among these predicates demonstrate low distance and high overlap. The *Person* domain has the highest at *86.66%*, whereas the *Drug* domain has the lowest at *42.39%* (**Answer to RQ1**). Moreover, predicates that have the greatest number of relationships, reaching a maximum of three, are more prevalent in the domains of *Person*, *Music*, and *Film*, whereas predicates in the *Drug* domain typically are related to only one predicate. It shows *60.6%* of predicates in *Drug* domain have only one relationship. *71.7%* of predicates in domain *Film* have

[3] TIB SPARQL endpoint for Wikidata.

relationships with three predicates, indicating that many properties convey the same meaning but may be represented by different names (**Answer to RQ2**).

5 Related Work

In recent years, KGs have become popular for representing large amounts of structured data. However, despite their usefulness, KGs based on open-world assumption (OWA) are quite incomplete [2], such as Wikidata [4]. In the absence of a pre-defined formal ontology, Wikidata consists of many properties, such as property constraints, to assist analyze and refine inconsistent data and errors [5]. Contributors may introduce new properties. These may overlap or be synonymous with existing properties in the KG. Several methods have been utilized to detect synonym predicates, including lexical and semantic methods. These methods involve analyzing the text of predicates and related entities [7], utilizing embedding techniques to identify similarities among predicates [10], and employing frequent item set mining [1]. Additionally, knowledge-based methods, which use external knowledge sources, e.g., ontologies, can also be used to identify synonym predicates [14]. One promising approach involves discovering relationships among properties, which can help uncover new relationships within the data and improve overall accuracy and completeness of the KGs.

6 Conclusions and Future Work

This study highlights the problem of discovering relationships among predicates in community-maintained knowledge graphs such as Wikidata. Knowledge graphs may contain numerous predicates that are similar in meaning but presented in different ways. The relationships among predicates are identified using a Class-based Relationship Discovery (CRD) concept, which employs diverse metrics to capture relations at the class level and measure the distance and overlap between predicates. The empirical evaluation validates the presence of relationships among predicates in Wikidata. As a result, there is at least one relationship among *86.66%* of predicates in the *Person* domain. We aim to help researchers recognize the presence of numerous relationships among predicates in Wikidata to improve the quality of their research. Future efforts will focus on leveraging these relationships to curate knowledge graph quality and downstream tasks, e.g., prediction tasks, negative sampling, and query processing.

Acknowledgement. This work is partially funded by EraMed project P4-LUCAT (GA No. 53000015).

References

1. Abedjan, Z., Naumann, F.: Synonym analysis for predicate expansion. In: Cimiano, P., Corcho, O., Presutti, V., Hollink, L., Rudolph, S. (eds.) ESWC 2013. LNCS, vol. 7882, pp. 140–154. Springer, Heidelberg (2013). https://doi.org/10.1007/978-3-642-38288-8_10

2. Abiteboul, S.: Querying semi-structured data. In: Afrati, F., Kolaitis, P. (eds.) ICDT 1997. LNCS, vol. 1186, pp. 1–18. Springer, Heidelberg (1997). https://doi.org/10.1007/3-540-62222-5_33
3. Auer, S., Bizer, C., Kobilarov, G., Lehmann, J., Cyganiak, R., Ives, Z.: DBpedia: a nucleus for a web of open data. In: Aberer, K., et al. (eds.) ASWC/ISWC -2007. LNCS, vol. 4825, pp. 722–735. Springer, Heidelberg (2007). https://doi.org/10.1007/978-3-540-76298-0_52
4. Balaraman, V., Razniewski, S., Nutt, W.: Recoin: relative completeness in Wikidata. In: WWW. ACM (2018). https://doi.org/10.1145/3184558.3191641
5. Ferranti, N., De Souza, J., Ahmetaj, S., Polleres, A.: Formalizing and validating Wikidata's property constraints using SHACL and SPARQL. SWJ (2024)
6. Haller, A., Polleres, A., Dobriy, D., Ferranti, N., Rodriguez Mendez, S.J.: An analysis of links in Wikidata. In: Groth, P., et al. (eds.) ESWC 2022. LNCS, vol. 13261, pp. 21–38. Springer, Cham (2022). https://doi.org/10.1007/978-3-031-06981-9_2
7. Han, Z., Feng, Y., Zhao, D.: Detecting synonymous predicates from online encyclopedia with rich features. In: Ma, S., et al. (eds.) AIRS 2016. LNCS, vol. 9994, pp. 111–122. Springer, Cham (2016). https://doi.org/10.1007/978-3-319-48051-0_9
8. Hernández, D., Hogan, A., et al.: Reifying RDF: what works well with Wikidata? In: ISWC (2015). https://ceur-ws.org/Vol-1457/SSWS2015_paper3.pdf
9. Hogan, A., Blomqvist, E., et al.: Knowledge Graphs (2021). https://kgbook.org/
10. Kalo, J.-C., Ehler, P., Balke, W.-T.: Knowledge graph consolidation by unifying synonymous relationships. In: Ghidini, C., et al. (eds.) ISWC 2019. LNCS, vol. 11778, pp. 276–292. Springer, Cham (2019). https://doi.org/10.1007/978-3-030-30793-6_16
11. Niazmand, E.: Enhancing query answer completeness with query expansion based on synonym predicat. In: WWW (2022). https://doi.org/10.1145/3487553.3524198
12. Ristoski, P., Rosati, J., et al.: RDF2Vec: RDF graph embeddings and their applications. Semant. Web (2019). https://doi.org/10.3233/SW-180317
13. Ruge, G.: Automatic detection of thesaurus relations for information retrieval applications. In: Freksa, C., Jantzen, M., Valk, R. (eds.) Foundations of Computer Science. LNCS, vol. 1337, pp. 499–506. Springer, Heidelberg (1997). https://doi.org/10.1007/BFb0052119
14. Suchanek, F.M., et al.: PARIS: probabilistic alignment of relations, instances, and schema. VLDB (2011). https://doi.org/10.14778/2078331.2078332
15. Vrandecic, D., Krötzsch, M.: Wikidata: a free collaborative knowledgebase. Commun. ACM (2014). https://doi.org/10.1145/2629489

Using a Spatial Grid Model to Interpret Players Movement in Field Sports

Valerio Antonini[1(✉)], Michael Scriney[1,2], Alessandra Mileo[1,2], and Mark Roantree[1,2]

[1] School of Computing, Dublin City University, Dublin 9 D09 V209, Ireland
valerio.antonini3@mail.dcu.ie,
{michael.scriney,alessandra.mileo,mark.roantree}@dcu.ie

[2] Insight SFI Research Centre for Data Analytics, School of Computing, Dublin City University, Dublin 9, Ireland

Abstract. The global sports analytics industry has a market value of USD 3.78 billion in 2023. The increase of wearables such as GPS sensors has provided analysts with large fine-grained datasets detailing player performance. Traditional analysis of this data focuses on individual athletes with measures of internal and external loading such as distance covered in speed zones or rate of perceived exertion. However these metrics do not provide enough information to understand team dynamics within field sports. The spatio-temporal nature of match play necessitates an investment in date-engineering to adequately transform the data into a suitable format to extract features such as areas of activity. In this paper we present an approach to construct Time-Window Spatial Activity Graphs (TWGs) for field sports. Using GPS data obtained from Gaelic Football matches we demonstrate how our approach can be utilised to extract spatio-temporal features from GPS sensor data.

1 Introduction

GPS data have emerged as a valuable resource for researchers and sports professionals to identify speed zones [1], define athletes' external load during competition [2], identify sequential and recurrent players movements [3], among other applications. Recently, machine learning approaches, such as ensemble learning [4], graph clustering [5], and neural-network based [6], have been applied to extract new insights. GPS analytics have traditionally focused on individual players to improve performance and prevent injuries, neglecting team dynamics. Complex analyses of player relationships, movement direction, and intensity are currently lacking. Player profiles have been studied without considering teammate interactions. To enhance individual and collective metrics, analyzing a network of players, locations, and movements is necessary, understanding

This work was funded by Science Foundation Ireland through the Centre for Research Training in Machine Learning (18/CRT/6183) and Grant Number SFI/12/RC/2289_P2, co-funded by the European Regional Development Fund.

R. Wrembel et al. (Eds.): DaWaK 2024, LNCS 14912, pp. 395–400, 2024.
https://doi.org/10.1007/978-3-031-68323-7_36

its evolution over time. Spatio-temporal data, with interconnected dependencies, suit graph representation, where nodes are locations and edges are events between them. We introduce a graph-based framework to represent Gaelic Football players' movements during games, extracting patterns and insights. This helps analyze high-action areas, compare games and players, and enable data-driven decisions by sports scientists and coaches based on real-time tracking data. An extended version of this work is available online [7].

2 Related Research

Spatio-temporal graphs analyze GPS data by representing spatial locations and temporal dynamics, exploring movement patterns and interactions. In sports analytics, graphs represent players or pitch areas as nodes and actions as edges, aiding strategy and performance analysis. Authors in [8] use betweenness centrality to identify key players, while our study focuses on pitch areas and movement speeds. In a different research, [9] introduce uPATO to generate adjacency matrices of player interactions for team performance assessment. Our research also uses centrality scores but differs in purpose and topology. A successive study [10] links passing networks and positioning to match outcomes, focusing on static positions. Researchers in [11] use a spatio-temporal graph with player nodes and distance-based edges for possession change prediction. Our study examines the importance of pitch areas over time. In conclusion, [12] analyze pitch zones in scoring actions using nodes for zones and edges for ball movements. Our study differs in data sources, pitch division, and focus on temporal dynamics.

3 Methodology

This work presents a novel framework which provides graph transformations from spatiotemporal datasets of team-based sports. Through this framework we are able to perform graph-based analytics which permits the extraction of insights about players' activity during games.

The framework is composed of four main steps:

- **Step 1: Data collection, cleaning & Annotation**. This step involves the initial export of data from the GPS sensing devices, cleaning and annotation of additional features (such as average speed) to the data to produce a dataset of player actions.
- **Step 2: Spatial Grid Mapping**. For each game, a spatial grid is constructed using the Tesspy library [13], which divides space into non-overlapping subspaces. The spatial grid mapping process involves:
  1. *Specification of a Point of Interest (POI)*: Tesspy takes a POI (e.g., 'Wembley') and creates a Tessellation object.
  2. *Spatial Grid Generation*: An initial cell size resolution is defined, and adaptive square grids are generated based on the POI data. The centroids of these grids are returned as latitude and longitude coordinates.

3. *Spatial Grid Mapping*: players' coordinates during the game are assigned to the nearest centroid. Centroids without player data are excluded, representing unoccupied areas.

  Each centroid represents a cell on the pitch, which is used to construct a spatio-temporal graph. After the cells are mapped, a domain expert specifies a time interval to create snapshots of spatial activity within each cell using a rolling window. This results in the Time-Window Spatial Activity Graph.

- **Step 3: Spatio-Temporal Graphs Construction**. This process builds the spatio-temporal graphs through an iterative process. The Time-Window Spatial Activity Graph TWG is a set of spatial-activity graphs where each graph is a snapshot of activity for a timepoint w within a cell a. Each Spatial Activity Graph is a directed graph $TWG_{a,w} = (N, E)$ where N is the set of nodes representing an area of activity denoted a tuple containing the latitude and longitude for said area $n = <lat, lng>, n \in N$. E denotes the set of edges within the graph where each edge e represents a four-tuple $e = <n_{from}, n_{to}, P, \bar{Speed}>, e \in E$ where, n_{from} and n_{to} are the *to* and *from* nodes for the directed edge, P is the distinct list of players who traversed between two areas of activity and $\bar{Speed}$ is the average speed for all players during that transition.
- **Step 4: Graph-based Analysis**. This step implements graph theory techniques to analyze patterns of behaviors of players during games or time windows of the same game.

4 Results

The dataset consists of raw GPS data from 11 competitive Gaelic Football games between 2019 and 2021. Gaelic Football, originating in Ireland, blends rugby and soccer. Players wore 10 Hz GPS sensors (STATSports Apex 10 Hz), recording latitude, longitude, and speed (m/s) at 10 observations per second. The data, filtered and pre-processed by STATSports software (v4.5.19), is accurate within 1–2% error [14]. The dataset was further processed using methods described in previous work [15]. All players, except goalkeepers, were included. The case study is a 78-min county-level male Gaelic Football match with 16 players, resulting in 13,586 records.

4.1 Spatial Mapping of the Pitch

The process starts by specifying the POI (stadium name) and a cell resolution. The Tesspy library provides grid coordinates for the pitch. Player coordinates from GPS data are then associated with the nearest grid points. The result is shown in Fig. 1.

4.2 TWG Generation

Once the spatial grid mapping is complete, the next step is generating the Time-Window Graph (TWG). This involves defining a rolling time window, with size

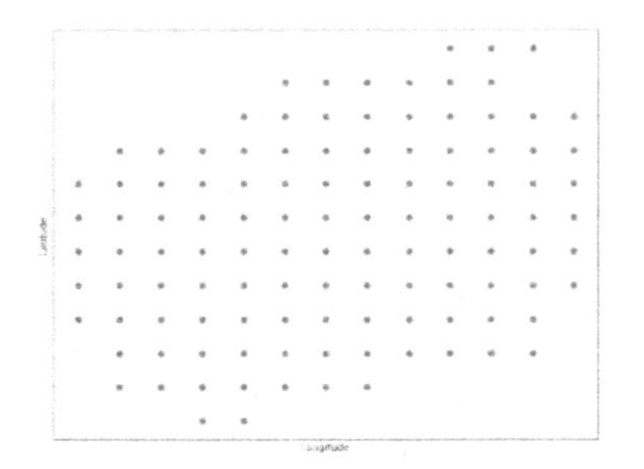

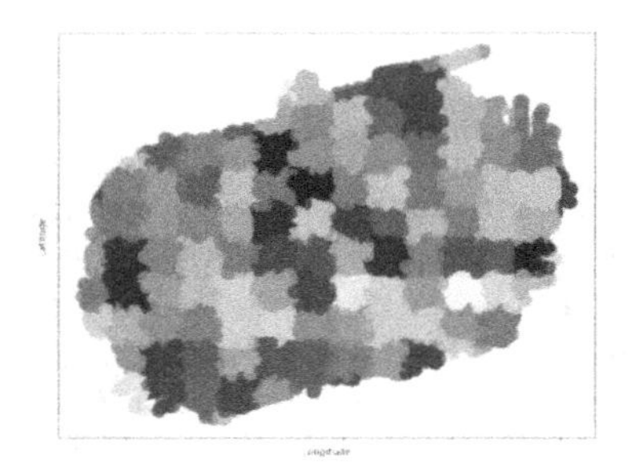

Fig. 1. Spatial grid mapping of the pitch. **Left**: grid points returned by the library. **Right**: result of the association of players' coordinates to their closest grid point.

set by a domain expert. Larger windows create more connected graphs but lose detail, while narrower windows capture fewer actions and provide less insight.

For this paper, we use a 5-min rolling window: [0,5), [1,6), etc., generating a TWG of size 74 for a 78-min game. Table 1 presents metrics for each 5-min window, including the number of nodes (pitch cells), edges (movements between cells), average edge weight (speed of actions), graph density, average edges per node, clustering degree, and average shortest path length

Table 1. Descriptive statistics regarding the characteristics of the TWG within each time window.

T.W.	Nodes	Edges	Avg. Weight E.	Density	Avg. Degree	Avg. Clust. Coeff.	Avg. Shortest Path
[0, 5)	118	952	3.2	0.07	11.4	0.26	3.5
[1,6)	118	882	3.1	0.06	10.7	0.21	3.8
[2,7)	118	872	3.0	0.06	10.8	0.23	3.8

4.3 Analysis of Areas of Activity

Using betweenness centrality, we can compare cell centrality across time windows, highlighting activity areas as the game progresses. Figure 2 exhibits the betweenness centrality for two different 5-min rolling windows. The locations displayed are the ones covered by players only in the specific time windows and can change over time. The most important locations (in terms of speed and connectivity) are frequently placed in the central area of the pitch. The dynamic of the game shifts as we move across windows, highlighting strategic alterations in the distribution of the scores, reflecting a change in the importance of locations. The importance of the cells is given by their connectivity. This means that these cells are reached by a higher number of different cells. Players tend to move from or in these specific cells more than others.

Similar to the fluctuation observed in centrality scores, we employed the Louvain community detection algorithm to identify distinct communities across

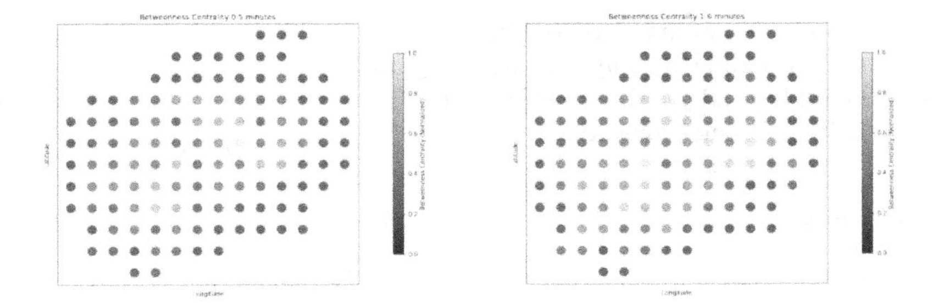

Fig. 2. Betweenees centrality scores per pitch cell during 5-min rolling windows: [0, 5) (**left**), [1, 6) (**right**).

different time windows in the TWG. Figure 3 details the community detection result across two consecutive time windows (black points represent coordinates not covered by players in the time window). These communities change shapes and internal composition over time. This behaviour may derived from a wider spread of players across the pitch, leading to a greater number of less populated communities. The detection of communities formed mainly by adjacent cells suggests that players tend to traverse short distances before changing speed zones.

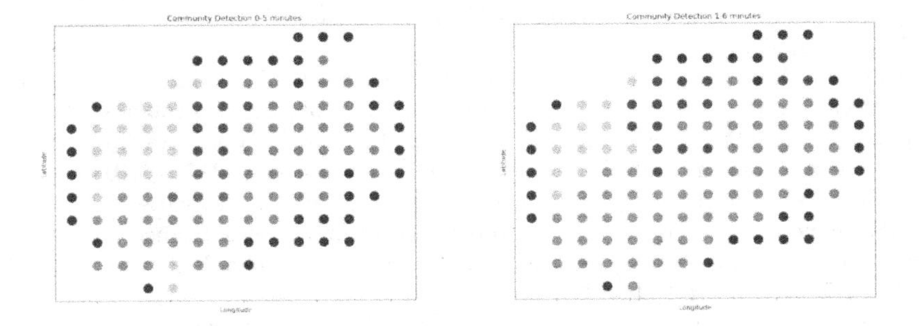

Fig. 3. Communities formed by pitch cells during 5-min rolling windows: [0, 5) (**left**), [1, 6) (**right**).

5 Conclusion and Future Work

As sports analytics evolve, more teams and organizations are adopting data-driven approaches to training and strategy. This shift necessitates investment in data engineering to transform and structure data for team-focused analytics during matches. We present a framework to construct a Time Window Spatial Activity Graph, a collection of spatio-temporal graphs across rolling time windows, focused on player movement. Our evaluation shows how graph-analytics, like betweenness centrality and community detection, can provide insights into team dynamics, aiding in player positioning and team strategies. Future work will explore additional graph-analytics methods, such as link prediction, to forecast graph state changes over time.

References

1. Coutts, A.J., Quinn, J., Hocking, J., Castagna, C., Rampinini, E.: Match running performance in elite Australian Rules Football. J. Sci. Med. Sport **13**(5), 543–548 (2010)
2. Malone, S., Solan, B., Collins, K.D., Doran, D.A.: Positional match running performance in elite Gaelic football. J. Strength Conditioning Res. **30**(8), 2292–2298 (2016)
3. White, R., Palczewska, A., Weaving, D., Collins, N., Jones, B.: Sequential movement pattern-mining (SMP) in field-based team-sport: a framework for quantifying spatiotemporal data and improve training specificity? J. Sports Sci. **40**(2), 164–174 (2022)
4. Sheridan, D., Brady, A.J., Nie, D., Roantree, M.: Predictive analysis of ratings of perceived exertion in elite Gaelic football. Biol. Sport **41**(4), 61–68 (2024)
5. Brady, A.J., Scriney, M., Moyna, N.M., McCarren, A.: Identification of movement categories and associated velocity thresholds for elite Gaelic football and hurling referees. Int. J. Perform. Anal. Sport **21**(5), 741–753 (2021)
6. Kim, J., Kim, H., Lee, J., Lee, J., Yoon, J., Ko, S.K.: A deep learning approach for fatigue prediction in sports using GPS data and rate of perceived exertion. IEEE Access **10**, 103056–103064 (2022)
7. Antonini, V., Scriney, M., Mileo, A., Roantree, M.: A Framework for Spatio-Temporal Graph Analytics in Field Sports (2024). https://doras.dcu.ie/30059/1/
8. Gama, J., et al.: Network analysis and intra-team activity in attacking phases of professional football. Int. J. Perform. Anal. Sport **14**(3), 692–708 (2014)
9. Silva, F.G., Gomes, A.J., Nguyen, Q.T., Martins, F.M., Clemente, F.M.: A new tool for network analysis on team sports the ultimate performance analysis tool. In: 2017 International Conference on Engineering, Technology and Innovation (ICE/ITMC), pp. 439–445. IEEE (2017)
10. Gonçalves, B., Coutinho, D., Santos, S., Lago-Penas, C., Jiménez, S., Sampaio, J.: Exploring team passing networks and player movement dynamics in youth association football. PLoS ONE **12**(1), e0171156 (2017)
11. Raabe, D., Nabben, R., Memmert, D.: Graph representations for the analysis of multi-agent spatiotemporal sports data. Appl. Intell. **53**(4), 3783–3803 (2023)
12. Mclean, S., Salmon, P.M., Gorman, A.D., Stevens, N.J., Solomon, C.: A social network analysis of the goal scoring passing networks of the 2016 European Football Championships. Hum. Mov. Sci. **57**, 400–408 (2018)
13. https://tesspy.readthedocs.io/en/latest/index.html
14. Beato, M., Coratella, G., Stiff, A., Iacono, A.D.: The validity and between-unit variability of GNSS units (STATSports Apex 10 and 18 Hz) for measuring distance and peak speed in team sports. Front. Physiol. **9**, 411796 (2018)
15. Antonini, V., Mileo, A., Roantree, M.: Engineering features from raw sensor data to analyse player movements during competition. Sensors **24**(4), 1308 (2024)

Author Index

R. Wrembel et al. (Eds.): DaWaK 2024, LNCS 14912, pp. 401–402, 2024.
https://doi.org/10.1007/978-3-031-68323-7

Printed in the USA
CPSIA information can be obtained
at www.ICGtesting.com
CBHW061144110924
14384CB00005B/167

9 783031 683220